Dynamic Force Spectroscopy
and
Biomolecular Recognition

Dynamic Force Spectroscopy

and

Biomolecular Recognition

Edited by

Anna Rita Bizzarri • Salvatore Cannistraro

CRC Press
Taylor & Francis Group
Boca Raton London New York

CRC Press is an imprint of the
Taylor & Francis Group, an **informa** business

CRC Press
Taylor & Francis Group
6000 Broken Sound Parkway NW, Suite 300
Boca Raton, FL 33487-2742

First issued in paperback 2018

© 2012 by Taylor & Francis Group, LLC
CRC Press is an imprint of Taylor & Francis Group, an Informa business

No claim to original U.S. Government works

ISBN-13: 978-1-4398-6237-7 (hbk)
ISBN-13: 978-1-138-37452-2 (pbk)

Library of Congress Cataloging-in-Publication Data

Dynamic force spectroscopy and biomolecular recognition / [edited by] Anna Rita
 Bizzarri and Salvatore Cannistraro.
    p. cm.
  ISBN 978-1-4398-6237-7 (hardback)
  1. Molecular spectroscopy. 2. Molecular recognition. I. Bizzarri, Anna Rita. II.
Cannistraro, Salvatore.

QD96.M65D96 2012
572'.36--dc23   2011045051

Visit the Taylor & Francis Web site at
http://www.taylorandfrancis.com

and the CRC Press Web site at
http://www.crcpress.com

Contents

Preface

The ability of a biomolecule to attract and bind to another molecule is commonly referred to as biomolecular recognition and plays a fundamental role in life. Through specific recognition processes, biomolecules, such as proteins, DNA, ribonucleic acid, lipids, and so on, can build reversible, or irreversible, complexes and aggregates able to perform a variety of functions, for example, cell adhesion, genome replication and transcription, signaling, immune responses, maintaining of the cell architecture, and so on. In view of the tremendous number of potential vital interactions occurring in the biological environment, the mechanism underlying the formation of molecular complexes is a key aspect that needs to be rigorously described.

Until recently, the conventional concepts of affinity and standard kinetic constants, used to describe chemical equilibria, have been assumed to provide a suitable framework to deal with biomolecular recognition processes. However, during the last decades, it appeared that these concepts were insufficient to provide a comprehensive description of the totality of the mechanisms regulating biomolecular interactions. Indeed, the outcomes of these interactions were found also to depend on other factors that are not currently accounted for, such as molecular crowding, bond sensitivity to disruptive forces, distance and orientation among the molecules, interaction restrictions between molecules at surfaces, the necessity to reach rapidity and efficiency in the interaction, and so on. For example, the immune system, whose task consists in detecting foreign and potentially harmful particles or molecules to destroy them, must couple specificity and rapid response to foreign materials; immune defense would be uneffective if an excessive amount of time were required to trigger a response. Biomolecular recognition is strongly influenced by the properties of the environment in which the biomolecules are embedded; for example, biomolecule interactions may involve not only soluble molecules (3D interactions) but also surface-attached molecules (2D interactions). Furthermore, cell adhesion to a solid surface, a fundamental process that influences many steps of cell function, may depend on the mechanical force to which the system undergoes. Finally, the knowledge of aspects inherent to individual molecules, for example, rare events, transient phenomena, crowding effects, population heterogeneity, and so on, generally hidden in bulk measurements, is a key requirement for a deeper understanding of biological processes. On such a basis, the necessity to develop both new experimental approaches and theoretical frameworks to reach a deeper comprehension of biomolecule recognition processes, even at the single molecule details, has strongly emerged.

Dynamic force spectroscopy (DFS) is a remarkable example of an innovative approach, stemming from a number of different experimental and theoretical advances able to shed a new light on important and unresolved issues related to biorecognition. DFS, which is experimentally based on the extension of widespread atomic force microscopy (AFM), allows to measure with a picoNewton, or even lower, sensitivity, the unbinding force between two individual biomolecular partners, one firmly bound to the AFM tip and the other to the substrate, under the application of an external force. Since these DFS measurements are carried out in nonequilibrium conditions, information on kinetics and thermodynamical properties at equilibrium can be extracted from the data by applying suitable theoretical models, based on the transition rate theory describing the escape over an energy barrier under the influence of an external perturbation.

Besides providing detailed information on the kinetics and thermodynamics of a single pair of biomolecular interactions, complementing, in this way, traditional biochemical approaches, DFS constitutes a great promise also for the knowledge of nonconventional aspects of biorecognition processes. This is especially due to the almost unique capability of DFS to investigate biomolecular interactions at single molecular level, in almost real time, in near-physiological conditions and without labeling.

The acronym DFS generally indicates the method by which the unbinding force is measured as a function of the loading rate, which is the rate at which the force is applied. Although the word spectroscopy could be misleading since no matter-radiation interaction is involved in DFS experiments, it finds a widespread use in the literature. In other cases, atomic force spectroscopy is used instead of DFS to emphasize that an AFM equipment is employed to perform the experiments. For the sake of completeness, it should be mentioned that DFS has also been used when other experimental approaches, such as biomembrane force probe, laminar flow chambers, optical tweezers, have been used to measure the unbinding force between two biomolecular partners.

This book is devoted to the application of DFS to investigate biorecognition processes between biological partners and is a collection of seven chapters written by experts in specific fields covering the main aspects correlated to biorecognition processes, the basic principles and instrumentation of DFS, and its application to investigate biological complexes with an extensive treatment of the used theoretical models and of the data analysis procedures.

In Chapter 1, the basic and novel aspects of biorecognition processes are reviewed and discussed in the perspective of exploiting the emerging capabilities of single molecule techniques to disclose kinetic properties and molecular mechanisms usually hidden in bulk measurements. A brief introduction on the computational methods finalized to study biorecognition processes is also presented.

In Chapter 2, the basic principles of AFM and DFS are described with a particular attention to those instrumental and theoretical aspects more strictly related to the study of biomolecules. Along this direction, a description of some innovative technical improvements of AFM equipments is also included.

In Chapter 3, the theoretical background within which experimental data taken in nonequilibrium measurements of biomolecular unbinding forces are extrapolated to equilibrium conditions is overviewed. While the Bell-Evans model is amply treated, some novel emergent theoretical models to analyze DFS data are also reviewed and discussed.

In Chapter 4, the most common and efficient strategies adopted in DFS experiments to immobilize the interacting biomolecules to the AFM tip and to the substrate are reviewed with a particular emphasis to the use of molecular linkers to bind the biomolecules to the surfaces. The experimental and theoretical criteria to discriminate among both specific and nonspecific events and single and multiple bond ruptures are widely discussed.

In Chapter 5, the most representative aspects related to the analysis of DFS data and the challenges of integrating well-defined criteria to calibrate force distance data, peak detection, histogram plotting, and so on in automatic routinary procedures are presented and discussed.

In Chapter 6, an overview of the most relevant DFS applications to study biorecognition processes is presented. Starting from the biotin–avidin pair, which represents a benchmark for investigating biorecognition processes, selected results on various biological complexes, including antigen–antibody, proteins–DNA, complexes involved in adhesion processes, and so on, are reviewed and discussed. Some emphasis is given to both the capability of DFS to elucidate biorecognition details that cannot be disclosed by standard bulk techniques and even to the possible application of DFS to nanobiosensing. Finally, in Chapter 7, the main results obtained by DFS applied to the study of biorecognition processes are briefly summarized along with the most debated and forthcoming theoretical and experimental advances. The topics of this book were also conceived within the networking research activity of the EU COST Action on Applications of Atomic Force Microscopy to NanoMedicine and Life Sciences (AFM4NanoMed&Bio).

MATLAB® is a registered trademark of The MathWorks, Inc. For product information, please contact:

The MathWorks, Inc.
3 Apple Hill Drive
Natick, MA 01760-2098 USA
Tel: 508 647 7000
Fax: 508-647-7001
E-mail: info@mathworks.com
Web: www.mathworks.com

Editors

Anna Rita Bizzarri received her degree in physics in 1987 from the University of Rome. She obtained a PhD in biophysics in 1992 from SISSA in Trieste. After postdoctoral fellowships in Perugia and Mainz, she joined the Science Faculty, Tuscia University (Italy), as a research assistant. In 2000, she became associate professor of physics, and in 2006, she received the position of full professor.

Her scientific interests include spectroscopic investigations and MD simulations of electron transfer metalloproteins. More recently, she is focused on single molecule level detection by surface-enhanced Raman spectroscopy and scanning probe microscopies for both fundamental and applicative aims.

Salvatore Cannistraro obtained a degree in physics in 1972 from Pisa University. He received a PhD in biophysics at Liegi University (Belgium). In 1977, he became reader of biophysics at Calabria University. He moved to Perugia University in 1981 as associate professor of molecular physics. In 1991, he became a full professor of physics, biophysics, and nanoscience at Tuscia University, leading the Biophysics and Nanoscience Centre.

His scientific interests include optical, magnetic, neutron spectroscopies, and modeling of amorphous and biological systems. More recently, he is focusing his activity on the application of AFM, STM, and Raman SERS to single biomolecule detection and nanobiomedicine.

Contributors

Boris B. Akhremitchev
Department of Chemistry
Florida Institute of Technology
Melbourne, Florida

Anna Rita Bizzarri
Biophysics and Nanoscience Centre
Università della Tuscia
Largo dell'Università
Viterbo, Italy

Pierre Bongrand
Laboratoire 'Adhesion et Inflammation'
INSERM UMR600
Marseille, France

Salvatore Cannistraro
Biophysics and Nanoscience Centre
Università della Tuscia
Largo dell'Università
Viterbo, Italy

Shu-wen W. Chen
CEA, iBEB
Service de Biochimie et Toxicologie
 Nucléaire
Bagnols-sur-Cèze, France

Yannick Delcuze
CEA Marcoule
Service de Biochimie et Toxicologie
 Nucléaire
DSV/iBEB/SBTN
Bagnols-sur-Cèze, France

Raymond W. Friddle
Sandia National Laboratories
Livermore, California

Hendrik Hölscher
Karlsruhe Institute of Technology (KIT)
Institute of Microstructure
 Technology (IMT)
Eggenstein-Leopoldshafen, Germany

Michael Odorico
CEA Marcoule
Service de Biochimie et Toxicologie
 Nucléaire
DSV/iBEB/SBTN
Bagnols-sur-Cèze, France

Pierre Parot
CEA Marcoule
Service de Biochimie et Toxicologie
 Nucléaire
DSV/iBEB/SBTN
Bagnols-sur-Cèze, France

Jean-Luc Pellequer
CEA Marcoule
Service de Biochimie et Toxicologie
 Nucléaire
DSV/iBEB/SBTN
Bagnols-sur-Cèze, France

Jean-Marie Teulon
CEA Marcoule
Service de Biochimie et Toxicologie
 Nucléaire
DSV/iBEB/SBTN
Bagnols-sur-Cèze, France

1 Biomolecular Recognition: The Current Challenge

Pierre Bongrand

CONTENTS

1.1 INTRODUCTION

Life relies on myriads of interactions between the molecular components of living systems. Proteins are a remarkable example in view of their diversity (the very name of proteins stems from Proteus, a Greek god known for his capacity to change shape). Several decades ago, the author of a well-known treatise on proteins [45] wrote that "... the biological function of proteins almost invariably depends on their direct physical interaction with other molecules." More recently, systematic use of powerful techniques such as yeast double hybrid or mass spectrometry was the basis for a large-scale attempt to build exhaustive databases of protein interactions, the so-called *interactome* [16]. Over 250,000 interactions between about 22,000 proteins were recorded in the Unified Interactome Database in the year 2008 [30].

Until recently, it seemed that the conventional concepts and methods used to study chemical equilibria provided a suitable framework to deal with biomolecular recognition. As was reckoned two decades ago [195], the concepts of specificity and affinity had seemed sufficient to deal with biological phenomena for many years, and only conventional kinetic constants had to be added to explain some recent findings. However, a number of reports supported the importance of forces in biological interactions [28] [94] and theoretical models of cell functions such as adhesion have included mechanical parameters [13] [127]. This was an incentive to devise experimental methods allowing us to study the response of biomolecules to forces with high temporal and spatial resolution up to the single molecule level. Simultaneously, continuous progress in molecular dynamics allowed computer scientists to report on simulations of the response of biomolecules to external forces [83] [93] [160], thus allowing deeper interpretation of experimental results [71] [167]. These advances were also facilitated by the tremendous increase of structural data on biomolecules, based on X-ray crystallography and nuclear magnetic resonance (NMR), and the use of genetic engineering techniques to relate structural and functional data, as exemplified by alanine scanning that consists of systematically replacing amino acids with alanin in protein–protein interaction areas to obtain a direct estimate of their contribution to binding energy [47]. The development of *dynamic force spectroscopy* (DFS) is a remarkable example of an innovative approach stemming for a number of different advances and yielding a new kind of information that might shed new light on important and unresolved issues.

The goal of this chapter is to present as palatably as possible a number of biological processes and recent methodological advances that played an important role in the development of DFS and may benefit from this growing domain. The first section includes selected examples of biological situations that are heavily dependent on biomolecular recognition. This will be the basis for defining the questions we need to ask. The next section is a brief outline of recent progress done in the study of molecular interactions, particularly at the single bond level, which shaped the present state of the art. The next section is intended to define and analyze the parameters required to provide an adequate account of biomolecule interactions, that is, to include the pieces of information that are needed to predict the behavior of a given ligand–receptor couple under physiological conditions. The last section gives a brief description of the application of conventional physical–chemical knowledge and newer computer simulation methods to the study of links between biomolecule structure and association properties. Admittedly, the field of biomolecule interactions is too vast to be exhaustively discussed in the limited space available. Also, it is unavoidable that the topics selected in this chapter should reflect the limitations of the author's fields of competence and interest. Therefore, I apologize for the omission of many key references that would certainly have enriched this presentation.

1.2   WHAT IS THE USE OF BIOMOLECULAR INTERACTIONS?

The goal of this section is to describe several important biological processes to illustrate the role of biomolecule interactions and the constraints that must be met.

1.2.1   CELL STRUCTURE: STATICS AND DYNAMICS

Clearly, any living cell or organism would fall into pieces in absence of the molecular interactions linking their components. It is important to emphasize that both qualitative and quantitative properties of these interactions are essential. Thus, it is well recognized that cell formation requires an autoorganization capacity of biomolecules that must be able to bind to each other with sufficient *specificity* to avoid durable presence of potentially harmful molecular interactions [188]. In addition, the rheological properties of cells are considered to be driven by the properties of the underlying cytoskeletal elements, which are themselves dependent on the *kinetic* and *mechanical* properties of intermolecular associations [192]. These points are important in view of the recently recognized importance of cell mechanics in situations of medical interest such as cancer cell metastasis [79] [165] or lethal inflammatory processes such as the acute respiratory disease syndrome [135]. Cell shape is considered to be highly dependent on the dynamic organization of a network of rod-like structures including actin microfilaments, tubulin microtubules, and intermediate filaments. These are highly plastic structures whose growth or retraction is determined by a variety of interaction events, and particularly polymerization/depolymerization as a consequence of tunable *kinetics* of monomer association or dissociation. Other important events are movements driven by the so-called *motor molecules* such as myosin or kinesin that are able to generate force-dependent displacements. Much effort was recently done to investigate the mechanisms of association/dissociation and *force* generation by these molecules.

1.2.2　CELL DIFFERENTIATION

A remarkable feature of living cells composing complex organisms is their capacity to acquire different structural and functional capacity, whereas they share a common set of genes. While the mechanisms of differentiation are not yet fully understood, a primary process is the selective synthesis of particular proteins as a consequence of gene activation by a combination of over 100 DNA binding proteins with a specificity for a number of regulatory sites on the DNA. Such a complex set of interactions remains incompletely known, but an extensive network of DNA/protein interactions clearly plays an important role in differentiation [9].

1.2.3　CELL ADHESION

As previously reviewed [151], cell adhesion is a fundamental process that influences nearly all steps of cell function. Thus, cell *survival* and *proliferation* are often dependent on a strong attachment to solid surfaces, a phenomenon known as *anchorage dependence* [31] [72] . An attractive interpretation of experimental findings was that cell adhesion might be required to induce marked cell flattening and spreading on the surface, and that cell behavior might be shape sensitive [134] [153].

Cell migration on a surface is also highly dependent on the qualitative and quantitative properties of cell surface interactions. It has long been shown that efficient cell migration required that *binding strength*, that is, the mechanical force required to detach adherent cells, fell within a particular range [140]. Cells that are too strongly adherent are expected to remain stuck on a fixed place [97]. In contrast, a minimal adhesion efficiency is probably required in order that a lamellipodium sent forward by a motile cell be able to remain stuck on the surface and drag forward the cell body with concomitant detachment of the rear part of the cell [141]. More recently, it was reported that moving cells were able to probe the rigidity of underlying surfaces and move toward more rigid regions, a phenomenon called *durotaxis* [118].

Cell differentiation is also strongly influenced by the properties of underlying surfaces. While this well-known phenomenon has long been interpreted by hypothesizing that cells were essentially sensitive to the biochemical structures of ligands exposed by surrounding surfaces and recognized by their receptors [105], more recent experiments showed that cell responses were also dependent on the stiffness of these surfaces [59]. The mechanisms allowing cells to measure surface stiffness remain poorly understood, but it is likely that this involves the *response to forces* of surface biomolecules adhering to nearby ligands.

Indeed, cells continually probe their environment to adapt their shape, motion, and other functions such as proliferation or mediator release. Environment sensing may result from the uptake of soluble ligands by membranes. However, a more accurate and less noise-sensitive way of probing cellular environment may result from mechanical exploration through continual formation and retraction of protrusions such as lamellipodia [54] and finger-like filopodia [14] [69] or through transverse membrane undulations [156] [203], thus inducing transient contacts between membrane receptors and fixed ligands, which may provide a powerful way of rapidly gathering information [158]. The outcome of interactions is heavily dependent on the *kinetics* of bond formation between surface-attached ligands and receptors and the *strength* of attachments. These phenomena are highly dependent on the kinetics and mechanics of receptor–ligand interactions. Arguably, cells use DFS to probe their environment (Figure 1.1).

Inflammation is an ubiquitous process used by multicellular organisms to cope with various forms of aggression, and particularly infection. A key step is the adhesion of flowing blood leukocytes to the vessel walls, with subsequent transmigration through these walls and entry into tissues containing infectious agents or damaged cells. Unraveling the mechanisms of leukocyte interaction with endothelial cells coating the vessel walls was a major task during the last two decades, and this provided a model of prominent biophysical interest (Figure 1.2).

It has been known for more than a century that locally activated endothelial cells are able to bind to flowing leukocytes, which undergo a nearly 100-fold velocity decrease (typically from 1 mm/s to 10 μm/s). Leukocytes then display a characteristically jerky motion called *rolling*. During the rolling phase, leukocytes remain sufficiently close to the wall to detect specific molecules with a capacity to activate

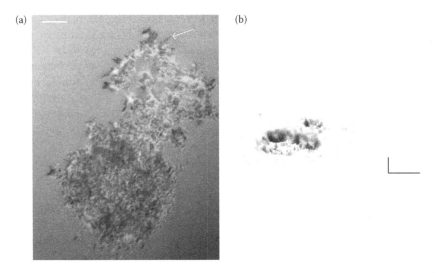

(a)   (b)

FIGURE 1.1 Cells probe their environment. (a) A monocytic THP-1 cell was deposited on a surface coated with fibronectin and was examined with interference-reflection contrast microscopy (IRM). Short filopodia (white arrow) appear as black lines. (b) The image shows the underside of a lymphocyte falling on an activating surface, and tridimensional shape was derived from IRM images. A dynamic study revealed undulations of a few nm amplitude and Hz frequency. Horizontal bar length is 5 μm and vertical bar length is 100 nm. See reference 46 for details.

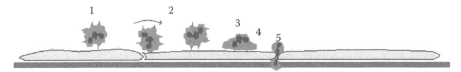

FIGURE 1.2 Leukocyte arrest on activated endothelium. Rapidly flowing leukocytes are first tethered by receptors such as P-selectin, which appear on the membrane of activated endothelial cells (1). Then, they begin rolling with a nearly 100-fold velocity decrease (2), which allows them to detect activating mediators such as chemokines on endothelial surfaces. These molecules activate leukocyte integrins, which results in firm cell adhesion (3). Then, cells migrate toward endothelial cell junctions (4) and undergo impressive deformation that allows them to traverse the endothelial junctions. Finally, after crossing the basal membrane, they accede to inflamed tissues (5).

strong leukocyte attachment and arrest. Displacement toward interendothelial junctions and migration to the peripheral tissues then follow. The progress of molecular biology and monoclonal antibody technology allowed identification of the adhesion molecules involved in leukocyte/endothelial interaction during the 1980s, and the overall mechanisms of rolling and firm adhesion were disclosed in 1991 [115] [189]. Briefly, proper stimulation of endothelial cells was shown to generate rapid expression of so-called selectin molecules on the membranes of endothelial cells. Thus, P-selectin that is stored in specialized granules may be externalized within minutes. P-selectin is a long (about 40 nm) molecule the distal extremity of which bears a binding sites specific for characteristic structures bearing the *sialyl − Lewis^x* tetrasaccharide. This ligand is exposed on molecules borne by leukocyte membranes such as PSGL-1 (a 40-nm carbohydrate-rich molecule). The jerky rolling motion may be accounted for by a rapid formation and dissociation of P-selectin/PSGL-1 bonds as shown with model systems [24]. During the rolling phase, leukocyte adhesion receptors belonging to the integrin family get activated by molecules linked to the endothelial cell pericellular matrix. These integrins then strongly bind to their ligand and induce a durable arrest. As an example, leukocyte integrin LFA-1 (which means lymphocyte function associated-1, also called CD11a/CD18) will bind to ICAM-1 (intercellular cell adhesion molecule-I, CD54) on endothelial cell surfaces. Flow chambers (see below) were used to study leukocyte arrest on planar surfaces coated with endothelial cell monolayers or molecules. A question that rapidly emerged consisted of understanding why P-selectin/PSGL-1 interaction resulted in rolling, whereas integrin/ligand association could not occur in the absence of rolling, even if leukocyte integrins were activated before the experiment. Since the affinity of P-selectin/PSGL-1 and integrin/ligand interaction fell into the same range, it was soon suggested that P-selectin/PSGL-1 interaction might display peculiar physical properties, with *high association and dissociation rates* (allowing rapid cell attachment and detachment) and *high mechanical strength* to resist hydrodynamic forces (otherwise, these interactions would not generate any detectable cell arrest).

A general conclusion of these studies is that cell function is dependent on precise kinetic and mechanical properties of their adhesion receptors and tight regulation of these parameters.

1.2.4 IMMUNE RECOGNITION

The immune system provides particularly important models of biological recognition. The task of immune cells consists of detecting foreign and potentially harmful particles or molecules to destroy them. Foreign particles may be pathogens, cancer cells, or damaged cells that may release harmful metabolites. Immune recognition is of utmost importance and failure may entail devastating consequences. Indeed, a marked immune deficiency is known to result in lethal infection within a few days or even hours after birth. Conversely, excessive immune activation may result in death as may be observed in allergic conditions or autoimmune diseases. It is probably because of this utmost importance that three complementary recognition mechanisms evolved and remain active in higher vertebrates.

Antibodies are protein molecules that may be generated by injecting animals with foreign substances that are consequently called *antigens* (which means antibody generators). Antibodies share remarkable structural properties shared by the plasma proteins called *immunoglobulins*. Each antibody molecule possesses between 2 and 10 identical antigen binding sites called *paratopes*. There seems to be no limit to the recognition capacity of antibodies: They can specifically bind to proteins, carbohydrates, lipids, nucleic acids, and even totally artificial structures such as dinitrophenol. Further, antibody efficiency is dependent on quantitative properties of binding sites such as affinity constant or association kinetics [53] [73], as explained below. Antibodies bind antigens with an affinity constant that may be as high as $10^{10} - 10^{12}M^{-1}$ and their specificity is illustrated by their capacity to discriminate between antignenic sites (called *epitopes*) differing by a single amino acid. The study of antibodies was long made difficult by the high heterogeneity of antibodies raised after injecting animals with a given antigen. However, monoclonal antibodies provide a highly efficient basis for studying molecular recognition [131].

The specific antigen receptors born by T lymphocytes (*T cell receptors* or TCRs) represent a different recognition system. As exemplified in Figure 1.3, a major task of T lymphocytes consists of detecting cells containing foreign material such as viral proteins. The recognition principle is remarkable: Most cells express on their surface on the order of 10,000 oligopeptides of 10–15 amino acids nearly randomly sampled from the proteins they synthesize. Each oligopeptide appears as a few units bound to specialized membrane molecules encoded by genes belonging to the *major histocompatibility complex* [131].

It is remarkable that a T lymphocyte can detect a few or even a single foreign oligopeptide on a cell after scanning its membrane for 5–10 minutes [19]. Another remarkable point is that a number of studies strongly supported the hypothesis that the outcome of the recognition of a foreign oligopeptide by a T lymphocytes is dependent on the physical properties of TCR/ligand interaction. Indeed, the lifetime of individual TCR/ligand bonds might be a key determinant of T lymphocyte

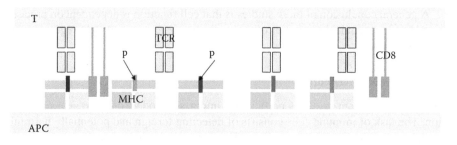

FIGURE 1.3 (See color insert.) T cell activation. A CD4 or CD8 (shown here) T lymphocyte (T) interacting with an antigen-presenting cell (APC) is endowed with several tens of thousands of *identical* T cell receptors (TCR) specific for a unique combination of a MHC molecule and an oligopeptide resulting from the degradation of a particular protein. There may be only a few tens, or less, of specific ligands of a TCR on an APC.

activation since a too-short interaction might result in cell paralysis rather than activation of effector functions [126]. Thus, quantifying these interactions between membrane-bound receptors and ligands is a current challenge of prominent importance [89] [90].

While the aforementioned two recognition mechanisms have been a focus of intense investigation during the last three or four decades, it is well recognized that the immune function also requires a set of so-called *innate recognition mechanisms* that are able to detect foreign microorganisms or damaged cells. Thus, a variety of receptors such as scavenger receptors [82] or toll-like receptors [91] can detect remarkable structures such as double-stranded ribonucleic acids, which are not expressed by eukaryotic cells or denatured proteins and altered lipids that appear in damaged cells. The exquisite specificity of antibodies and TCRs may be responsible for the necessity of an additional recognition mechanism: Since a given lymphocyte bears receptors of a single specificity (this is a basic tenet of the so-called clonal theory), due to the high amount of receptor specificities, the probability that a foreign particle entering a multicellular organism be recognized by a lymphocyte it had just encountered is very low. Since immune defenses would be uneffective if an excessive amount of time was required to initiate an immune response, there is a need for rapid ways of detecting the presence of foreign material with limited specificity. Understanding the involved recognition mechanisms is a challenge of high current interest.

1.2.5 SIGNAL GENERATION

A general consequence of biorecognition events is the selective binding of specific molecules by cell membrane receptors and subsequent generation of intracellular signals that drive cell function. It has long been considered that this phenomenon was fully accounted for by the specificity of intermolecular recognition events. Also, signal generation was usually ascribed to two prominent mechanisms: (1) in many cases, ligand–receptor association was found to result in a change of receptor conformation with concomitant acquisition of signaling sites. G protein-coupled receptors are a general example, and more than 700 of these receptors appear to be encoded in the human genome out of about 25,000 genes [1]. (2) Another general mechanism of signal generation is the surface aggregation of membrane receptors following association with multivalent ligands. This aggregation may result in conformational changes, or in encounter between enzymes and substrata bound to the intracellular part of receptors. Thus, T lymphocyte activation often involves a clustering of tyrosine kinases such as p56lck, which are constitutively bound to cytoplasmic domains of so-called coreceptors. Coreceptor clustering may thus trigger the phosphorylation of tyrosines borne by the cytoplasmic chains of nearby molecules. These phosphorylated tyrosines will then become ligands for intracellular scaffolding proteins bearing cognate SH2 domains [180]. However, while the importance and frequency of aforementioned mechanisms are well established, recent reports supported the view that a number of membrane receptors might behave as force sensors and generate signals through different kinds of mechanisms. Binding of surface-attached ligands might result in force generation, thus generating conformational changes and appearance of binding sites that might nucleate signaling scaffold. Thus, recent data suggested that TCR signaling might indeed be influenced by forces [119]. Also, it was recently reported that a force of only a few piconewtons applied on molecule talin might result in the appearance of new reactive sites [51]. Thus, the *effect of forces on molecules* involved in recognition events is of direct functional significance.

In conclusion, most aspects of cell function are dependent on specific interactions between biomolecules. The outcome of interaction depends not only on affinity but also on association and dissociation kinetics and bond sensitivity to disruptive forces. Further, in view of the tremendous number of potential interactions occurring in the biological environment, the *specificity* of binding molecules is a key property that needs to be rigorously evaluated. Finally, biomolecule interactions may involve not only soluble molecules but also surface-attached receptors. As a consequence of this situation, it appeared during the last decades that the conventional theoretical framework developed during the last century to account for soluble phase (also called 3D) interactions was insufficient to deal with cell function. This was an incentive to develop new methods of studying interactions between surface-attached molecules (i.e., 2D interactions). These methods gave accurate information on bond formation and dissociation at the single molecule level. This exquisite sensitivity provided investigators with a direct grasp on specific aspects of molecular behavior such as random thermal fluctuations. Data interpretation thus required a reexamination of older theoretical models. These recent developments will be rapidly sketched in the following section.

1.3 BRIEF HISTORICAL OUTLINE OF RECENT INVESTIGATIONS MADE ON BIOMOLECULE RECOGNITION AT THE SINGLE BOND LEVEL

The purpose of this section is to give a brief account of a series of investigations essentially performed during the last two decades to analyze interactions between surface-bound molecules at the single bond level. Indeed, the kind of understanding brought by these studies proved highly relevant to biomolecule function, and this was an incentive to reexamine theoretical frameworks elaborated more than a century ago to account for the basic mechanisms of molecule association and separation [66], [85], [111]. It is hoped that this brief outline will help the reader grasp more easily the rationale of more recent work.

1.3.1 STUDYING BOND RUPTURE AT THE SINGLE MOLECULE LEVEL

A theoretical paper authored by George Bell [12] may be considered as a starting point to all recent work on the force sensitivity of single bonds. The purpose of Bell's paper was to find a relationship between the function of cell membrane receptors responsible for adhesive phenomena and the properties of soluble forms of these molecules. Two main points of this paper consisted (1) of separating the encounter phase of interaction that was supposed to be different under 2D and 3D conditions, and the second phase of complex formation that was postulated to be similar in free and surface-anchored molecules, and (2) of suggesting a simple model to account for the effect of disruptive forces on dissociation rates, leading to the so-called Bell's law:

$$k_{\text{off}}(F) = k_{\text{off}}(0)\exp(Fd/k_BT) = k_{\text{off}}(0)\exp(F/F^0) \qquad (1.1)$$

where $k_{\text{off}}(F)$ is the dissociation rate of a bond subjected to force F, as shown in Figure 1.4, x_β is a parameter with the dimension of a length that was interpreted as the distance between the equilibrium distance and the transition state of the ligand–receptor complex as observed on a one-dimensional energy landscape, and k_B and T are Boltzmann's constant and the absolute temperature. F is a parameter with the dimension of a force that may be viewed as an indicator of bond mechanical strength.

This formula is now denominated as Bell's law. A theoretical justification based on Smoluchowski's equation was elaborated a few years later by Evans [62]. Also, while it seemed reasonable to expect that a disrupting force should reduce the lifetime of a bond, rigorous thermodynamic reasoning lead Dembo and colleagues [52] to notice that a disrupting force should reduce the affinity of a bond, but since the affinity constant is the ratio between the association and dissociation rates, it was conceivable that a pulling force might somewhat paradoxically increase bond lifetime. The authors dubbed *slip bonds* "normal" bonds displaying decreased lifetime in presence of forces, and *catch bonds* "strange" bonds displaying increased lifetime in presence of force.

Remarkably, within a few years, several complementary methods [18] allowed a number of investigators to test the theoretical predictions that had recently been reported. Goldsmith used a moving capillary tube to monitor the rupture of doublets made between osmotically sphered red cells coated with a minimal amount of antibodies [184] and subjected to shear flow. The normal force at separation ranged between 60 and 197 pN. Assuming that binding involved a few or even one antibody molecule, this order of magnitude was consistent with Bell's prediction. A few years later, Evans used a dual pipette apparatus to monitor the rupture of attachments

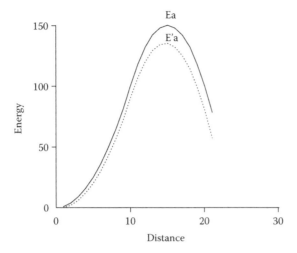

FIGURE 1.4 Bell's law. Bell interpreted bond rupture as the exit of an energy well on a unidimensional energy landscape. Assuming that the frequency of particle attempts at crossing the barrier was a constant, the probability of success was estimated at $\exp(Ea/k_BT)$, where Ea is the activation energy. The effect of a force is to lower the energy curve in proportion to the distance (broken line) [12].

between red cells bound by a minimal amount of antibodies [60]. He estimated at a few tens of piconewtons the rupture force and ascribed it to the uprooting of membrane molecules, a possibility already suggested by Bell [12]. Soon thereafter, *laminar flow chambers* (Figure 1.5) were used to monitor the formation and rupture of attachments between moving particles and surfaces coated with receptor and ligand molecules.

This approach proved a highly sensitive way of observing single bond formation and dissociation, since a cell size sphere subjected to a wall shear rate on the order of a few s^{-1} displays a translational velocity of a few μm/s and is subjected to a distractive force on the order of a piconewton, which is sufficiently low to permit a single weak bond to maintain a particle at rest during a detectable amount of time. The lifetime of single bonds formed between E-selectin molecules and ligands borne by flowing neutrophils was estimated at about 2.4 s [106]. During the following years, flow chambers were used to estimate Bell's F^0 coefficient for the force dependence of dissociation rates, yielding about 90 pN for P-selectin/PSGL-1 couple [5]. However, it was soon reported that single bond rupture was more complex than that predicted with Bell's law, since ligand-receptor association behaved as a multiphasic reaction [146], [147]. Another problem that was later emphasized was the difficulty of ensuring that single bonds were indeed observed [201]. This difficulty may provide an explanation for the discrepancy found between different estimates of parameter F^0 [5], [65].

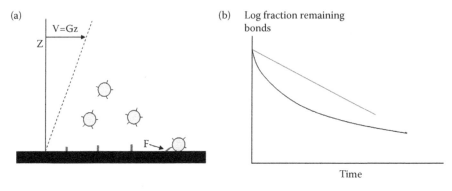

FIGURE 1.5 **(See color insert.)** Studying molecular interactions with a flow chamber. (a) Optimal information can be obtained by studying the motion of receptor-coated microspheres near ligand-coated surfaces in presence of a wall shear rate of a few s^{-1}. Using microspheres of a few μm diameter, trajectories can be monitored with an accuracy of several tens of nm and time resolution of 20 ms with standard video equipment. The force exerted on a particle of 1.4 μm radius may be a fraction of a piconewton, and the force on the bond may be estimated at a few pN when the wall shear rate is on the order of several s^{-1}, which provides high sensitivity. The possibility to scan extensive contact areas is well suited to the use of low surface density coatings and determination of association rates. (b) Bond rupture may be studied by plotting the logarithm of the fraction of surviving bond versus time after initial arrest. In some cases, curves are straight lines and the slope represents the off rate, which may depend on the wall shear rate (red line). In many cases, the curve is more complex (black line) and may be analyzed to estimate some quantitative properties of energy landscapes. The capacity of the flow chamber to measure the kinetic and mechanical properties of weak bonds is described in a recent review [156].

Atomic force microscopy (AFM) provided another way of tackling with single bonds. Initial studies [71] [117] were performed on the avidin/biotin interaction, that is known for its high affinity constant on the order of $10^{15}M^{-1}$. Gaub and colleagues reported on the rupture of association between cantilever tips derivatized with avidin and agarose beads coated with biotin or analogs. The avidin/biotin separation forces appeared as integer multiples of 160 pN, which was interpreted as the strengh of a single bond [71]. As previously reviewed [20], during the following years, different authors used flow chambers, AFM, and also optical tweezers [136] [186] to measure the rupture force of a number of ligand–receptor couples. A major advance came from Evans' laboratory when he markedly enhanced the power of his micromechanical approach by developing the so-called *biomembrane force probe* (BFP) [61] [129]: He glued a latex microbead on an erythrocyte that was used as a tunable cantilever. He used pipettes mounted on a piezoelectric system, allowing computer-controlled displacement with high velocity and subnanometer accuracy. Finally, a rapid video camera allowed excellent time resolution. This device allowed Evans to convince the scientific community that the *unbinding forces* commonly reported in AFM-based studies were not intrinsic parameters of a given ligand–receptor couple. Indeed, even with an interaction as strong as the avidin–biotin bond, spontaneous rupture will occur in absence of force if observation is performed for a sufficiently long time (that may be centuries!). When individual bonds were subjected to a pulling force increasing at constant rate (the so-called *loading rate*, expressed as pN/s) the force at the moment of rupture was linearly dependent on the logarithm of the loading rate. When the loading rate was varied over an impressive range of six orders of magnitude and the rupture force was plotted versus the logarithm of the loading rate, the curve appeared as a sequence of straight lines that could be related to the localization of barriers in the energy landscape (see below). This method dubbed DFS [64] provided a powerful way of analyzing ligand–receptor interactions. At this stage, bond rupture might be viewed as the serial passage of a series of barriers in a 1D energy landscape that could be analyzed with DFS [64]. Each barrier was crossed with a frequency that seemed to increase in presence of forces following Bell's law. Flow chambers and AFM or BFP appeared as consistent and complementary methods. Thus, while BFP gave accurate information on a notable part of energy landscapes, flow chambers operated at several wall shear rates allowed direct visualization of the random character of bond rupture (as illustrated by the distribution of bond lifetimes). The multiplicity of bound states, corresponding to the multiplicity of energy barriers, was an early finding [146]. Results obtained on a same molecular model such as homotypic cadherin association with a flow chamber [144] and BFP [145] appeared fairly consistent. A general finding was that flow chambers were better suited to probe weak interactions or to analyze the outer part of energy landscapes of strong interactions such as avidin/biotin association [154], while AFM and BFP provided more information on the inner part of these energy landscapes. While Bell's law was considered to account quite satisfactorily for many experimental models [32] as a convenient *zeroth-order phenomenological theory* [55], experimental results obtained with flow chambers [123], [185], and AFM [123] on a bacterial model of lectin-mediated adhesion and the P-selectin/PSGL-1 interaction strongly supported the view that the catch bonds fancied by Dembo and colleagues actually existed.

This was an incentive to reexamine the theoretical framework used to analyze experimental data. Further details will be provided in the following chapters.

1.3.2   MEASURING BOND FORMATION AT THE SINGLE MOLECULE LEVEL

As previously emphasized [149], studying bond formation between surface-attached molecules appeared more difficult than the aforementioned investigations on bond dissociation for several reasons: First, while it is relatively easy to compare the frequencies of bond rupture under 2D and 3D conditions since both are expressed in the same units (i.e., s^{-1}), rates of molecular association are respectively expressed in $M^{-1}s^{-1}$ and in *molecule*$^{-1}\mu m^2 s^{-1}$ under 3D and 2D conditions, respectively [57], [152]. Secondly, while it is relatively simple to exert a force on a bond until it breaks, studying bond formation requires to bring two molecules into close contact, wait for a given amount of time, then exert a force to determine whether molecules are bound. However, the choice or force or waiting time is quite arbitrary and many combinations must be tried. Thirdly, while bond strength is mainly dependent on the molecular properties of interaction sites but see [63], the properties of linkers between molecules and surfaces may play a dominant role in binding kinetics [150]. The earliest determination of biomolecule association rate at the single bond level was performed with AFM by Hinterdorfer [86] who studied the interaction between a mica surface coated with bovine serum albumin (BSA) and a cantilever tip coated with anti-BSA antibodies connected through a 8-nm-long polyethyleneglycol linker. The association rate was derived from the binding frequency, assuming free motion of the antibody site (paratope) in a half sphere. The association rate k_{on} was estimated at $5 \times 10^4 \, \text{M}^{-1}\text{s}^{-1}$, which was deemed comparable to values reported on several antibody/peptide couples. Soon thereafter, a laminar flow chamber was used to measure the binding frequency of beads and planar surfaces coated with fragments of C-cadherin, an homotypic adhesion molecule [149]. The binding frequency was estimated on the basis of computer simulations, yielding a quantitative estimate of the actual interaction time between beads and surfaces, as a consequence of vertical Brownian motion. The estimate of about $1.2 \times 10^{-3}\text{s}^{-1}$ for the binding frequency would yield an association rate of about $0.2 \, \text{M}^{-1}\text{s}^{-1}$ as estimated with a similar reasoning as that suggested by Hinterdorfer. This value is much lower than an estimate of cadherin association rate obtained with AFM [10]. On the same year, a clever way of estimating association rates was reported in Zhu's laboratory [35]. This consisted of generating numerous transient encounters between erythrocytes coated with immunoglobulin G (IgG) and transfected CHO cells expressing IgG receptors. This was achieved with two micropipettes and a piezoelectric system was used to vary the encounter time in the several second range. Binding events were revealed by transient deformation of softer erythrocytes, which allowed exquisitely sensitive count. Concomitant determination of the surface density of ligands and receptors allowed quantitative determination of the product between the conventional association rate k_{on} and contact area, which was estimated at $2.6 \times 10^{-7}\mu\text{m}^4\text{s}^{-1}$. The molecular contact area was estimated at a few percent of a macroscopic contact area of $3 \, \mu\text{m}^2$. More recently, Zhu et al. improved this method by using a more sensitive way of detecting attachments. Instead of looking for a deformation of the softer cell membrane, they recorded the thermal fluctuations of the BFP [33] [34]. This allowed them to measure bond formation and dissociation with better than 100-ms resolution. During the same period of time, flow chambers went on being used to study bond formation kinetics. Careful analysis of the relationship between contact duration between surfaces and binding probability led to the intriguing finding that the probability of bond formation was not proportional to encounter duration, but rather that a minimum contact time was required to allow binding [169]. This finding might cast a doubt on the suitability of the association rate parameter to account for different experimental models. Note that this conclusion does not mean that the use of an association rate is incorrect. Only, if a binding reaction is highly multiphasic, that is, if it involves numerous sequential reactions with a number of association rate parameters, it was found that association kinetics might be approximated with a simple law involving a single parameter. This point will be further discussed in the following section.

1.4   WHICH PARAMETERS DO WE NEED TO ACCOUNT FOR BIOMOLECULE RECOGNITION?

The examples provided in Section 1.2 show that a quantitative description of biomolecule recognition is required to understand how these biomolecules fulfill their function. The historical outline given in Section 1.3 shows that a new kind of knowledge is now available concerning biomolecule interaction. On the basis of this progress, it is now warranted to reexamine the suitability of older parameters used to investigate molecular interactions. It is important to notice that there is a certain degree of freedom in the choice of basic parameters. As an example, either forces or energies might be chosen as primitive parameters for developing theoretical mechanics. However, it is important to understand that an improper choice might lead to conceptual limitation and seriously hamper a quantitative interpretation of experimental data. This point may be illustrated with the following two examples.

(1) When the force-induced rupture of molecular bonds began being studied at the single molecule level, a natural parameter might be the rupture frequency k_{off}, as was naturally chosen when laminar flow chambers were used as experimental devices [5] [106] [147]. However, the unbinding force was chosen by investigators using AFM [10] [71] [86]. Theoretical [62] and experimental [129] advances were needed to show that unbinding forces were not intrinsic parameters and were strongly dependent on loading rates. This new understanding may be considered as the starting point for DFS [62] [129].

(2) A common theoretical procedure initiated by Eyring consisted of modeling bond formation and dissociation as consequences of time-dependent evolution of a single coordinate in a unidimensional energy landscape, which was interpreted as a valley in a multidimensional hypersurface [66]. Deeper analysis was needed to understand that the choice of a reaction coordinate is by no means straightforward since an improperly chosen coordinate cannot provide an "intrinsic" description of a system if it is dependent on the system history [15].

Thus, a good set of interaction parameters must satisfy the following criteria: (1) it must be sufficiently exhaustive to predict biomolecule behavior when numerical values of parameters are known, (2) it should be liable to experimental determination, (3) it must be sufficiently "intrinsic" to be independent of a particular experimental setup used for experimental study, and (4) it should be feasible to relate each parameter to molecular structure.

Commonly used interaction parameters (or coordinates) will now be rapidly considered.

1.4.1   THE AFFINITY CONSTANT

As previously acknowledged [195], the concept of affinity still dominated most thinking about complex biological reactions only two decades ago. Starting from the standard equation

$$A + B \rightleftarrows (AB) \; ; \; K_a = \frac{[AB]}{[A][B]} \tag{1.2}$$

where A and B are a ligand and a receptor molecule, $[A]$, $[B]$, and $[AB]$ are, respectively, the molar concentrations of isolated molecules A and B and of the molecular complex AB, and K_a is the affinity constant, we can, in principle, calculate the amount of complex if we know the total amounts of molecules A and B. Further, determining the affinity constant between soluble receptors and ligands may be easily achieved with powerful and widely available methods such as those based on optical biosensors [174] (some caution is however warranted [166]). Finally, the thermodynamic relationship

$$K_a = \exp(-\Delta G^0 / RT) \tag{1.3}$$

allows us to relate the affinity constant to the free enthalpy of reaction under standard conditions (see standard treatises or [20] for more details). However, there are two problems with this formalism.

Firstly, while Equation 1.2 is useful under equilibrium conditions, life works out of equilibrium. As an example, the affinity constant may conveniently account for the amount of occupied receptors on the cell membrane in a stable environment, but it is certainly insufficient to account for the evolution of rapid signaling cascades.

Secondly, while Equation 1.2 can be used to deal with two soluble reactants, or a cell receptor interacting with soluble ligands, it cannot account for interactions between surface-attached molecules. A major problem is related to the reaction entropy. As emphasized by Page and Jencks, the standard free enthalpy ΔG^0 is the sum of an "intrinsic term" that represents the intrinsic binding energy and a connecting term that represents the loss of entropy generated by complex formation [132] [139]. The problem is that both terms are of comparable order of magnitude and they may be quite different when interacting molecules are bound to surfaces, which may dramatically restrict their motion and number of degrees of freedom.

Reasoning with kinetic parameters instead of affinity constants may suffice to deal with out-of-equilibrium processes. As was emphasized, dealing with surface-attached molecules will result in the replacement of two *numbers*, the reaction on-rate and off-rate, with two *functions*, namely $k_{off}(F)$, that is, the dissociation rate as a function of applied force, and $k_{on}(d)$, that is, the association frequency of two molecules maintained at a fixed distance d [148]. The suitability of these functions will be considered below. Unfortunately, this additional complexity remains insufficient to deal with all situations of biological interest.

1.4.2 KINETIC CONSTANTS: THE ON-RATE AND THE OFF-RATE

The kinetic description of molecular interaction may seem more intuitive than the thermodynamic description. It makes use of two parameters as follows:

$$A + B \underset{k_{\text{off}}}{\overset{k_{\text{on}}}{\rightleftarrows}} AB \quad ; \quad d[AB]/dt = k_{\text{on}}[A][B] - k_{\text{off}}[AB] \tag{1.4}$$

Accounting for the kinetics of molecular interactions certainly contributed a major advance to the study of many biological phenomena. Thus, kinetics certainly plays a major role in determining the respective role of selectin and integrin adhesion receptors in leukocyte interaction with blood vessels. When a cell briefly encounters a foreign surface, only kinetic information can tell us whether contact will be durable enough to allow bond formation provided suitable receptors and ligands are expressed on surfaces. The remarkable treadmilling phenomenon [1] reported on cell cytoskeletal elements is understandable only on the basis of kinetic data. Actin microfilaments are oriented, and while the thermodynamics of monomer association/dissociation are similar on both ends, there is a 10-fold difference between kinetic constants. Also, signaling cascades generated by membrane receptors require the rapid formation of multimolecular scaffolds that are strongly influenced by interaction kinetics, as well as molecular localization. Finally, recent methodological advances such as the use of surface plasmon resonance technology allowed rapid increase of available data on the kinetics of a number of ligand–receptor couples [174] and experimental progress was an incentive to consider more thoroughly the significance of kinetic rates. For the sake of clarity, bond formation and dissociation will be considered separately. Bond dissociation will be first considered in view of its greater simplicity and historical order.

1.4.2.1 The Force-Dependent Dissociation Rates

Since the principles of bond dissociation with AFM and BFP and theoretical interpretations are described with much detail in Chapters 2 and 3, only some key points will be mentioned.

- First, dissociation rates are highly relevant to important experimental situations. As mentioned above, the outcome of interactions between a ligand and a receptor is certainly dependent on interaction lifetime. Prominent examples are (1) cell adhesion, since an essential factor of adhesion efficiency is the capacity of a single bond to maintain a cell in contact with an adhesive surface until a second bond occurred. This is the critical step to the formation of a firm adhesion that will be maintained by hundreds or thousands of bonds [17] [151], and (2) signaling, since in some cases exemplified by the TCR the duration of interaction will shape the cell response

[126]. It is therefore of obvious interest to determine dissociation rates. As illustrated by the many studies on leukocyte–endothelium interaction, the force dependence of interaction plays a dominant role in some situations.

- Second, as described above and in other chapters, some techniques allow experimental determination of $k_{off}(F)$, that is, the rupture frequency of a given bond in the presence of a disruptive force F. This may be achieved with a flow chamber that yields direct determination of $k_{off}(F)$ [157] or with AFM or BFP since theoretical models allow us to relate constant-force binding frequencies and unbinding forces measured at constant pulling speed [56] [75]. An important point is that the force-free dissociation rate and mechanical resistance may behave as different parameters: Thus, when the ligand CD34 of L-selectin was subjected to mild periodate oxidation, the force-free dissociation rate $k_{off}(0)$ was not substantially altered, in contrast with dissociation rates measured in presence of disruptive forces, as evidenced with a flow chamber [162]. This example supports the use of considering the force dependence of dissociation rates.

- Third, an important question is to know whether $k_{off}(F)$ may be viewed as an intrinsic property of a given ligand–receptor complex AB. While a positive answer might have appeared obvious a few years ago, two recent papers [124] [159] reflected the feeling that bond lifetime and dissociation rates were not intrinsic parameters since they depended on the *history* of studied complexes. This apparent paradox is indeed a consequence of a clear approximation in our language. It is only an approximation to refer to a complex AB since it is well known that AB may span a number of states that appear as local minima in a multidimensional energy landscape or even in a 1D reaction path [129] [146] [204]. Therefore, if the amount of time required to reach equilibrium is higher than the period of time between complex formation and dissociation rate determination, measured parameters will depend on the initial state of the molecular complex and on the time allowed for equilibration between different substates before beginning measurements. An additional point is that the dissociation probability of a molecular complex subjected to a time-dependent disruptive force is dependent on the history of force application [124] [193] and possibly, as suggested by molecular dynamics simulation, on the precise location of atoms at the moment of force application [193].

There are other properties that hamper the universality of function $k_{off}(F)$. Firstly, dissociation may depend not only on the intensity of a disruptive force but also on its direction [8] [202]. This may be important if free rotation is not allowed between binding molecules and surfaces. Secondly, dissociation is not only dependent on the properties of binding sites but also on linker molecules connecting these sites to surfaces [63] [193].

Fourth, several authors developed theoretical models to relate dissociation frequencies under constant load or loading rates to the location and depth of energy landscapes. The next step would be to relate these geometric and energy parameters to structural properties of binding molecules. This point will be rapidly considered in the next section.

In conclusion, while it might appear for some years that Equation 1.1 provided a tractable way of describing the force dependence of molecular bonds [32], more recent work showed that (1) $k_{off}(F)$ was often more complicated than suggested by Bell's law due to the existence of multiple barriers and possibly multiple dissociation pathways, as suggested to interpret catch bond behavior [143], and (2) a function such as $k_{off}(F)$ may not exist, even with a more complicated form than Equation 1.1 due to the effect of history and dependence on the properties of linker molecules. These points will be discussed with more detail in a following chapter of this book.

1.4.2.2 Distance-Dependent Association Rates

The importance and significance of association rates (i.e., k_{on} parameter) will now be discussed.

First, there are many important examples supporting the prominent biological importance of association rates. As indicated above, the efficiency of selectin molecules was ascribed to their capacity to tether rapidly flowing leukocytes to endothelial cells, which required a particularly high association rate. Also, experimental data supported the view that the association rate of antibodies progressively increased during the so-called maturation of immune responses, a finding that was interpeted as *a premium on the capacity to bind target rapidly* [73]. Finally, the cell capacity to probe its environment is dependent on the capacity of membrane receptors to bind to their ligand during a transient approach of a receptor-bearing membrane protrusion toward a ligand-bearing surface. In all these case, it seems that the efficiency of bond formation should be calculable if we knew a function $k_{on}(d)$ defined as the frequency (per unit of time) of bond formation between a ligand and a receptor molecules located at distance d. Such a function would include sufficient information to account for interactions between soluble molecules (i.e., 3D conditions) and surface-attached molecules (i.e., 2D conditions). Unfortunately, the determination and even the very definition of such a function are fraught with difficulties for at least two complementary reasons.

(1) If the association between molecules A and B is a multiphasic reaction involving a high number of interaction states, the discrimination between free and bound states may be somewhat arbitrary. Indeed, if bond formation is not an all-or-none phenomenon but require a progressive strengthening, it is not obvious to chose a threshold to discriminate between free and bound states. Thus, while the streptavidin–biotin interaction might have been considered as strong enough to allow easy detection of bound states, several investigators reported on the time-dependent maturation of this interaction [159] and existence of a number of weak association states [153]. Indeed, if the number of intermediate states is high, the concept of association constant becomes meaningless. This point was recently demonstrated in a quantitative study made on the binding efficiency of antibody-coated microspheres encountering antigen-bearing surfaces in a laminar flow chamber [169]. The probability of bond formation scaled as a power of encounter duration that was

significantly higher than 1, and under a number of conditions this probability varied as $erfc[(t/t_0)^{1/2}]$, where t was the contact time and t_0 was a constant on the order of 10 ms. It was further shown that this formula could be derived from a simple model where the reaction landscape was modeled as an unidimensional curve with a rugged segment (Figure 1.6). It may be useful to emphasize that this problem could not have been detected in studies of 3D interactions since in this case the encounter time is determined by the laws of diffusion and is not expected to display substantial variations between different experimental setups.

(2) AFM and micropipette-based methods that met with impressive success in analyzing bond rupture may be less well suited to the study of bond formation because they do not allow easy control of contact duration since contact is usually difficult to detect. Also, the contact area is often difficult to estimate since it is difficult to observe [35] and it may be markedly altered by forces exerted by the apparatus to induce molecular contact [186]. Finally, while hundreds or thousands of approach/retraction cycles can be performed on a given contact area, it is more difficult to sample extensive areas, which may be useful if low ligand and receptor densities are used to ensure that binding events are representative of single bond formation and dissociation.

Further, when ligands and receptors are attached to surfaces, association rates are less "intrinsic" parameters than dissociation rates because bond formation is highly dependent on the properties of linker molecules [98]. Indeed, if molecules are rigid, association will be impossible if ligands and receptors are not suitably oriented to allow proper match between interacting areas. In contrast, association will be strongly enhanced if ligands and receptors are forced against each other with binding configuration. Also, the microtopology of surfaces bearing ligands and receptors

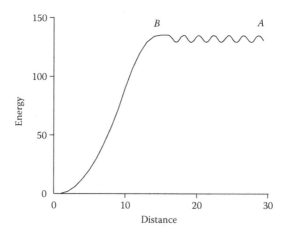

FIGURE 1.6 Model of bond formation. Bond formation is modeled as a passage through a rough segment of an energy landscape, represented as passing from A to B. This was found to match experimental findings obtained with a flow chamber [169].

may strongly influence association rates. As an example, the association rate between capsules bearing immunoglobulins and immunoglobulin receptors displayed 50-fold decrease when a smooth erythrocyte was replaced with a rough nucleated cell [196].

In conclusion, the new kind of information that was recently obtained on biomolecule interactions by studying single bond formation and dissociation in the presence of forces is directly relevant to a number of important biological processes. However, connecting this information to structural data still requires significant theoretical and experimental progress. In addition, accounting for biological processes still requires to consider other less well-defined parameters than k_{off} and k_{on}. Thus, we shall briefly discuss the frequently used concepts of *avidity* and *specificity*.

1.4.3   AVIDITY OF BIOMOLECULE INTERACTIONS: AN INCOMPLETELY DEFINED PARAMETER

While aforementioned development might convey the view that ligand–receptor interaction are liable to rigorous quantification, it has long been recognized that the affinity constant or association rates did not fully account for biological phenomena. Antigen recognition by antibodies provides a suitable example in view of the huge diversity of interactions and number of applications in hospital and research laboratory. As written in a standard treatise several decades ago [80] "In the literature, affinity and avidity commonly are used synonymously ... However, it is now accepted that the term affinity is a thermodynamic expression ... Avidity also involves other contributing factors such as antibody valence, antigen valence." A similar opinion remains in use today [131] "... The total binding strength of a molecule with more than one binding site is called the avidity." Thus, although it is accepted that avidity is not defined as accurately as affinity, a general concept is that this may be related to the capacity of forming multivalent associations. Indeed, many situations suggest that a most common way of forming strong associations involves the formation of multiple bonds. The following examples are intended to support the importance of the concept of avidity and the complexity associated to the multivalency of molecular interactions.

Many biological interactions need to be multivalent. There are many examples suggesting that a single noncovalent interaction between a ligand and a receptor may be too transient to be significant. Cell adhesion is driven by a number of membrane receptors that often require multivalent interactions. Cadherins, which are thought to play a dominant role in the stability of epithelia, are an important example. The importance of lateral clustering was very elegantly demonstrated [199] by studying the adhesion of cells expressing engineered cadherins, which could be oligomerized at will by bridging the intracellular domains with a drug. Similarly, integrins play a prominent role in cell adhesion to extracellular matrix components. It has long been known that cell surface integrins are often in an inactive state, and events including clustering or conformational changes are required to enable these integrins to bind their ligands. Some recent examples clearly demonstrated that clustering integrins could directly enhance the binding to multivalent, not monovalent ligands without any affinity change [26]. As another example, ICAM-1, a ligand of integrin LFA-1, was reported to bind to immobilized LFA-1 with high avidity (dissociation constant was 8 nM) after dimerization, while no measurable interaction was observed with monomeric ICAM-1 [130]. It was further checked that ICAM-1 monomer expressed a complete LFA-1 binding surface [102].

It is difficult to relate the properties of divalent and monovalent interactions. As emphasized above, it is because of this difficulty that single molecule studies revolutionized our understanding of biomolecule interactions. This difficulty is due to several reasons. Firstly, the rupture frequency of multivalent attachments may be drastically decreased by the possibility of rebinding events. Indeed, while a monovalent attachment is expected to break spontaneously as a consequence of thermal fluctuations, a multivalent attachment may need an external force for rupture if rebinding occurs [178]. Also, the force sensitivity of multivalent attachments is strongly dependent on force sharing between different bonds, and unbinding forces may follow a number of different laws depending on forces and bond arrangement [177] [182] [183].

In conclusion, while single molecule studies essentially provided an accurate description of the interaction between binding sites exposed by biomolecules, we need to better understand the requirement for multivalent association. Clearly, multivalency is dependent on the topographical relationship between different binding sites and molecular flexibility. When interactions involve surface-attached molecules, other additional factors are important, including static and dynamic length and flexibility of linkers between surfaces and binding sites, as well as rugosity of the surface region surrounding molecules, and lateral mobility of molecules. These points will be briefly listed in a later section.

1.4.4 SPECIFICITY OF BIOMOLECULE INTERACTIONS: AN ESSENTIAL PROPERTY THAT IS DIFFICULT TO DEFINE ACCURATELY

Obviously, biomolecules must bind specifically to adequate targets to fulfil their task. Specificity seems easy to define qualitatively. A ligand–receptor interaction is the more specific as the interaction between the same receptor and a "slightly" different ligand is "weaker." However, there is no general way of defining the similarity of two molecules or the strength of an interaction.

First, two molecules may differ according to their shape (e.g., as mentioned above for ortho- or para-dinitrophenol), their electric charge, their hydrogen bonding capacity, or their hydrophobicity. As indicated in the next section, all these properties are involved in biomolecule recognition, but their relative importance may be different in varying situations. The similarity (or dissimilarity) between two molecules is not an absolute quantitative concept. This arbitrariness was indeed pointed out many years ago [96].

Second, an interaction may be considered as "weaker" than another one if it occurs less often under physiological conditions. Thus, the affinity constant may be the dominant parameter if we are interested in the proportion of receptor molecules that are occupied by their ligand at equlilibrium, for example, the number of insulin receptors at a given moment. However, if we are interested in the detection of an immobilized ligand on a surface dynamically explored by a cell protrusion, the kinetic rate of bond formation may be more important. Finally, if we are interested in the specificity of cell tethering on a surface, the interaction strength may be the dominant parameter. Interestingly, these parameters are not necessarily correlated. Thus, when mutant streptavidin molecules were made to bind to biotin, the rupture forces were different, but they were correlated to the thermodynamic enthalpy rather than free energy of reaction [36], which is tightly related to the affinity constant, as recalled in Equation 1.2. Also, as mentioned above, the zero-force dissociation rate may not be correlated to the force resistance as represented with Bell's distance parameter [155] [162].

Third, the difficulty of defining specificity is further illustrated by the so-called *promiscuous* receptors that may bind specifically to a number of very different ligands, while a slight alteration of a given ligand may abolish the interaction. As an example, a monoclonal antibody was reported to bind specifically to 2,4 dinitrophenol, $K_d = 20$ nM, with a negligible affinity for the close analogs 2-nitrophenol and 2-nitro-4-iodophenol ($K_d > 100\,\mu$M), but which also bound unrelated compounds such as furazolidone with high affinity ($K_d = 1.2\,\mu$M) [95].

In conclusion, while it is recognized that both affinity and specificity are essential properties of ligand–receptor interactions [96] [194], the latter may remain difficult to define ambiguously. Specificity cannot be considered as an intrinsic parameter. A receptor may be considered as specific for its ligand if it does not interact with other molecules *that it is liable to encounter under biologically relevant situations*. The significance of interaction specificity will be discussed more precisely in a further section devoted to the structural basis of biomolecule interactions.

1.4.5   LIGAND–RECEPTOR INTERACTIONS ARE INFLUENCED BY PARAMETERS THAT ARE EXTRINSIC TO BOTH LIGAND AND RECEPTOR MOLECULES

In addition to the parameters we have just mentioned, it is important to recall that molecular associations occurring in the biological milieu may be deeply influenced by a number of external parameters that may obscure the intrinsic properties of interacting sites. Receptor-mediated cell adhesion provides many examples as shown below. We shall give selected examples to illustrate this point.

Presence of repellers on receptor-bearing surfaces. It is well known that the surfaces of living cells are coated with a carbohydrate-rich layer with a thickness of several tens of nanometers or more, called the *glycocalyx* or *pericellular matrix*. Much experimental evidence supports the view that (1) the glycocalyx may substantially impair the receptor capacity to bind to ligands, particularly during short encounters as occuring in a laminar flow chamber [142] [171], (2) under some circumstances, cells may increase their receptor capacity by rapid removal of glycocalyx components, thus increasing the accessibility of membrane receptors [172]. This inhibitory effect of the pericellular matrix is an example of the so-called steric repulsion [151].

Lateral mobility of binding molecules. It seems obvious that the probability of encounter between surface-attached binding molecules may be strongly enhanced if molecules can move freely on surfaces. This has long been demonstrated experimentally. As an example, when cells bearing CD2 surface molecules were micromanipulated into contact with surfaces coated with CD58, a ligand of CD2, either in immobilized form or freely diffusing in a supported lipid bilayer, adhesion efficiency was strongly increased when ligand molecules were mobile, and this effect was more apparent when the ligand density was decreased [29]. As another example, the adhesive efficiency of cell surface integrins was reported to increase in parallel with lateral mobility, as measured with enhanced video microscopy and single particle tracking [113].

Localization of binding molecules on surfaces. As previously indicated, surface roughness may strongly decrease the accessibility of surface receptors [196]. It is understandable that this phenomenon might depend on the localization of binding molecules as suggested by some experimental evidence. Thus, the capacity of selectin molecules to mediate binding of rapidly flowing leukocytes to the vessel walls was found to require the localization of these selectins on the tip of cell surface protrusions. Indeed, when this localization was prevented by changing the transmembrane domain of adhesion molecules, the dynamic binding capacity was abolished although binding sites were intact [27] [190].

Interactions between soluble biomolecules are also environment-sensitive. Recently, the kinetics of DNA hybridation was studied in living cells transfected with *fluorescence resonance energy transfer* (FRET)-labeled double-strand DNA [173]. Different kinetics were observed within cells and in the extracellular milieu, and differences were dependent on the length of strands. Further, the authors did not observe any direct effect of molecular crowding *in vitro*. Other authors concluded on the basis of experiments and computer simulation that the molecular crowding observed in the cell interior might change protein conformation [87].

In conclusion, the function of biomolecules involved in recognition events is dependent on a wide spectrum of parameters that are not all determined by the structure of binding sites or event of linker parts of binding molecules. It is certainly warranted to devote much attention to all these parameters in the forthcoming years.

1.5 RELATIONSHIP BETWEEN BIOMOLECULE STRUCTURE AND RECOGNITION EVENTS

As shown in the previous sections, the efficiency and selectivity of biomolecule interactions are dependent on a number of thermodynamic, kinetic, and mechanical parameters that can be determined experimentally with exquisite sensitivity, using a number of recently developed methodologies. These advances increase our need for a theoretical framework, allowing us to relate these quantitative binding parameters to structural properties. In addition to a mere intellectual appeal, such a framework would be useful (1) to help us integrate a daunting amount of available data, (2) to take advantage of increasingly available structural data to predict the interaction behavior of important molecules, and (3) to facilitate the rational design of molecules with desired interaction properties, for example, to act as drugs. In this section, three points related to this goal will be considered: (1) The main intermolecular forces responsible for biomolecule recognition will be rapidly listed. (2) We shall describe some experiments aimed at determining which forces are involved in the interactions between molecules of known structure. (3) Lastly, we shall rapidly discuss the interest of computer simulations as a practical way of increasing our understanding of biomolecule interactions. All these points were sketched in a previous review [20].

1.5.1   INTERMOLECULAR FORCES IN THE BIOLOGICAL MILIEU

An obvious prerequisite to relate biomolecule structure and recognition events is to know the basic physical forces involved in these interactions. As written more than half a century ago in a celebrated treatise on quantum chemistry [67], "In so far as quantum mechanics is correct, chemical questions are problems in applied mathematics." Unfortunately, the complexity of proteins makes it totally unfeasible to derive biomolecule behavior from basic equations and there is a need for some clear guidelines to help us build an intuitive view of molecular interactions. Thus, many authors looked for a classification of intermolecular forces and it must be understood that there has to be some arbitrariness is such an endeavor. Indeed, van Oss [187] found 17 so-called "primary interactions" after compiling a number of contemporary papers. Here, we shall only mention some characteristic features of four sets of forces, and we refer the reader to a standard treatise [92] or a review from our laboratory [20] for more details.

1.5.1.1   Electrostatic Forces

Electrostatic forces clearly constitute the basis for noncovalent interactions between biomolecules. In vacuum, the free energy of interaction between two charges q and q' separated by a distance r is simply

$$F = \frac{q \times q'}{4\pi\varepsilon_0 r} \tag{1.5}$$

where ε_0 is the dielectric constant of vacuum. Since we are interested in orders of magnitude, it is interesting to indicate that the interaction energy of two unit charges of 1.6×10^{-19} Coulomb separated by a distance r expressed in Angström is about $552\, k_B T/r$, where k_B is the Boltzmann's constant and T is the temperature. This means that the interaction energy of two unit charges in contact in vacuum would be more than 20-fold higher than any standard ligand–receptor interaction such as were described in the first section of this review.

However, in a material medium, the electric field generated by any charge results in a polarization and orientation of surrounding molecules, which generates a counter-field tending to decrease the total electric field. As a consequence, parameter ε_0 of Equation 1.5 must be multiplied by the relative dielectric constant ε_r, which is close to 78 in water. This high value is a consequence of two specific features of water: water molecules have a permanent dipole moment and the dipole moments are highly correlated due to hydrogen bonds. This results in a particularly efficient alignment along the surrounding electric fields [58]. Unfortunately, it is difficult to use Equation 1.5 together with the experimental value of ε_r to estimate the energy of electrostatic interactions in water when charges are at short distance since (1) water can no longer be approximated as a continuous medium, (2) water molecules may be expelled by steric effects, (3) water structure is expected to be altered near surfaces, and (4) the relative dielectric constant may decrease in presence of a very high electric field, a phenomenon known as dielectric saturation. The effective dielectric constant may thus be much lower than 78.

In addition to the charge screening by water molecules, the interaction between two fixed charges is decreased by surrounding ions. Indeed, according to Boltzmann's law, it is expected that free ions will get concentrated around ions of opposite charge. This phenomenon is well accounted for by Poisson–Boltzmann's equation:

$$\Delta V + \left[\left(\rho + \sum_i c_i q_i \exp(-q_i V / k_B T) \right) \right] / \varepsilon \tag{1.6}$$

where ρ is the volume density of charges other than soluble ions, c_i and q_i are the concentration and charge of ionic species i. A notable simplification is achieved if the terms $q_i V / k_B T$ are low enough to use a linearized form of this equation. The interaction energy between two charges q and q' at distance r may then be written as

$$F = \frac{q \, q' \, \exp(-\kappa r)}{4 \pi \varepsilon r} \tag{1.7}$$

In a 1:1 electrolyte solution such as physiological saline, parameter κ is equal to

$$\kappa = \left[\frac{2 c q^2}{k_B T \varepsilon} \right]^{1/2} \tag{1.8}$$

In a 0.15 M NaCl solution, the *Debye – Hückel* length $1/\kappa$ is about 8 Å at room temperature. Unfortunately, the linear approximation is not always fully valid in the biological milieu. The increase of computer power led to a revival of interest in the classical equations of electrostatic and numerical solution of Poisson–Boltzmann's equations allowed investigators to build maps of the electrostatic potential of protein surfaces, based on the surface distributions of charged amino acids such as glutamic acid, aspartic acid, lysine, or arginine. This allowed clear visualization of active sites, thus demonstrating the importance of electrostatic interactions in biomolecule recognition [88]. The effective interaction energy between two opposite charges such as a COO^- and a NH_3^+ group in a protein–protein interface was estimated at about 1.6 $k_B T$ in an experimental study made on the high affinity interaction between thrombin and hirudin, which involves four electrostatic bonds [181].

1.5.1.2 The Hydrogen Bond

In addition to charged ionic groups born by acidic amino acids, such as aspartic or glutamic acid, or basic amino acids such as lysine or arginine the surface of any protein bears local charges resulting from the differential distribution of atomic nuclei and electronic charges. These charges are expected to generate attractive or repulsive interactions between approaching protein surfaces. A prominent example is provided by hydrogen bonds. As a consequence of the. high electronegativity of atoms such as oxygen or nitrogen, the electronic cloud involved in a bond such as O-H or N-H is asymmetric, resulting in a net negative charge on oxygen or nitrogen and a net positive charge on the proton H. As a consequence, an atom with a net negative charge such as oxygen, will be attracted by H atoms bearing a net positive charge. While this simple mechanism may be considered as the basis of the hydrogen bond, it must be emphasized that this is only an approximation, and a quantum mechanical approach is required to achieve a more accurate description of this interaction. A notable consequence is that the hydrogen bond is dependent on the orientation of interaction molecules, not only on the distance between positive and negative sites, and there is still an interest in quantitative modeling of this interaction [37]. Hydrogen bonds such as O-H-O play a major role in protein or nucleic acid organization. Also, they are thought to account for the highly particular structure of water [58]. The H_2O molecule may be viewed as a tetrahedron with two positive and two negative charges on the four vertices, which may allow extensive clustering of these molecules. This extensive hydrogen bonding capacity of water is responsible for its high boiling point and dielectric constant.

As a consequence, protein–protein association often results in the replacement of protein–solvent hydrogen bonds with protein–protein hydrogen bonds, which makes it difficult to predict the total contribution of hydrogen bonds to biomolecule interactions. While the energy of a typical hydrogen bond may be on the order of about 8 k_BT, the contribution of an hydrogen bond to a protein–protein interaction may not be higher than k_BT [43].

1.5.1.3 Different Timescale: Electrodynamic Interactions

In addition to the aforementioned rather static view of electrical forces between fixed charges in a biological environment, other interactions are dependent on different timescales and deserve a separate treatment. The basis is the presence of a dipole of moment $\vec{p}$ that may be permanent or induced by the presence of a surrounding electric field $\vec{E}$ according to Equation 1.9

$$\vec{p} = \alpha \, \vec{E} \tag{1.9}$$

where α is called the *polarizability*. Further, the interaction energy between a dipole and an electric field is simply the scalar product $-\vec{p}.\vec{E}$, and the electric field generated at point $\vec{r}$ (starting from the dipole) is

$$\vec{E} = (1/4\pi\varepsilon)\vec{grad}(\vec{p}.\vec{r}/r^3) \tag{1.10}$$

These equations are the basis of two interactions:

The so-called *Keesom* interaction term represents the interaction between two freely rotating dipoles p_1 and p_2. Following Boltzmann's law, the relative orientations leading to a decreased free energy are favored. After integrating over all orientations and weighting with Boltzmann's factor, one obtains

$$F_K = -(p_1^2 p_2^2 / 24\pi^2 \varepsilon^2 k_B T)/r^6 \qquad (1.11)$$

For two water molecules, the numerator is about $4300\, k_B T$ when r is expressed in Angström. The timescale of molecular rotations ranges between picoseconds and nanoseconds depending on molecule size. The effective dielectric constant is expected to decrease according to rotation velocity.

The interaction between a dipole and polarizable molecule (see Equation 1.9) is called *Debye* interaction. While this is dependent on temperature, there is a nonzero high temperature limit since the net interaction between a dipole and a polarizable molecule is attractive whatever the dipole orientation. This limit is proportional to r^{-6}, similar to Keesom interaction.

Finally, it is known that even in the absence of permanent dipole moments, two polarizable molecules located at distance r exert a mutual attraction. This is called *dispersion* or *London* force since the first calculation of the dispersion interaction between hydrogen atoms was calculated by London (1930). This is proportional to r^{-6}. The numerical coefficient for two water molecules is about $740\, k_B T$, when r is expressed in Angström.

Thus, there is some theoretical support for the concept that two freely interacting molecular groups will exert a mutual attraction with an energy proportional to r^{-6}. This is often denominated as van der Waals attraction. Further, it is well known that in addition a short-distance repulsion will prevent a collapse of molecule pairs. This repulsion was sometimes called *Born repulsion*. It displays very rapid variation with distance, which led to represent it empirically either as a step function (this is the simplest "hard wall" model), or as a r^{-12} function, leading to the empirical 6–12 or *Lennard Jones* potential:

$$F_{LJ} = -4\varepsilon[(\sigma/r^6) - (\sigma/r^{12})] \qquad (1.12)$$

1.5.1.4 Using the Formalism of Surface Physical–Chemistry: Hydrophobic Bonds

The sharp energy distance relationship illustrated by Equation 1.12 is an incentive to view molecular interactions as contact forces occurring at the interface between rigid bodies. Clearly, this is not a rigorous model. A major problem is that molecular interactions are not fully additive [122] [198]. However, this view is simple enough to be felt useful, at least as a first approximation leading us to describe biomolecule interactions as a set of contact interactions between contacting groups. This concept provides a convenient way of accounting for the so-called hydrophobic bond. The formalism of surface chemistry provides a convenient framework to discuss this point. As already indicated, the free energy of interaction between two molecules numbered 1 and 2 embedded in a medium 3 results from a balance between bond formation and bond rupture as follows:

$$F_{12}^{(3)} = F_{12} - F_{13} - F_{23} + F_{33} \qquad (1.13)$$

since the formation of an interface between surfaces 1 and 2 (Equation 1.12) results in the destruction of interfaces Equations 1.13 and 1.23 represent the region of biomolecules surfaces that will be involved in interaction and release of solvent molecules that will exert a mutual attraction. Now, a convenient classification consists of discriminating between dispersion forces and polar interactions generated by charged groups, permanent dipoles and hydrogen bonds. Considering water, the surface tension is about 72 mJ/m^2, as compared to about 20 mJ/m^2 for many apolar substances. In a first approximation, it may be assumed that the difference of 52 mJ/m^2 represents polar interactions and that apolar components do not display important variations in different amino acids. According to Equation 1.13, the interaction between two apolar bodies will therefore amount to about 52 mJ/m^2. This interaction is the so-called hydrophobic interaction. On the basis of the very crude estimate mentioned above, this is expected to be on the order of $0.13\,k_BT$ per $Å^2$.

A practical way of applying this framework to biomolecule interactions is based on the concept of *accessible surface area* [116], which is defined as the area of the surface spanned by the center of a solvent molecule remaining in contact with the surface of a given molecule. A water molecule is usually modeled as a sphere of 1.4 Å radius (Figure 1.7). On the basis of this definition and of known free energies of transfer of amino acids from an organic solvent to water, Chothia estimated at about $0.04\,k_BT$ per $Å^2$ the free energy required to expose hydrophobic residues on protein surfaces [38].

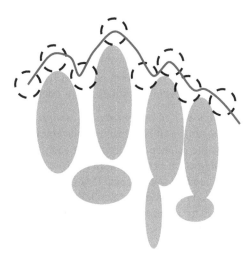

FIGURE 1.7 (See color insert.) Accessible area. The accessible surface (red line) may be defined as the surface spanned by the center of a sphere representing a water molecule (broken contour) moving in contact with atoms constituting the protein and modeled as hard spheres with a known van der Walls radius (dark blue areas).

Other authors looked for more detailed estimates of the transfer free energy of individual chemical groups from an apolar to an aqueous environment to obtain a more accurate estimate of the energy of protein folding. The basic equation was

$$\Delta F = \sum_i g_i A_i \qquad (1.14)$$

Where the summation extends over all groups i of the solute, and A_i is the conformation-dependent accessible surface area. The hydration free energy of chemical groups such as CH_3, OH, or C=O was estimated at about 0.03, 0.29, and 0.72 $k_B T / \mathring{A}^2$, respectively [138].

In conclusion, in this very brief review, we summarized some basic concepts that are used as a crude framework to describe intermolecular forces. This will be the starting point for a discussion of the basic mechanisms of ligand–receptor association.

1.5.2   GENERAL PROPERTIES OF LIGAND-RECEPTOR ASSOCIATION

While it would certainly be quite naive to look for universal properties of biomolecule interactions, it is useful to describe selected examples to help the reader get a reasonably realistic feeling for involved mechanisms. As indicated above, we shall essentially consider proteins.

1.5.2.1   A Static View of Ligand–Receptor Complexes

The contact area between two interacting proteins may be defined as the region that was accessible in isolated molecules and that is no longer accessible in complexes (assuming that the interaction did not trigger major morphological changes, which is usually the case). In many cases involving membrane receptors or antibodies, molecular association may result in a loss of accessible area comprised between about 800 and 1600 $\mathring{A}^2$ [39] [44] [101]. The interaction may involve between 10 and 30 side chains from each protein [42]. Several questions are of interest:

Do all residues located in contact areas contribute a similar part to binding affinity? The answer is probably negative, and extensive replacement of individual residues was used to try and assess the relative importance of each interaction. In a pioneering study made on the interaction between human growth hormone and its receptor [42] [47], it was concluded that more than 75% of the total binding energy was accounted by a central hydrophobic region essentially involving two trp residues: the replacement of each of these trp reduced the binding free energy by more than 7 $k_B T$ [42]. It was thus concluded that a few *hot spots* contributed most of the interaction energy. More recently, when a recombinant TCR was made to interact with a series of nonapeptides bound by a same MHC molecule, the replacement of a single leucine with a valine was sufficient to increase the binding free energy by 2 $k_B T$ and provoke a 3-fold decrease of the dissociation rate, and this difference resulted in nearly 10-fold change of T lymphocyte activation potency [4]. In a patient with a bleeding disease, a replacement of a lysine with a valine resulted in 2.2-fold decrease of the dissociation rate between von Willebrand factor and *GPIbα* ligand under flow [114]. In another study, the replacement of an alanine with a valine in a TCR chain resulted in a binding free energy change of about 3 k_BT [198]. The authors emphasized that the affinity changes displayed cooperative rather than additive behavior when they compared the effect of separate or concomitant changes of amino acids at several positions [198]. An important point is that in some cases binding properties were changed following an alteration of an amino acid residue located out of the contact area as identified on the basis of crystallographic studies.

Is a particular type of interaction favored in biologically relevant ligand–receptor couples? As recently emphasized [22], the role of electrostatics on binding affinities remains controversial, but a study of nearly 300 protein complexes led the authors to conclude that in most cases electrostatic forces did not strongly contribute to binding affinity. Indeed, there is a balance between desolvation energies required for molecular contact and attraction energies if there is a match between charges born by opposite surfaces. There are, however, some examples where electrostatic charges contribute significantly to binding efficiency. Electrostatic interactions contributed by four glutamic acid residues was estimated to account for 32% of the binding energy in the thrombin–hirudin couple [181]. Also, electrostatic interactions might be useful to give a proper orientation to approaching ligands and receptors, a phenomenon denominated as *electrostatic steering* [109]. More recently, alanine scanning was used to study the influence of individual aminoacids on the binding to a talin oligopeptide of a 25-mer sequence of the membrane-proximal tail of several integrins. A general finding was that alanine replacement of acidic groups increased binding efficiency, while alanine replacement of basic groups slightly reduced this interaction [78], thus supporting the influence of charges in some cases of molecular interactions.

Further, there is much evidence supporting the view that *hydrophobic bonds* are responsible for an important part of binding affinities [39] [44] [50]. In a compilation of 75 ligand–receptor complexes of known atomic structure, the fraction of apolar atoms was about 56%. Hydrophobic bonds might thus contribute on the order of several tens of k_BT to the binding free energy, based on the aforementioned estimates of contact areas and solvation free energies.

The average surface density of *hydrogen bonds* was estimated as one per 170 $\mathring{A}^2$ in the series of 75 complexes mentioned above [44]. A recent study of the effect of a number of mutations on the interaction between a triacylated lipopeptide and a pattern recognition receptor led the authors to emphasize the importance of a network of hydrogen bonds in binding [104].

In conclusion, despite a number of studies based on a growing number of 3D structures, recognition interfaces between ligands and receptors did not display universal specific features, and it seems accepted that binding is a variable blend of polar and apolar interactions that will result in sufficient binding energy provided there is a proper match in the shape and charge distribution of interacting surfaces.

1.5.2.2 Dynamics of Ligand–Receptor Interaction

A static view of the structure of biomolecule interfaces is not sufficient to provide a mechanistic understanding of molecular association. During the last decades, many investigators attempted to chose between two general concepts.

According to the *lock and key* hypothesis, proteins behave as rigid bodies and recognition events require that two regions of the surfaces of interacting molecules be complementary enough to allow the formation of a closely packed interface. The main consequences of this hypothesis would be as follows: (1) The 3D structures of free and bound receptors should be similar. This prediction can be tested more and more readily since structural databases contain a growing number of complex structures involving molecules that have already been studied in isolation. (2) Thus, the major requirement for bond formation would be that biomolecules encounter each other with a convenient orientation. The loss of translational entropy associated with bond formation would be rather limited and might be compensated for by the entropy increase associated to the formation of hydrophobic bonds.

The *induced fit* hypothesis suggests that biomolecule association should involve an adaptation of at least one of the interacting molecules to match the structure of the opposing surface. In this case, bond formation would be expected to involve a number of intermediate states.

The *conformation selection* hypothesis is only quantitatively different from the previous one: It is suggested that any protein will display a number of different conformations. Association between a receptor and a ligand would require that molecular encounter occurred between complementary conformations. In this case, a high proportion of encounters could not be conducive to association, and the rate of bond formation would be significantly lower.

It must be emphasized that the difference between the induced fit and conformation selection are only quantitative. In the latter case, the energies of different conformations are sufficiently similar to make each conformation detectable in the absence of association. In the former case, the free energy of the bound conformation is too high to be detectable in the absence of stabilization. During the last decades, a number of investigations disclosed the complexity of protein behavior. While crystallographic studies conveyed the feeling that many proteins displayed a fairly rigid conformation, with a precise localization of atoms at the Å scale, dynamic studies showed that proteins were in fact dynamic objects with an enormous number of states and substates [6]. Thus, the discrimination between aforementioned views may seem somewhat naive. However, this is often a basis for interpreting many experimental data. This point will now be clarified by some examples.

T cell receptor association with pMHC ligand has already been mentioned in the first section of this review as an important model for biomolecule recognition and it was felt that structural studies should illuminate the recognition process. Early studies suggested that the TCR exhibited extremely poor shape complementarity with the peptide, as exemplified by the interaction between 2C TCR and $H - 2K^b - dEV8$ ligand [77]. In a later thermodynamic study, it was indeed suggested that the induced fit of TCR to a peptide could allow the required discriminatory power [21]. A few years later, a structural study made on another TCR (2B4) led the investigators to the conclusion that the initial TCR/pMHC interaction essentially involved an interaction between the TCR and MHC and that a later phase involved an interaction with the TCR and the peptide bound to the MHC [197]. The authors emphasized that different MHCs could display different behavior. However, a few years later, the authors of a review based on 24 TCR/pMHC complexes acknowledged that crystal structures did not allow to define a common mechanism for the interaction and ongoing signaling phenomena [170]. Finally, kinetic studies of pMHC/TCR association with surface plasmon resonance led to the conclusion that binding was a multiphasic reaction, suggesting either induced fit or conformation selection [76]. The need for TCR to recognize a variety of peptides might require a particular role for molecule flexibility. However, there is also experimental evidence for the importance of molecular flexibility in less-variable models of biomolecular recognition. Thus, the interaction between IgE, the antibodies with a major role in allergic reactions, and IgE receptors was reported to involve alterations of receptor structure, as revealed with NMR spectra and release of bound fluorescent probes [161]. Other NMR-based studies yielded similar conclusions. Calmodulin is a cytoplasmic molecule that plays a major role in mediating functional consequences of intracellular calcium changes by interacting with a number of targets. It was reported that (1) ligand binding resulted in marked changes of calmodulin conformational dynamics, and (2) apparent changes of conformational entropy were linearly related to the change of overall binding entropy, showing that binding-associated changes of protein conformation significantly contributed to binding free energy [74]. In another study made on the interaction between cytokine interleukin-18 (IL-18) and specific antibodies, crystallographic studies revealed important displacements, up to $10\,\text{Å}$, of some specific residues [7]. Another direct proof of the importance of receptor flexibility was obtained with the surface forces apparatus, a device allowing to quantify the distance between decorated mica surfaces with $1\,\text{Å}$ resolution. This was used to study the interaction between DC-SIGN, a pattern recognition receptor, and oligosaccharide ligands. When surfaces bearing receptors and ligands were approached and retracted, a 2.4 nm alteration of receptor length was observed [128]. Note that in this case geometric changes might involve binding sites or protein domains linking them to the surface.

In contrast to the TCR/pMHC interaction, the recognition of a number of ligands by immunoreceptor NKG2D was claimed to involve only rigid adaptation, as suggested by higher binding rate and a more favorable binding entropy [125]. This supports the previous report that association involved only limited changes of conformation in 75 complexes studied with crystallography [44].

In conclusion, there is a growing evidence that proteins are highly flexible objects, and increasingly sensitive analytical methods are expected to reveal more and more structural or dynamical changes associated to bond formation. The relative role of these changes in binding free energy may, however, display wide variations when different biological systems are studied.

1.5.2.3   Mechanisms Influencing the Specificity of Biomolecule Interaction

As indicated above, specificity is an essential property of biomolecule association. The information described in the preceding section will be a basis to describe more precisely the meaning of molecular specificity. Indeed, there are several potential mechanisms that might account for the capacity of a given receptor to bind different ligands. We shall list them, then some informative examples will be given to support the significance of these mechanisms.

Possible mechanisms for cross reactivity.   (1) Different molecules may share a common binding region. This may be a consequence of a common origin since it is known that proteins share a limited number of basic folds. Alternatively, this may be a consequence of some kind of convergent evolution [50]. Finally, cross reactivity may be due to a random similarity of unrelated molecules. As an example, Kabat reported a similarity of 3D models of purin-6-oylglycine and 2'-deoxyadenylic acid, which might provide an explanation for observed cross reactions between antisera to purin-6-oyl-bovine serum albumin and DNA [103]. (2) In a given contact region between a receptor and different ligands, different hot spots might account for the binding of different molecules. (3) A biomolecule might display sufficient flexibility to bind to different molecules. (4) A given receptor might be endowed with a number of unrelated binding sites specific for different molecules. So-called adapters are but an example.

Examples of promiscuous receptors.   A crystallographic study revealed that a monoclonal antibody (CB4-1) could bind five unrelated peptides on the same binding region (paratope), but these peptides formed different contacts [110].

Other authors reported on a single antibody that could bind several unrelated antigens. They performed both X-ray crystallography and kinetic studies and provided strong proof of the occurrence of conformational diversity and induced fit [95]. The hypothesis that antibody promiscuity might be due to high flexibility is also reported by a comparison between two mannopyranoside antibodies, 1H7 and 2D10. The antibody with a narrow specificity, 1H7, displayed much lower flexibility than promiscuous antibody 2D10 [112].

In another model, the use of spectroscopic methods such as circular dichroism and fluorescence measurements revealed that hyaluronan-binding protein 1, an oligomeric protein with a capacity to bind different ligands such as C1q and hyaluronan, displayed marked structural changes when the ionic strength was altered concomitantly with a change of ligand specificity [97]. The importance of conformal diversity was also advocated in a study made on peptide recognition by a chicken MHC molecule [108].

The involvement of shared structures in some cases of recognition multiplicity is supported by a study of the $\alpha X \beta 2$ integrin, a leukocyte receptor with a capacity to detect danger signals. Recognition seemed to involve carboxylic groups that seemed to appear more frequently on damaged proteins [191]. Potential mechanisms of receptor promiscuity are displayed on Figure 1.8.

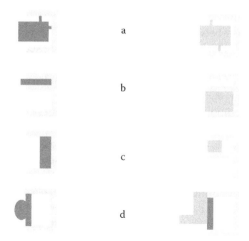

FIGURE 1.8 **(See color insert.)** Possible mechanisms of cross reactivity. Several different mechanisms were shown to result in receptor promiscuity. (a) Different ligands may form different elementary bonds in a same binding site [110]. (b) A molecule may display several unrelated binding sites on its surface. (c) A binding site may be flexible and may accommodate different ligands [95]. (d) Two unrelated molecules may display some local similarity [103].

In conclusion, while specificity is a hallmark of biomolecule interactions, promiscuity is also a common finding, as testified by the multiplicity of interactions found in databases.

In addition, the data presented in this brief review illustrate the diversity of situations and the complexity of the binding mechanisms that involve multiple conformational changes and may depend on structural details of molecular surfaces at the nanometer scale. Therefore, there is clearly a need for a more accurate description of protein behavior to interpret more and more refined experimental data. A tentative way of approaching this goal consists of using computer simulation to try and correlate subtle behavioral patterns to structural details. This point will be considered in the following section.

1.5.3 INFORMATION YIELDED BY COMPUTER SIMULATION

A conclusion of the results described in previous sections is that (1) experimental dissection of bond formation and dissociation at the single molecule level brought an accurate description of many models of biomolecule recognition as highly multiphasic processes with complex force dependence, (2) experimental study of isolated protein molecules or molecular complexes with techniques such as X-ray crystallography or NMR yielded a wealth of structural data with nearly Å accuracy, (3) currently available knowledge of intermolecular forces yielded a relatively intuitive understanding of the relationship between protein structure and binding behavior. However, this remains too approximative to allow accurate prediction of

outcomes in well-defined situations, and (4) current theoretical work aims at finding more detailed links between structure and function. However, it is well known
that "... In the real world, exact solutions are the notable exception" [164]. Further,
analytic solutions may be too complicated to be of real use.

During the past decades, the increasing availability and power of computers triggered the development of simulation as a general tool for trying to bring insights into
models that are too complex to allow simple mathematical description. The basic
idea consists of observing the behavior of a model system containing enough ingre-
dients to mimic a realistic situation. We shall now give a few examples to illustrate
the kind of information that can be obtained with this approach. However, a word of
caution as found in the introduction of a recent treatise of molecular dynamics simulation may be useful: "... Simulations must be kept honest, because seing is believing,
and animated displays can be very convincing irrespective of their veracity" [164].
We shall sequentially consider two approaches of growing complexity.

Simulations as numerical solutions of simple equations. Even fairly simple models cannot be described with plain mathematical formula. As an example, when
flow chambers are used to study the motion of receptor-coated microspheres near
ligand-bearing surfaces, a quantitative determination of the rate of bond formation
requires a knowledge of the frequency and duration of molecular encounters. The
distance between spheres and surfaces must therefore be known with millisecond
and nanometer accuracy, which is difficult to achieve with present day techniques.
This was an incentive to mimic the motion of microspheres near surfaces in the presence of hydrodynamic shear and account for Brownian motion and hydrodynamic
interactions between the sphere and the wall [149] [152]. This approach was necessary to obtain quantitative information on the kinetic rate of bond formation [149].
The validity of simulations was subjected to theoretical [152] and experimental [168]
check. In a later study, numerical simulations were found useful to demonstrate that
the bond formation between a ligand and a receptor might be viewed as the progres-
sion of the complex along a linear segment of a rugged energy landscape rather than
the passage across an energy barrier, as was usually considered [169].

Simulating complex systems: molecular dynamics and related approaches. The basic
idea of molecular dynamics was reported by Alder and Wainwright [2] in the late
50s. They modeled a system made of hard spheres in a box, starting with a random configuration defined as positions and velocities. Then, they applied the law of
mechanics to follow the displacements of all spheres through a series of ultrashort
steps, and the new positions and velocities of spheres were recorded at each step.
Fourteen years later, the motion of several tens of model water molecules interacting
with more realistic potentials was similarly simulated [163]. The continual growth
of computer power allowed investigators to study models of increasing complexity,
until they were able to simulate interactions between proteins molecules, including
tens of thousands of atoms, and surrounded by hundreds of water molecules. The
most detailed approach consisted of modeling interactions between pairs of atoms
or chemical groups with semiempirical potential functions derived from quantum

chemical considerations and experimental data. Using femtoseconds time steps, this method currently allows the observation of objects as complex as proteins during less than a microsecond, which is still insufficient to account for molecular interactions. Calculations are performed with arrays of computers, using a few standard software packages such as CHARMM [23], Gromos [40], Rosetta [48], or the freely available Xplor-NIH package [176]. In the following sections, we shall first discuss rapidly some limitations of the method and tricks suggested to progress. Then, we shall describe a few selected examples to illustrate the kind of information that can be obtained. More information can be found in aforementioned papers and standard treatises [164].

1.5.3.1 Limitations and Technical Advances

As shown above, there is a need to extend the range of accessible domains to take real advantage of computer simulations. Here are some noticeable points.

Long distance cut-off of interactions. Since most interatomic forces exhibit a rapid decay with distance, it seemed reasonable to neglect interactions at distance higher than some arbitrary cut-off. A useful point is that artefacts may be avoided by using, for example, a smoothened cut-off [23].

Poisson-Boltzmann statistics. Due to the complexity of solvent effect, it was tempting to account for water dielectric constant by considering individual molecules. Unfortunately, this is highly time-consuming. It was suggested to use numerical solution of Poisson–Boltzmann equation to calculate the potential on the surface of proteins [88].

Protein flexibility. A major difficulty is the formidable number of configurations available for proteins. Initial docking software thus treated proteins as rigid objects [68] [179], which greatly increased calculation speed, but it is also an important limitation, as discussed above. This was an incentive to elaborate clever research algorithm to sample a sufficient variety of conformations within a reasonable amount of time. Examples are the use of the so-called "genetic search algorithms," which allowed successful simulation of association of couples such as methotrexate and dihydrofolate reductase, or galactose and L-arabinose-binding protein [100].

Coarse-grained potential. A major limitation of molecular dynamics is that the passage across significant energy barriers may require an excessive amount of time. This difficulty may be overcome by smoothing barriers to allow a more rapid exploration of more extensive configurations. In a later step, a more refined potential may be used [133].

Brownian dynamics and hybrid simulation. Another way to overcome the time limitation of molecular simulations consists of combining deterministic equations accounted for ultrashort step with possibly random equations accounting for a high number of steps. For example, the displacement of a water molecule during a femtosecond interval would be a few thousandths of $\mathring{A}$ if the molecule velocity is on the order of a few hundreds of meter/s, corresponding to $(k_B T/m)^{1/2}$, where m is the molecule mass. It may be more rapid to generate random diffusive displacements according to the laws of diffusion [137]. In a later study, to simulate the dissociation of a streptavidin–biotin bond with an AFM, the cantilever motion was described with equations from classical mechanics, while the molecular deformations and motion were simulated with standard molecular dynamic approach. This approach was rightly denominated as hybrid simulation [200].

Reliability check. A major problem with any powerful method is that it is difficult to check its reliability when it yields data that cannot be obtained with any alternative approach. This difficulty is of major importance with computer simulation due to the number of required assumptions. A marked progress consisted of organizing systematic checks of predictions by asking investigators to study selected model systems the structure of which was currently studied. It was thus possible to compare the predictions yielded by different softwares to the experimental data that were available soon thereafter. This approach is the basis of CASP (critical assessment of structure prediction) and CAPRI (critical assessment of predicted interactions) [70] [175].

1.5.3.2   New Information on Molecular Association May Be Provided by Computer Simulations

Simulation of force-driven bond rupture. Soon after the report of experimental rupture of individual avidin–biotin bonds with an AFM [71] [117], computer simulations were performed to simulate unbinding over periods of up to 500 ps [83] [93]. Interestingly, when the BFP allowed a tentative dissection of the unbinding path [129], the authors could compare results from simulation [93] to the tentative energy landscape they were able to build on the basis of their experimental data (see Fig. 4 of [129]). As another example, Shulten and Leckband investigated the force-induced rupture of the CD2-CD58 complex [11]. This association might play a role in the interaction between T lymphocytes and APC. The authors found that two different dissociation mechanisms might occur depending on the loading rate. At a high loading rate of 35 or 70 pN/ps, the proteins began unfolding before being separated. As a 10-fold lower loading rate, the proteins separated before unfolding. Further, the authors concluded that salt bridges were the primary determinants of tensile strength.

A few years later, computer simulations [84] [120] were used to investigate the behavior of so-called *catch bonds*. These bonds were predicted to display a paradoxical force sensitivity with lifetime increase in presence of a disruptive force [52], and they were evidenced with flow chambers [123] [185] and AFM [123]. Soon thereafter, a theoretical report showed that this behavior was consistent with a two-pathway model of bond rupture [143]. The following year, a report based on crystallographic studies and computer simulation suggested a relationship between

L-selectin behavior and the existence of two conformations, with a possible effect of forces on transition between these two states [120]. In a later paper [121], the authors suggested a more precise mechanism.

Dissecting the kinetics of molecular interactions. The interaction between acetyl-cholinesterase and tetramethylammonium (TMA), a molecule resembling the bulkiest part of acetylcholine, was studied with molecular dynamics. The authors concluded that local conformational fluctuations were required to allow the ligand passage [25]. More recently, molecular dynamics was used to investigate the mechanisms of interactions between tyrosine kinases and a monoclonal antibody [3]. The authors concluded that binding preferences were determined by conformational selection.

Predicting molecular interactions. A major goal of molecular simulation is to predict molecular interactions. As already mentioned, molecular simulations have been used for years to try and predict molecular associations [49] [100] in addition to other knowledge-based approaches [41] [107]. As indicated above, it would be most useful to be able to predict molecular interactions and the present state of the art is reviewed in CAPRI rounds [70]. Successes go on being reported in present day literature [49] [81].

1.6   CONCLUSION

Biomolecule recognition is a process of outstanding biological importance that has been studied for decades with standard physical–chemical methods and concepts, based on the determination of affinity constants and kinetic association and dissociation rates. Results were interpreted within the framework of thermodynamics, statistical physics, and physical chemistry as elaborated nearly a century ago. More recently, the experimental dissection of molecular interactions at the single bond level, increasing availability of structural data with nearly Angström resolution, and development of simulation methods with a power matching protein complexity made it conceivable to describe molecular interactions with much improved accuracy, with a deeper understanding of the force sensitivity and transition between a multitude of substates. In parallel, the progress of our understanding of all the complexity of biological systems is an incentive to use this new kind of knowledge to achieve a better understanding of cell function. It is hoped that the development of DFS will be a substantial factor of progress along this line.

REFERENCES

1. Alberts, B., A. Johnson, J. Lewis, M. Raff, K. Roberts, and P. Walter. 2008. *Molecular Biology of the Cell* 5th edition, New York: Garland.
2. Alder, B. J., and T. E. Wainwright. 1959. Studies in molecular dynamics. I. General methods. *J. Chem. Phys.* 31:459–466.

3. Aleksandrov, A., and T. Simonson. 2010. Molecular dynamics simulations show that conformational selection governs the binding preferences of Imatinib for several tyrosine kinases. *J. Biol. Chem.* 285:13807–13815.

4. Aleksic M., O. Dushek, H. Zhang, E. Shenderov, J-L. Chen, V. Cerundolo, D. Coombs, and P. A. van der Merwe. 2010. Dependence of T cell antigen recognition on T cell receptor-peptide MHC confinement time. *Immunity* 32:163–174.

5. Alon, R., D. A. Hammer, and T. A. Springer. 1995. Lifetime of P-selectin-carbohydrate bond and its response to tensile force in hydrodynamic flow. *Nature* 374:539–542. (Correction: *Nature* 377:86, 1995).

6. Ansari, A., J. Berendzen, S. F. Bowne, H. Frauenfelder, I. E. T. Iben, T. B. Sauke, E. Shyamsunder, and R. T. Young. 1985. Protein states and proteinquakes. *Proc. Natl. Acad. Sci. USA* 82:5000–5004.

7. Argiriadi, M. A., T. Xiang, C. Wu, T. Ghayur, and D. W. Borhani. 2009. Unusual water-mediated antigenic recognition of the proinflammatory cytokine Interleukin-18. *J. Biol. Chem.* 284:24478–24489.

8. Astrof, N. S., A. Salas, M. Shimaoka, J. F. Chen, and T. A. Springer. 2006. Importance of force linkage in mechanochemistry of adhesion receptors. *Biochemistry* 45:15020–15028.

9. Badis, G., M. F. Berger, A. Anthony, A. A. Philippakis, S. Talukder, A. R. Gehrke, S. A. Jaeger, E. T. Chan, G. Metzler, A. Vedenko, X. Chen, H. Kuznetsov, C-F. Wang, D. Coburn, D. E. Newburger, Q. Morris, T. R. Hughes, and M. L. Bulyk. 2009. Diversity and complexity in DNA recognition by transcription factors. *Science* 324:1720–1723.

10. Baumgartner, W., P. Hinterdorfer, W. Ness, A. Raab A, D. Vestweber, H. Schindler, and D. Drenckhahn. 2000. Cadherin interaction probed by atomic force microscopy. *Proc. Natl. Acad. Sci. USA* 97:4005–4010.

11. Bayas, M. V., K. Schulten, and D. Leckband. 2003. Forced detachment of the CD2-CD58 complex. *Biophys. J.* 84:2223–2233.

12. Bell, G. I. 1978. Models for the specific adhesion of cells to cells. *Science* 200:618–627.

13. Bell, G. I., M. Dembo, and P. Bongrand. 1984. Cell adhesion: competition between nonspecific repulsion and specific bonding. *Biophys. J.* 45:1051–1064.

14. Bentley, D., and A. Toroian-Raymond. 1986. Disoriented pathfinding by pioneer neurone growth cones deprived of filopodia by cytochalasin treatment. *Nature* 323:712–715.

15. Best, R. B., and G. Hummer. 2005. Reaction coordinates and rates from transition paths. *Proc. Natl. Acad. Sci. USA* 102:6732–6737.

16. Blow, N. 2009. Untangling the protein web. *Nature.* 460:415–418.

17. Bongrand, P., and G. I. Bell. 1984. Cell-cell adhesion: Parameters and possible mechanism. In *Cell Surface Dynamics. Concepts and Models.* eds A. S. Perelson, C. DeLisi, and F. W. Wiegel. 459–493. New York: Marcel Dekker.

18. Bongrand, P., P. M. Claesson, and A. S. G. Curtis (eds). 1994. *Studying cell adhesion.* Berlin: Springer-Verlag.

19. Bongrand, P., and B. Malissen. 1998. Quantitative aspects of T-cell recognition: From within the antigen-presenting cell to within the T cell. *Bioessays* 20:412–422.

20. Bongrand, P. 1999. Ligand-receptor interactions. *Rep. Prog. Phys.* 62:921–968.

21. Boniface, J. J., Z. Reich, D. S. Lyon, and M. M. Davies. 1999. Thermodynamics of T cell receptor binding to peptide-MHC: Evidence for a general mechanism of molecular scanning. *Proc. Natl. Acad. Sci. USA* 96:1446–11451.

22. Brock, K., K. Talley, K. Coley, P. Kundrotas, and E. Alexov. 2007. Optimization of electrostatic interactions in protein-protein complexes. *Biophys. J.* 93:3340–3352.

23. Brooks, B. R., C. L. Brooks III, A. D. MacKerell Jr., L. Nilsson, R. J. Petrella, B. Roux, Y. Won, G. Archontis, C. Bartels, S. Boresch, A. Caflisch, L. Caves, Q. Cui, A. R. Dinnr, M. Feig, S. Fishcher, J. Gao, M. Hodoscek, W. Im, K. Kuczera, T. Lazaridis, J. Ma, V. Ovchinnikiv, E. Paci, R. W. Pastor, C. B. Post, J. Z. Pu, M. Schaefer, B. Tidor, R. M. Venable, H. L. Woodcock, X. Wu, W. Yang, D. M. York, and M. Karplus. 2009. CHARMM: The biomolecular simulation program. *J. Comput Chem.* 30:1545–1614.

24. Brunk, D. K., D. J. Goetz, and D. A. Hammer. 1996. Sialyl Lewisx/E-selectin-mediated rolling in a cell free system. *Biophys. J.* 71:2902–2907.

25. Bui, J. M., R. H. Henchman, and J. A. McCammon. 2003. The dynamics of ligand barrier crossing inside the acetylcholinesterase gorge. *Biophys. J.* 85:2267–2272.

26. Bunch, T. A. 2010. Integrin alpha-IIb beta-3 activation in chinese hamster ovary cells and platelets increases clustering rather than affinity. *J. Biol. Chem.* 285:1841–1849.

27. Buscher, K., S. B. Riese, M. Shakibaei, C. Reich, J. Dernedde, R. Tauber, and K. Ley. 2010. The transmembrane domains of L-selectin and CD44 regulate receptor cell surface positioning and leukocyte adhesion under flow. *J. Biol. Chem.* 285:13490–13497.

28. Capo, C., P. Bongrand, A. M. Benoliel, and R. Depieds. 1978. Dependence of phagocytosis on strength of phagocyte-particle interaction. *Immunology* 35:177–182.

29. Chan, P. Y., M. B. Lawrence, M. L. Dustin, L. M. Ferguson, D. E. Golan, and T. A. Springer. 1991. Influence of receptor lateral mobility on adhesion strengthening between membranes containing LFA-3 and CD2. *J. Cell Biol.* 115:245–255.

30. Chaurasia, G., S. Malhotra, J. Russ, S. Schnoegl, C. Hnig C, E. E. Wanker, and M. E. Futschik. 2009. UniHI 4: New tools for query, analysis and visualization of the human protein-protein interactome. *Nuc. Acids Research.* 37:D657-D660. Epub 2008.

31. Chen C. S., M. Mrksich, S. Huang, G. M. Whitesides, and D. E. Ingber. 1997. Geometric control of cell life and death. *Science* 276:1425–1428.

32. Chen, S., and T. A. Springer. 2001. Selectin receptor-ligand bonds: Formation limited by shear rate and dissociation governed by the Bell model. *Proc. Natl. Acad. Sci. USA.* 98:950–955.

33. Chen, W., E. A. Evans, R. P. McEver, and C. Zhu. 2008. Monitoring receptor-ligand interactions between surfaces by thermal fluctuations. *Biophys. J.* 94:694–701.

34. Chen W, V. I. Zarnitzyna, K. K. Sarangapani, J. Huang, and C. Zhu. 2008a. Measuring receptor-ligand binding kinetics on cell surfaces: From adhesion frequency to thermal fluctuation methods. *Cell. Mol. Bioeng.* 1:276–288.

35. Chesla, S. E., P. Selvaraj, and C. Zhu. 1998. Measuring two-dimensional receptor-ligand binding kinetics by micropipette. *Biophys. J.* 75:1553–1572.

36. Chilkoti, A., T. Boland, B. D. Ratner, and P. S. Stayton. 1995. The relationship between ligand binding thermodynamics and protein-ligand interaction forces measured by atomic force microscopy. *Biophys. J.* 69:2125–2130.

37. Choi, H., H. Kang, and H. Park. 2010. New angle-dependent potential energy function for backbone-backbone hydrogen bond in protein-protein interactions. *J. Comput. Chem.* 31:897–903.

38. Chothia, C. 1974. Hydrophobic bonding and accessible surface area in proteins. *Nature* 248:338–339.

39. Chothia, C., and J. Janin. 1975. Principles of protein-protein recognition. *Nature* 256:705–708.

40. Christen, M., P. H. Hnenberger, D. Bakowies, R. Baron, R. Brgi, D. P. Geerke, T. N. Heinz, M. A. Kastenholz, V. Krutler, C. Oostenbrink, C. Peter, D. Trzesniak. 2005. The GROMOS software for biomolecular simulation: GROMOS05. *J. Comput. Chem.* 26:1719–1751.

41. Chuang, G-Y., D. Kozakov, R. Brenke, S. R. Comeau, and S. Vajda. 2008. DARS (Decoys As the Reference State) potentials for protein-protein docking. *Biophys. J.* 95:4217–4227.

42. Clackson T., and J. A. Wells. 1995. A hotspot of binding energy in a hormone-receptor interface. *Science* 267:383–386.

43. Connelly, P. R., R. A. Aldape, F. J. Bruzzese, S. P. Chambers, M. J. Fitzgibbon, M. A. Fleming, S. Itoh, D. J. Livingston, M. A. Navia, J. A. Thomson, and K. P. Wilson. 1994. Enthalpy of hydrogen bond formation in a protein-ligand binding reaction. *Proc. Natl. Acad. Sci. USA* 91:1964–1968.

44. Conte, L. L., C. Chothia, and J. Janin. 1999. The atomic structure of protein-protein recognition sites. *J. Mol. Biol.* 285:2177–2198.

45. Creighton, T. E. 1983. *Proteins – structure and molecular properties*, 1st edition, New York: Freeman.

46. Cretel, E., D. Touchard, A. M. Benoliel, P. Bongrand, and A. Pierres. 2010. *J. Phys. Condensed matter* 22:194107.

47. Cunningham, B. C., and J. A. Wells. 1989. High-resolution epitope mapping of hGH-receptor interactions by alanine scanning mutagenesis. *Science* 244:1081–1085.

48. Das, R., and D. Baker. 2008. Macromolecular modeling with Rosetta. *Ann. Rev. Biochem.* 77:363–382.

49. Das, R., I. Andr, Y. Shen, Y. Wu, A. Lemak, S. Bansal, C. H. Arrowsmith, T. Szyperski, and D. Baker. 2009. Simultaneous prediction of protein folding and docking at high resolution. *Proc. Natl. Acad. Sci. USA.* 106:18978–18983.

50. DeLano, W. L., M. H. Ultsch, A. M. de Vos, and J. A. Wells. 2000. Convergent solutions to binding at a protein-protein interface. *Science* 287:1279–1283.

51. Del Rio, A., R. Perez-Jimenez, R. Liu, P. Roca-Cusachs, J. M. Fernandez, and M. P. Sheets. 2009. Stretching single talin rod molecules activates vinculing binding. *Science* 323:638–641.

52. Dembo, M., D. C. Torney, K. Saxman, and D. Hammer. 1988. The reaction-limited kinetics of membrane-to-surface adhesion and detachment. *Proc. Roy. Soc. Lond. B* 234:55–83.

53. Diz, R., S. K. McCray, and S. H. Clarke. 2008. B cell receptor affinity and B cell subset identity integrage to define the effectiveness, affinity threshold, and mechanism of anergy. *J. Immunol.* 181:3834–3840.

54. Dobereiner, H. G., B J. Dubin-Thaler, J. M. Hofman, H. S. Xenias, T. N. Sims, G. Giannone, M. Dustin, C. H. Wiggins, and M. P. Sheetz. 2006. Lateral membrane waves constitute a universal dynamic pattern of motile cells. *Phys. Rev. Let.* 97:038102.

55. Dudko, O. K., G. Hummer, and A. Szabo. 2006. Intrinsic rates and activation free energies from single-molecule pulling experiments. *Phys. Rev. Letters* 96:108101.

56. Dukdo, O.K., G. Hummer, and A. Szabo. 2008. Theory, analysis and interpretation of single-molecule force spectroscopy experiments. *Proc. Natl. Acad. Sci. USA* 105:15755–15760.

57. Dustin, M. L., S. K. Bromley, M. M. Davis, and C. Zhu. 2001. Identification of self through two-dimensional chemistry and synapses. *Ann. Rev. Cell Dev. Biol.* 17:133–157.

58. Eisenberg, D., and W. Kauzmann. 1969. *The Structure and Properties of Water.* Oxford: Oxford University Press.

59. Engler, A. J., S. Sen, H. L. Sweeney, and D. E. Discher. 2006. Matric elasticity directs stem cell lineage specification. *Cell* 126:677–689.

60. Evans, E., D. Berk, and A. Leung. 1991. Detachment of agglutinin-bonded red blood cells. I - Forces to rupture molecular-point attachments. *Biophys. J.* 59:838–848.

61. Evans, E., R. Merkel, K. Ritchie, S. Tha, A. Zilker. 1994. Picoforce method to probe submicroscopic actions in biomembrane adhesion. In *Studying Cell Adhesion*, eds P. Bongrand, P. M. Claesson, and A. S. G. Curtis. 125–139. Berlin: Springer-Verlag.

62. Evans, E., and K. Ritchie. 1997. Dynamic strength of molecular adhesion bonds. *Biophys. J.* 72:1541–1555.

63. Evans, E., and K. Ritchie. 1999. Strength of a weak bond connecting flexible polymer chains. *Biophys. J.* 76:2439–2447.

64. Evans, E. 2001. Probing the relation between force - lifetime - and chemistry in single molecular bonds. *Annu. Rev. Biophys. Biomol. Struct.* 30:105–128.

65. Evans E., A. Leung, D. Hammer, and S. Simon. 2001. Chemically distinct transition states govern rapid dissociation of single L-selectin bonds under force. *Proc. Natl. Acad. Sci. USA.* 98:3784–3789.

66. Eyring, H. 1935. The activated complex in chemical reactions. *J. Chem. Phys.* 3:107–115.

67. Eyring, H., J. Walter, and G. E. Kimball. 1944. *Quantum chemistry.* New York: Wiley.

68. Fahmy, A., and G. Wagner. 2002. TreeDock: A tool for protein docking based on minimizing van der Waals energies. *J. Am. chem. Soc.* 124:1241–1250.

69. Faix, J., and K. Rottner. 2006. The making of filopodia. *Curr. Opin. Cell Biol.* 18:18–25.

70. Fernandez-Recio, J., and M. J. E. Sternberg. 2010. The 4th meeting on the Critical Assessment of PRedicted Interaction (CAPRI) held a the Mare Nostrum, Barcelona. *Proteins.* 78:3065–3066.

71. Florin, E. L., V. T. Moy, and H. E. Gaub. 1994. Adhesion forces between individual ligand-receptor pairs. *Science* 264:415–417.

72. Folkman, J., and A. Moscona. 1978. Role of cell shape in growth control. *Nature* 273:345–349.

73. Foote, J., and C. Milstein. 1991. Kinetic maturation of an immune response. *Nature* 352:530–532.

74. Frederick, K. K., M. S. Marlow, K. G. Valentine, and A. J. Wand. 2007. Conformational entropy in molecular recognition by proteins. *Nature* 448:325–330.

75. Freund, L. B. 2009. Characterizing the resistance generated by a molecular bond as it is forcibly separated. *Proc. Natl. Acad. Sci. USA* 106:8818–8823.

76. Gakamsky, D. M., I. F. Luescher, and I. Pecht. 2004. T cell receptor-ligand interactions: a conformational preequilibrium or an induced fit. *Proc. Natl. Acad. Sci. USA* 101:9063–9066.

77. Garcia, K. C., M. Degano, L. R. Pease, M. Huang, P. A. Peterson, L. Teyton, and I. A. Wilson. 1998. Structural basis of plasticity in T cell receptor recognition of a self-peptide-MHC antigen. *Science* 279:1166–1172.

78. Gingras, A. R., W. H. Ziegler, A. A. Bobkov, M. G. Joyce, D. Fasci, M. Himmel, S. Rothemund, A. Ritter, J. G. Grossmann, B. Patel, N. Bate, B. T. Goult, J. Emsley, I. L. Barsukov, G. C. K. Roberts, R. C. Liddington, M. H. Ginsberg, and D. R. Critchley. 2009. Structural determinants of integrin binding to the talin rod. *J. Biol. Chem.* 284:8866–8876.

79. Glinsky, V. V., G. V. Glinsky, O. V. Glinskii, V. H. Huxley, R. Turk JR, V. V. Mossine, S. L. Deutscher, K. J. Pienta, and T. P. Quinn. 2003. Intravascular metastatic cancer cell homotypic aggregation at the sites of primary attachment to the endothelium. *Cancer Res.* 63:3805–3811.

80. Glynn, E., and M. V. Steward (eds). 1977. *Immunochemistry: An Advanced Textbook.* 2nd edition. New york: Wiley. 240.

81. Goley, E. D., A. Rammohan, E. A. Znameroski, E. N. Firat-Karalar, D. Sept, and M. D. Welch. 2010. An actin-filament-binding interface on the Arp2/3 complex is critical for nucleation and branch stability. *Proc. Natl. Acad. Sci.* USA 107:8159–8164.

82. Greaves, D. R., and S. Gordon. 2009. The macrophage scavenger receptor at 30 years of age: Current knowledge and future challenges. *J. Lipid. Res.* 50:S282–S286.

83. Grubmuller, H., B. Heymann, and P. Tavan. 1996. Ligand binding: Molecular mechanics calculation of the streptavidin-biotin rupture force. *Science* 271:997–999.

84. Gunnerson, K. N., Y. V. Pereverzev, O. V. Prezhdo. 2009. Atomistic simulation combined with analytic theory to study the response of the P-selectin/PSGL-1 complex to an external force. *J. Phys. Chem. B* 113:2090–2100.

85. Hanggi, P., P. Talkner, and M. Borkovec. 1990. Reaction-rate theory: Fifty years after Kramers. *Rev. Mod. Phys.* 62:251–341.

86. Hinterdorfer, P., W. Baumgartner, H. J. Gruber, K. Schilcher, and H. Schindler. 1996. Detection and localization of individual antibody-antigen recognition events by atomic force microscopy. *Proc. Natl. Acad. Sci. USA* 93:3477–3481.

87. Homouz, D., M. Perham, A. Samiotakis, M. S. Cheung, and P. Wittung-Stafshede. 2008. Crowded, cell-like environement induces shape changes in aspherical protein. *Proc. Natl. Acad. Sci. USA* 105:11754–11759.

88. Honig, B., and A. Nicholls. 1995. Classical electrostatics in biology and chemistry. *Science* 268:1144–1149.

89. Huang, J., V. Zarnitsyna, B. Liu, L. J. Edwards, N. Jiang, B. D. Evavold BD, and C. Zhu. 2010. The kinetics of two-dimensional TCR and pMHC interactions determine T-cell responsiveness. *Nature* 464:932–936.

90. Huppa, J. B., M. Axmann, M. A. Mortelmaier, B. F. Lillemeier, E. W. Newell, M. Brameshuber, L. O. Klein, G. J. Schutz, and M. M. Davis. 2010. TCR-peptide-MHC interactions in situ show accelerated kinetics and increased affinity. *Nature* 463:963–970.

91. Imler, J-L., and J. A. Hoffmann. 2001. Toll receptors in innate immunity. *Trends Cell Biol.* 11:304–311.

92. Israelachvili, J. N. 1991. *Intermolecular and surface forces.* 2nd edition. New York: Academic Press.

93. Izrailev, S., S. Stepaniants, M. Balsera, Y. Oono, and K. Schulten. 1997. Molecular dynamics study of unbinding of the avidin-biotin complex. *Biophys. J.* 72:1568–1581.

94. Jaalouk, D. E., and J. Lammerding. 2009. Mechanotransduction gone awry. *Nature Rev. Mol. Cell Biol.* 10:63–73.

95. James, L. C., P. Roversi, and D. S. Tawfik. 2003. Antibody multispecificity mediated by conformational diversity. *Science* 299:1362–1367.

96. Janin, J. 1996. Quantifying biological specificity: The statistical mechanics of molecular recognition. *Proteins* 25:438–445.

97. Jay, P. Y., P. A. Pham, S. A. Wong, and E. L. Elson. 1995. A mechanical function of myosin II in cell motility. *J. Cell Sci.* 108:387–393.

98. Jeppesen, C., J. Y. Wong, T. L. Kuhl, J. N. Israelachvili, N. Mullah, S. Zalipsky, and C. M. Marques. 2001. Impact of polymer tether length on multiple ligand-receptor bond formation. *Science* 293:465–468.

99. Jha, B. K., D. M. Salunke, and K. Datta. 2003. Structural flexibility of multifunctinal HABP1 may be important for regulating its binding to different ligands. *J. Biol. Chem.* 278:27464–27472.

100. Jones, G., P. Willett, and R. C. Glen. 1995. Molecular recognition of receptor sites using a genetic algorithm with a description of desolvation. *J. Mol. Biol.* 245:43–53.

101. Jones, S., and J. M. Thornton. 1996. Principles of protein-protein interactions. *Proc. Natl. Acad. Sci. USA* 93:13–20.

102. Jun, C-D., M. Shimaoka, C. V. Carman, J. Takagi, and T. A. Springer. 2001. Dimerization and the effectiveness of ICAM-1 in mediating LFA-1-dependent adhesion. *Proc. Natl. Acad. Sci. USA* 98:6830–6835.

103. Kabat, E. A., S. M. Beiser, and S. W. Tanenbaum. 1966. Structural similarity as seen in 3-dimensional models between the N-[purin-6-oyl]-substituted amino acids and 2'-deoxyadenylic acid. *Cancer Res.* 26:459–460.

104. Kajava, A. V., and T. Vasselon. 2010. A network of hydrogen bonds on the surface of TLR2 controls ligand positioning and cell signaling. *J. Biol. Chem.* 285:6227–6234.

105. Kaplan, G., and G. Gaudernack. 1982. In vitro differentiation of human monocytes. Differences in monocyte phenotypes induced by cultivation on glass or on collagen. *J. Exp. Med.* 156:1101–1114.

106. Kaplanski G., C. Farnarier, O. Tissot, A. Pierres, A-M. Benoliel, M-C Alessi, S. Kaplanski, and P. Bongrand. 1993. Granulocyte-endothelium initial adhesion. Analysis of transient binding events mediated by E-selectin in a laminar shear flow. *Biophys. J.*, 64:1922–1933.

107. Keiser, M. J., V. Setola, J. J. Irwin, C. Laggner, A. I. Abbas, S. J. Hufensein, N. H. Jensen, M. B. Kuijer, R. C. Matos, T. B. Tran, R. Whaley, R. A. Glennon, J. Hert, K. L. H. Thomas, D. D. Edwards, B. K. Shoichet, and B. L. Roth. 2009. Predicting new molecular targets for known drugs. *Nature* 462:175–182.

108. Koch, M., S. Camp, T. Collen, D. Avila, J. Salomonsen, H-J Wallny, A. van Hateren, L. Hunt, J. P. Jacob, F. Johnston, D. A. Marston, I. Shaw, P. R. Dunbar, V. Cerundolo, E. Y. Jones, and J. Kaufman. 2007. Structures of an MHC class I molecule from B21 chickens illustrate promiscuous peptide binding. *Immunity* 27:885–899.

109. Kozack, R. E., M. J. d'Mello, and S. Subramanian. 1995. Computer modeling of electrostatic steering and orientational effects in antibody-antigen association. *Biophys. J.* 68:807–814.

110. Kramer, A., T. Keitel, K. Winkler, W. Stocklein, W. Hohne, and J. Schneider-Mergener. 1997. Molecular basis for the binding promiscuity of an anti-p24 (HIV-1) monoclonal antibody. *Cell* 91:799–809.

111. Kramers, H. A. 1940. Brownian motion in a field of force and the diffusion model of chemical reactions. *Physica* VII:284–304.

112. Krishnan, L., G. Sahni, K. J. Kaur, and D. M. Salunke. 2008. Role of antibody paratope conformational flexibility in the manifestation of molecular mimicry. *Biophys. J.* 94:1367–1376.

113. Kucik, D. F., M. L. Dustin, J. M. Miller, and E. J. Brown. 1996. Adhesion-activating phorbol ester increases the mobility of leukocyte integrin LFA-1 in cultured lymphocytes. *J. Clin. Invest.* 97:2139–2144.

114. Kumar, R. A., J. F. Dong, J. A. Thaggard, M. A. Cruz, J. A. Lopez, and L. V. McIntire. 2003. Kinetics of GPEbalpha-vVWF-A1 tether bond under flow: Effect of GPIbalpha mutations on the association and dissociation rates. *Biophys. J.* 85:4099–4109.

115. Lawrence, M. B., and T. A. Springer. 1991. Leukocytes roll on a selectin at physiologic flow rates: Distinction from and prerequisite for adhesion through integrins. *Cell* 65:859–873.

116. Lee, B., and F. M. Richards. 1971. The interpretation of protein structures: estimation of static accessibility. *J. Mol. Biol.* 55:379–400.

117. Lee, G. U., D. A. Kidwell, and R. J. Colton. 1994. Sensing discrete streptavidin-biotin interactions with atomic force microscopy. *Langmuir* 10:354–357.

118. Lo, C. M., H. B. Wang, M. Dembo, and Y. L. Wang. 2000. Cell movement is guided by the rigidity of the substrate. *Biophys. J.* 79:144–152.

119. Ma, Z., P. Janmey, and T. H. Finkel. 2008. The receptor deformation model of TCR triggering. *FASEB. J.* 22:1002–1008.

120. Lou, J., T. Yago, A. G. Klopocki, P. Mehta, W. Chen, V. I. Zarnitsyna, N. V. Bovin, C. Zhu, and R. P. McEver. 2006. Flow-enhanced adhesion regulated by a selectin interdomain hinge. *J. Cell Biol.* 174:1107–1117.

121. Lou, J., and C. Zhu. 2007. A structure-based sliding-rebinding mechanism for catch bonds. *Biophys. J.* 92:1471–1485.

122. Margenau, H., and N. R. Kestner. 1969. *Theory of Intermolecular Forces*. Oxford: Pergamon.

123. Marshall, B., M. Long, J. W. Piper, T. Yago, R. P. McEver, and C. Zhu. 2003. Direct observation of catch bonds involving cell-adhesion molecules. *Nature* 423:190–193.

124. Marshall, B. T., K. K. Sarangapani, J. Lou, R. P. McEver, and C. Zhu. 2005. Force history dependence of receptor-ligand dissociation. *Biophys. J.* 88:1458–1466.

125. McFarland, B. J., and R. K. Strong. 2003. Thermodynamic analysis of degenerate recognition by the NKG2D immunoreceptor: Not induced fit but rigid adaptation. *Immunity* 19:803–812.

126. McKeithan, T. W. 1995. Kinetic proofreading in T-cell receptor signal transduction. *Proc. Natl. Acad. Sci. USA* 92:5042–5046.

127. Mege, J. L., C. Capo, A. M. Benoliel, and P. Bongrand. 1987. Use of cell contour analysis to evaluate the affinity between macrophages and glutaraldehyde-treated erythrocytes. *Biophys. J.* 52:177–186.

128. Menon, S., K. Rosenberg, S. A. Graham, E. M. Ward, M. E. Taylor, K. Drickamer, and D. Leckband. 2009. Binding-site geometry and flexibility in DC-SIGN demonstrated with surface force measurements. *Proc. Natl. Acad. Sci. USA* 106:11524–11529.

129. Merkel R., P. Nassoy, A. Leung, K. Ritchie, and E. Evans. 1999. Energy landscapes of receptor-ligand bonds explored with dynamic force spectroscopy. *Nature* 397:50–53.

130. Miller, J., R. Knorr, M. Ferrone, R. Houdei, C. P. Carron, and M. L. Dustin. 1995. Intercellular adhesion molecule-1 dimerization and its consequences for adhesion mediated by lymphocyte function associated-1. *J. Exp. Med.* 182:1231–1241.

131. Murphy, K., P. Travers, and M. Walport. 2008. *Janeway's Immunobiology* 7th edition, New York: Garland Science.

132. Jencks, W. P. 1981. On the attribution and additivity of binding energies. *Proc. Natl. Acad. Sci. USA* 78:4046–4050.

133. Moritsugu, K., and J. C. Smith. 2007. Coarse-grained biomolecular simulation with REACH: realistic extension algorithm via covariance Hessian. *Biophys.* J. 93:3460–3469.

134. Neves, S. R., P. Tsokas, A. Sarkar, E. A. Grace, P. Rangamani, S. M. Taubenfeld, C. M. Alberini, J. C. Schaff, R. D. Blitzer, I. I. Moraru, and R. Iyengar. 2008. Cell shape and negative links in regulatory motivs together control spatial information flow in signaling networks. *Cell* 133:666–680.

135. Nishino, M., H. Tanaka, H. Ogura, Y. Inoue, T. Koh, K. Fujita and H. Sugimoto .2005. Serial changes in leukocyte deformability and whole blood rheology in patients with sepsis or trauma. *J. Trauma Injury Infection Critical Care* 59:1425–1431.

136. Nishizaka, T., H. Miyata, H. Yoshikawa, S'I. Ishiwata, and K. Kinosita. 1995. Unbinding force of a single motor molecule measured using optical tweezers. *Nature* 377:251–254.

137. Northrup, S. H., and H. P. Erickson. 1992. Kinetics of protein-protein association explained by brownian dynamics computer simulation. *Proc. Natl. Acad. Sci. USA* 89:3338–3342.

138. Ooi, T., M. Oobatake, G. Nemethy, and H. A. Scheraga. 1987. Accessible surface areas as a measure of the thermodynamic parameters of hydration of peptides. *Proc. Natl. Acad. Sci. USA* 84:3086–3090.

139. Page, M. I., and W. P. Jencks. 1971. Entropic contributions to rate accelerations in enzymic and intramolecular reactions and the chelacte effect. *Proc. Natl. Acad. Sci. USA* 68:1678–1683.

140. Palecek, S. P., J. C. Loftus, M. H. Ginsberg, D. A. Lauffenburger, and A. F. Horwitz. 1997. Integrin-ligand binding properties govern cell migration speed through cell-substratum adhesiveness. *Nature* 385:537–539.

141. Palecek, S. P., A. Huttenlocher, A. F. Horwitz, and D. A. Lauffenburger. 1998. Physical and biochemical regulation of integrin release during rear detachment of migratin cells. *J. Cell Sci.* 111:929–940.

142. Patel, K. D., M. U. Nollert, and R. P. McEver. 1995. P-selectin must extend a sufficient length from the plasma membrane to mediate rolling of neutrophils. *J. Cell Biol.* 131:1893–1902.

143. Pereverzev, Y. V., O. Prezhdo, W. E. Thomas, and E. V. Sokurenko 2005. Distinctive features of the biological catch bond in the jump-ramp force regime predicted by the two-pathway model. *Phys. Rev. E* 72:010903.

144. Perret, E. A. M. Benoliel, P. Nassoy, A. Pierres, V. Delmas, J. P. Thiéry, P. Bongrand, and H. Feracci, 2002. Fast dissociation kinetics of the recognition between individual E-cadherin fragments revealed by flow chamber analysis. *EMBO J.* 21:2537–2546.

145. Perret E., A. Leung, and E. Evans. 2004. Trans-bonded pairs of E-cadherin exhibit a remarkable hierarchy of mechanical strengths. *Proc. Natl. Acad. Sci. USA* 101:16472–16477.

146. Pierres, A., A. M. Benoliel, and P. Bongrand. 1995. Measuring the lifetime of bonds made between surface-linked molecules. *J. Biol. Chem.* 270:26586–26592.

147. Pierres A., A. M. Benoliel, P. Bongrand, and P. A. van der Merwe. 1996. Determination of the lifetime and force dependence of interactions of single bonds between surface-attached CD2 and CD48 adhesion molecules. *Proc. Natl. Acad. Sci. USA* 93:15114–15118.

148. Pierres, A., A. M. Benoliel, and P. Bongrand. 1996a. Measuring bonds between surface-associated molecules. *J. Immunological Methods* 196:105–120.

149. Pierres, A., H. Feracci, V. Delmas, A. M. Benoliel, J. P. Thiéry, and P. Bongrand. 1998. Experimental study of the interaction range and association rate of surface-attached cadherin 11. *Proc. Natl. Acad. Sci. USA* 95:9256–9261.

150. Pierres, A., A. M. Benoliel, and P. Bongrand. 1998a. Studying receptor-mediated cell adhesion at the single molecule level. *Cell Adhesion Commun.* 5:375–395.

151. Pierres, A., A. M. Benoliel and P. Bongrand P. 2000. Cell-cell interactions. In *Physical chemistry of biological interfaces*, eds. A. Baszkin, and W. Norde. 459–522, Amsterdam: Marcel Dekker Inc.

152. Pierres, A., A-M. Benoliel, C. Zhu, and P. Bongrand. 2001. Diffusion of microspheres in shear flow near a wall: Use to measure binding rates between attached molecules. *Biophys. J.* 81:25–42.

153. Pierres, A., A. M. Benoliel, and P. Bongrand. 2002. Cell fitting to adhesive surfaces: A prerequisite to firm attachment and subsequent events. *Eur Cell Materials* 3:31–45.

154. Pierres A, D. Touchard, A. M. Benoliel, and P. Bongrand. 2002. Dissecting streptavidin-biotin interaction with a laminar flow chamber. *Biophys. J.* 82:3214–3223.

155. Pierres, A. J. Vitte, A-M. Benoliel, and P. Bongrand. 2006. Dissecting individual ligand-receptor bonds with a laminar flow chamber. *Biophys. Rev. Letters* 1:231–257.

156. Pierres, A., A. M. Benoliel, D. Touchard, and P. Bongrand. 2008. How cells tiptoe on adhesive surfaces before sticking. *Biophys. J.* 94:4114–4122.

157. Pierres, A., A. M. Benoliel, and P. Bongrand. 2008. Studying molecular interactions at the single bond level with a laminar flow chamber. *Cell. Mol. Bioengineering* 1:247–262.

158. Pierres, A., V. Monnet-Corti, A-M. Benoliel, and P. Bongrand. 2009. Do membrane undulations help cells probe the world? *Trends Cell Biol.* 19:428–433.

159. Pincet, F., and J. Husson. 2005. The solution to the streptavidin-biotin paradox: The influence of history on the strength of single molecular bonds. *Biophys. J.* 89:4374–4381.

160. Puklin-Faucher, E., M. Gao, K. Schulten, and V. Vogel. 2006. How the headpiece hinge angle is opened: New insights into the dynamics of integrin activation. *J. Cell. Biol.* 175:349–360.

161. Price, N. E., N. C. Price, S. M. Kelly, and J. M. McDonnell. 2005. The key role of protein flexibility in modulating IgE interactions. *J. Biol. Chem.* 280:2324–2330.

162. Puri, K. D., S. Chen, and T. A. Springer. 1998. Modifying the mechanical property and shear threshold of L-selectin adhesion independently of equilibrium properties. *Nature* 392:930–933.

163. Rahman, A., and F. H. Stillinger. 1971. Molecular dynamics study of liquid water. *J. Chem. Phys.* 55:3336–3359.

164. Rapaport, D. C. 2004. *The Art of Molecular Dynamics Simulation.* 2nd edition, 549. Cambridge: Cambridge University Press.

165. Remmerbach, T. W., F. Wottawah, J. Dietrich, B. Lincoln, C. Wittekind, and J. Guck J. 2009. Oral cancer diagnosis by mechanical phenotyping. *Cancer Res.* 69:1728–1732.

166. Rich, R. L., and D. G. Myszka. 2006. Survey of the year 2005 commercial optical biosensor literature. *J. Mol. Recognit.* 19:478–534.

167. Rief, M., M. Gautel, F. Oesterhelt, J. M. Fernandez, and H. E. Gaub. 1997. Reversible unfolding of individual titin immunoglobulin domains by AFM. *Science* 276:1109–1112.

168. Robert, P., K. Sengupta, P. H. Puech, P. Bongrand, and L. Limozin. 2008. Tuning the formation and rupture of single ligand-receptor bonds by hyaluronan-induced repulsion *Biophys. J.* 95:3999–4012.

169. Robert, P., L. Limozin, A. Pierres, and P. Bongrand. 2009. Biomolecule association rates do not provide a complete description of bond formation. *Biophys. J.* 96:4642–4650.

170. Rudolph, M. G., R. L. Stanfield, and I. A. Wilson. 2006. How TCRs bind MHCs, peptides and coreceptors. *Annu. Rev. Immunol.* 24:419–466.

171. Sabri, S., A. Pierres, A. M. Benoliel, and P. Bongrand. 1995. Influence of surface charges on cell adhesion: Difference between static and dynamic conditions. *Biochem. Cell Biol.* 73:411–420.

172. Sabri, S., M. Soler, C. Foa, A. Pierres, A. M. Benoliel, and P. Bongrand. 2000. Glycocalyx modulation is a physiological means of regulating cell adhesion. *J. Cell Sci.* 113:1589–1600.

173. Schoen, I., H. Krammer, and D. Braun. 2009. Hybridization kinetics is different inside cells. *Proc. Natl. Acad. Sci* USA 106:21649–21654.

174. Schuck, P. 1997. Use of surface plasmon resonance to probe the equilibrium and dynamic aspects of interactions between biological macromolecules. *Ann. Rev. Biophys. Biomol. Structure* 26:541–566.

175. Schueler-Furman, O., C. Wang, P. Bradley, K. Misura, and D. Baker. 2005. Progress in modeling of protein structures and interactions. *Science* 310:638–642.

176. Schwieters, C. D., J. J. Kuszewski, N. Tjandra, and G. M. Clore. 2003. The Xplor-NIH NMR molecular structure determination package. *J. Magn. Res.* 160:65–73.

177. Seifert, U. 2000. Rupture of multiple parallel molecular bonds under dynamic loading. *Phys. Rev. Lett.* 84:2750–2753.

178. Seifert, U. 2001. Dynamic strength of adhesion molecules: Role of rebinding and self-consistent rates. *Europhys. Lett* 58:792–798.

179. Shoichet, B. K., R. M. Stroud, D. V. Santi, I. D. Kuntz, K. M. Perry. 1993. Structure-based discovery of inhibitors of thymidylate synthase. *Science* 259:1445–1450.

180. Smith-Garvin, J. E., G. A. Koretzky, and M. S. Jordan. 2009. T cell activation. *Ann. Rev. Immunol.* 27:591–619.

181. Stone, S. R., S. Dennis, and J. Hofsteenge. 1989. Quantitative evaluation of the contribution of ionic interactions to the formation of the thrombin-hirudin complex. *Biochemistry.* 28:6857–6863.

182. Sulchek, T. A., R. W. Friddle, K. Langry, E. Y. Lau, H. Albrecht, T. V. Ratto, S. J. DeNardo, M. E. Colvin, and A. Noy. 2005. Dynamic force spectroscopy of parallel individual mucin-1-antibody bonds. *Proc. Natl. Acad. Sci. USA* 102:16638–16643.

183. Tang, C-C., Y-P. Chu, and H-Y. Chen. 2007. Lifetime of ligand-receptor clusters under external force. *Phys. Rev. E* 76:061905.

184. Tha, S. P., J. Shuster, and H. L. Goldsmith. 1986. Interaction forces between red cells agglutinated by antibody. IV Time and force dependence of break-up *Biophys. J.* 50:1117–1126.

185. Thomas, W. E., E. Trintchina, M. Forero, V. Vogel, and E. Sokurenko. 2002. Bacterial adhesion to target cells enhanced by shear forces. *Cell* 109:913–923.

186. Thoumine, O., P. Kocian, A. Kottelat, and J. J. Meister. 2000. Short-term binding of fibroblasts to fibronectin: Optical tweezers experiments and probabilistic analysis. *Eur. Biophys. J.* 29:398–408.

187. van Oss, C. J. F. 1991. Interaction forces between biological and other polar entities in water: How many different primary forces are there? *J. Dispersion Sci. Technol.* 12:201–219.

188. Vavouri, T., J. I. Semple, R. Garcia-Verdugo, and B. Lehner. 2009. Intrinsic protein disorder and interaction promiscuity are widely associated with dosage sensitivity. *Cell* 138:198–208.

189. Von Andrian, U. H., J. D. Chambers, L. M. McEvoy, R. F. Bargatze, K. E. Arfors, and E. C. Butcher. 1991. Two-step model of leukocyte-endothelial cell interaction in inflammation: distinct roles for LECAM-1 and the leukocyte beta 2 integrins in vivo. *Proc. Natl. Acad. Sci. USA* 88:7538–7542.

190. Von Andrian, U. H., S. R. Hasslen, R. D. Nelson, S. L. Erlandsen, and E. C. Butcher. 1995. A central role for microvillous receptor presentation in leukocyte adhesion under flow. *Cell* 82:989–999.

191. Vorup-Jensen, T., C. V. Carman, M. Shimaoka, P. Schuck, J. Svitel, and T. A. Springer. 2005. Exposure of acidic residues as a danger signal for recognition of fibrinogen and other macromolecules by integrin alphaXbeta2. *Proc. Natl. Acad. Sci. USA* 102:1614–1619.

192. Wachsstock, D. H., W. H. Schwarz, and T. D. Pollard. 1994. Cross-linker dynamics determine the mechanical properties of actin gels. *Biophys. J.* 66:801–809.

193. Walton, E. B., S. Lee, and K. J. van Vliet. 2008. Extending Bell's model: How force transducer stiffness alters measured unbinding forces and kinetics of molecular complexes. *Biophys. J.* 94:2621–2630.

194. Wang, J., K. Zhang, H. Lu, and E. Wang. 2006. Quantifying the kinetic paths of flexible biomolecular recognition. *Biophys. J.* 91:866–872.

195. Williams, A. F. 1991. Out of equilibrium. *Nature* 352:473–474.
196. Williams, T. E., S. Nagarajan, P. Selvaraj, and C. Zhu. 2001. Quantifying the impact of membrane microtopology on effective two-dimensional affinity. *J. Biol. Chem.* 276:13283–13288.
197. Wu, L. C., D. S. Tuot, D. S. Lyons, K. C. Garcia, and M. M. Davis. 2002. Two-step binding mechanism for T-cell receptor recognition of peptide-MHC. *Nature* 418:552–556.
198. Yang, J., C. P. Swaminathan, Y. Huang, R. Guan, S. Cho, M. C. Kieke, D. M. Cranz, R. A. Mariuzza, and E. J. Sundberg. 2003. Dissecting cooperative and additive binding energetics in the affinity maturation pathway of a protein-protein interface. *J. Biol. Chem.* 278:50412–50421.
199. Yap, A. S., W. M. Brieher, M. Pruschy, and B. M. Gumbiner. 1997. Lateral clustering of the adhesive ectodomain: a fundamental determinant of cadherin function. *Current Biol.* 7:308–315.
200. Zhou, J., L. Zhang, Y. Leng, H-K. Tsao, Y-J. Sheng, and S. Jiang. 2006. Unbinding of the streptavidin-biotin complex by atomic force microscopy: A hybrid simulation study. *J. Chem. Phys.* 125:104905.
201. Zhu, C., M. Long, S. E. Chesla, and P. Bongrand. 2002. Measuring receptor/ligand interaction at the single-bond level: Experimental and interpretative issues. *Ann. Biomed. Engineering* 30:305–314.
202. Zhu, J., B-H. Luo, T. Xiao, C. Zhang, N. Nishida, and T. A. Springer. 2008. Structure of a complete integrin ectodomain in a physiologic resting state and activation and deactivation by applied forces. *Mol. Cell* 32:849–861.
203. Zidovska, A., and E. Sackmann. 2006. Brownian motion of nucleated cell envelopes impedes adhesion. *Phys. Rev. Letters* 96:048103.
204. Zwanzig, R. 1988. Diffusion in a rough potential. *Proc. Natl. Acad. Sci. USA* 85:2029–2030.

2 Atomic Force Microscopy and Spectroscopy

Hendrik Hölscher

CONTENTS

2.1 A SHORT INTRODUCTION INTO ATOMIC FORCE MICROSCOPY

The direct measurement of the force interaction between distinct molecules has been a challenge for scientists for many years. Only very recently these forces can be directly measured for single atomic and molecular bonds. Interestingly, the applied technique is surprisingly simple. A spring with a defined elasticity is elongated or compressed due to the weight of the object to be measured. The compression Δz of the spring (with spring constant k_{cant}) is a direct measure of the force F exerted, which in the regime of elastic deformation obeys Hooke's law:

$$F = k_{cant} \times \Delta z \qquad (2.1)$$

In atomic force microscopy (AFM) (Binnig et al. 1986), the "spring" is a bendable cantilever with a stiffness between 0.01 and 10 N/m. Since intra-atomic forces are in the range of some nanonewton, the cantilever will be deflected by 0.01–100 nm. Consequently, the precise detection of the cantilever bending is the key feature of AFM. If a sufficiently sharp tip is directly attached to the cantilever, we could then measure the interacting forces between the last atoms of the tip and the sample through the bending of the cantilever.

2.1.1 EXPERIMENTAL SETUPS

During the last years, many experimental setups have been developed and today commercial AFMs are offered from various manufactures. Although most of these instruments are designed for specific applications and environments, they are typically based on the following types of sensors, detection methods, and scanning principles.

Sensors. Cantilevers are produced by standard microfabrication techniques (Binnig et al., 1987; Wolter et al., 1991), mostly from silicon and silicon nitride as rectangular or V-shaped cantilevers. Spring constants and resonance frequencies of cantilevers depend on the actual mode of operation. For contact AFM measurements, they are about 0.01–1 N/m and 5–100 kHz, respectively. In a typical force microscope, cantilever deflections in the range from 0.1 Å to a few micrometers are measured. This corresponds to a force sensitivity ranging from 10^{-13} to 10^{-5} N.

Figure 2.1 shows two scanning electron microscope (SEM) images of a typical rectangular silicon cantilever. Using this imaging technique, the length (l), width (w), and thickness (t) can be precisely measured. The spring constant k_{cant} of the cantilever can then be determined from these values (Meyer et al., 2004)

$$k_{cant} = E_{Si} \frac{w}{4} \left(\frac{t}{l}\right)^3 \tag{2.2}$$

where $E_{Si} = 1.69 \times 10^{11}$ N/m^2 is the Youngs's modulus. Typical dimensions of silicon cantilevers are as follows: lengths of 100–500 μm, widths of 10–50 μm, and thicknesses of 0.3–5 μm.

The torsion of the cantilever due to lateral forces between tip and surface depends also on the height of the tip h. The torsional spring constant can be calculated from (Meyer et al., 2004)

$$k_{tor} = \frac{G}{3} \frac{wt^3}{lh^2} \tag{2.3}$$

where $G_{Si} = 0.68 \times 10^{11}$ N/m is the shear modulus of silicon.

Since the dimensions of cantilevers given by the manufacturer are only average values, high accuracy calibration of the spring constant requires the measurement of length, width, and thickness for each individual cantilever. The length and the width can be measured with sufficient accuracy using an optical microscope, but the thickness requires high-resolution techniques like SEM. To avoid this time- and cost-consuming measurement, one can determine the cantilever thickness from its eigenfrequency in normal direction (Lüthi et al., 1995; Meyer et al., 2004; Bhushan

(a)

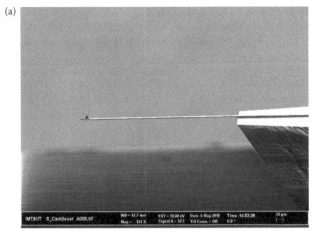

(b)

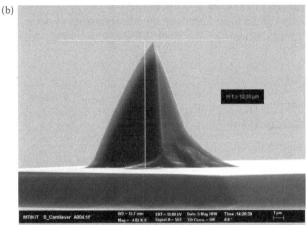

FIGURE 2.1 (a) Scanning electron micrograph of a rectangular silicon cantilever. (b) A closer view of the tip reveals its pyramidal shape obtained by the anisotropic etching of silicon. (Images courtesy of Michael Röhrig, Karlsruhe Institute of Technology.)

and Marti, 2005)

$$t = \underbrace{\frac{4\sqrt{3}}{0.596861^2\pi} \sqrt{\frac{\rho_{Si}}{E_{Si}}}}_{\approx 7.23\times10^{-4}\ \text{s/m}} l^2 f_0 \tag{2.4}$$

with the density of silicon $\rho_{Si} = 2330\,\text{kg/m}^3$.

The formulas presented above are only valid for rectangular cantilevers, but equations for V-shaped cantilevers were given by Neumeister and Ducker (1994) and Sader (1995). Since the calibration of the cantilever is a key issue in every AFM experiment, several researchers developed many other methods to calibrate the forces measured with an AFM (Schwarz et al., 1996; Varenberg et al., 2003; Bilas et al.,

2004; Green et al., 2004; Cook et al., 2006). Today, most commercial AFM setups include already software routines to calibrate the vertical spring constant via thermal noise analysis (Hutter and Bechhofer, 1993; Butt and Jaschke, 1995; Cook et al., 2006) in a convenient way.

Detection methods. Today, nearly all commercial AFMs use the so-called *laser beam deflection* scheme shown in Figure 2.2. The bending and torsion of cantilevers can be detected by a laser beam reflected from their backside (Alexander et al., 1988; Meyer & Amer, 1988), while the reflected laser spot is detected with a sectioned photodiode. The different parts are read out separately. Usually a four-quadrant diode is used to detect the normal and the torsional movements of the cantilever. With the cantilever at equilibrium, the spot is adjusted such that the upper and the lower sections show the same intensity. If the cantilever bends up or down, the spot moves, and the difference signal between upper and lower section is a measure of the bending. A detailed analysis of the optimal position where to focus the laser spot on the back side of the cantilever was given by Schäffer and Fuchs (2005).

The sensitivity can be even improved by interferometer systems adapted by several research groups (Rugar et al., 1989; Moser et al., 1993; Allers et al., 1998;

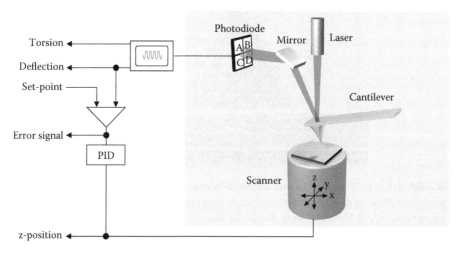

FIGURE 2.2 Principle of an atomic force microscope working with the laser beam deflection method. Deflection (normal force) and torsion (friction) of the cantilever are measured simultaneously by measuring the lateral and vertical deflection of a laser beam while the sample is scanned in the x-y plane. The laser beam deflection is determined using a four-quadrant photo diode. If A, B, C, and D are proportional to the intensity of the incident light of the corresponding quadrant, the signal (A+B)-(C+D) is a measure for the deflection and (A+C)-(B+D) is a measure for the torsion of the cantilever. A schematic of the feedback system is shown by solid lines. The actual deflection signal of the photodiode is compared with the set-point chosen by the experimentalist. The resulting error signal is fed into the PID controller, which moves the z-position of the scanner to minimize the deflection signal.

Kawakatsu et al., 2002). This detection scheme is very often used in low-temperature, ultrahigh vacuum (UHV) systems. It is also possible to use cantilevers with integrated deflection sensors based on piezoresistive films (Tortonese et al., 1993; Yuan et al., 1994; Linnemann et al., 1996). Since no optical parts are needed in the experimental setup of an AFM with self-sensing cantilevers, their design can be very compact (Stahl et al., 1994). However, since it is very difficult to produce piezoresistive cantilevers with a high, consistent quality, they are rarely used today.

During scanning of the surface, the deflection of the cantilever is kept constant by a feedback system, which controls the vertical movement of the scanner. A schematic of the feedback system is drawn in Figure 2.3. It works as follows: The current signal of the photodiode is compared with a preset value. The feedback system including a proportional, integral, and differential (PID) controller varies the z-movement of the scanner to minimize the difference. As a consequence, the tip-sample force is kept practically constant for an optimal setup of the PID parameters.

While the cantilever is moving relative to the sample in the x-y-plane of the surface by a piezoelectric scanner, the current z-position of the scanner is recorded as a function of the lateral x-y-position with (ideally) sub-Ångström precision. The

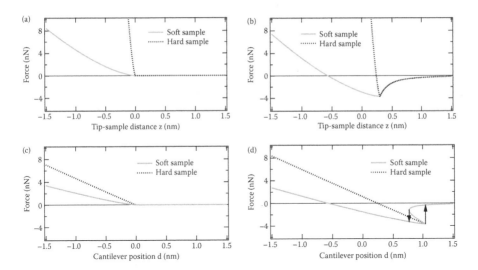

FIGURE 2.3 (**See color insert.**) Tip-sample forces using the (a) Hertz and (b) DMT-M model for a hard ($E_{hard} = 100\,\text{GPa}$) and soft ($E_{soft} = 1\,\text{GPa}$) sample assuming a tip radius of 10 nm. The other parameters are $z_0 = 0.3\,\text{nm}$, $\mu_t = \mu_s = 0.3$, $E_t = 130\,\text{GPa}$, $A_H = 0.2\,\text{aJ}$. (c) If these forces are measured with an atomic force microscope using a cantilever with a spring constant of 5 N/m, the resulting force versus curves show significantly reduced slope due to the elasticity of the cantilever. Without the presence of adhesion forces (Hertz model), the curves are continuous. (d) Adhesion results in a hysteresis between forward and backward movement of the cantilever as marked by the arrows. (See Section. 2.2 for more details on this effect.)

obtained data represents a map of equal forces, which is analyzed and visualized by computer processing.

2.1.2 TIP-SAMPLE FORCES

A large variety of sample properties related to tip-sample forces can be detected with an AFM. The obtained contrast depends on the operational mode and the actual tip-sample interactions. Before discussing details of the operational modes of AFM, we first specify the most important tip-sample interactions.

Figure 2.3b shows the typical shape of the interaction force curve the tip senses during an approach toward the sample surface. Upon approach of the tip toward the sample, the negative attractive forces, representing, for example, van der Waals or electrostatic interaction forces, increase until a minimum is reached. This turnaround point is due to the onset of repulsive forces, caused by Pauli repulsion, which will start to dominate upon further approach. Eventually, the tip is pushed into the sample surface and elastic deformation will occur.

In general, the effective tip-sample interaction force is a sum of different force contributions. They can be roughly divided into attractive and repulsive components. The most important forces are summarized in the following Israelachvili (1992) Sarid (1994), and Butt & Kappl (2010).

Van der Waals forces. These forces are caused by fluctuating induced electric dipoles in atoms and molecules and they are also sometimes named Casimir forces (Parsegian, 2006). The distance dependence of this force for two distinct molecules is attractive in most cases (Munday et al., 2009) and follows a $-1/z^7$ force law. For simplicity, solid bodies are often assumed to consist of many independent noninteracting molecules and the van der Waals forces of these bodies are obtained by simple summation. For example, for a sphere over a flat surface, the van der Waals force is given by

$$F_{vdW}(z) = -\frac{A_H R}{6z^2} \tag{2.5}$$

where R is the radius of the sphere, and A_H is the Hamaker constant, which is typically in the range of $\approx 0.1\,\text{aJ}$ (Israelachvili, 1992). This geometry is often used to approximate the van der Waals forces between tip and sample. Due to the $1/z^2$ dependency, van der Waals forces are considered long-range forces compared with other forces occurring in AFM.

Capillary forces. Water molecules condense at the sample surface (and also on the tip) under ambient conditions and cause the occurrence of an adsorption layer. Consequently, the AFM tip penetrates through this layer when approaching the sample surface. At the tip-sample contact, a water meniscus is formed that causes a very strong attractive force (Stifter et al., 2000). For soft samples such forces often lead to unwanted deformations of the surface. However, this effect can be circumvented by measuring directly in liquids. Alternatively, capillary forces can be avoided by performing the experiments in a glove box with dry gases or in vacuum.

Pauli or ionic repulsion. Repulsive forces are the most important forces in conventional contact mode AFM. The Pauli exclusion principle forbids that the charge clouds of two electrons showing the same quantum numbers can have some significant overlap; first, the energy of one of the electrons has to be increased. This yields a repulsive force. In addition, overlap of the charge clouds of electrons can cause an insufficient screening of the nuclear charge, leading to ionic repulsion of Coulombic nature. The Pauli and the ionic repulsion are nearly hard wall potentials. Thus, for tip and sample in intimate contact, most of the (repulsive) interaction is carried by the atoms directly at the interface. The Pauli repulsion is of purely quantum mechanical origin and semiempirical potentials are mostly used to allow an easy and fast calculation. A well-known model is the *Lennard-Jones* potential, which combines short-range repulsive interactions with long-range attractive van der Waals interactions:

$$V_{LJ}(z) = E_0 \left(\left(\frac{r_0}{z}\right)^{12} - 2\left(\frac{r_0}{z}\right)^6 \right) \tag{2.6}$$

where E_0 is the bonding energy and r_0 is the equilibrium distance. In this case, the repulsion is described by an inverse power law with $n = 12$. The term with $n = 6$ describes the attractive van der Waals potential between two atoms/molecules.

Elastic forces. If the tip is in contact with the sample, elastic deformations can occur. Since this deformation affects the effective contact area the knowledge about the elastic forces and the corresponding deformation mechanics of the contact is an important issue in AFM. The repulsive forces occurring during elastic indentation of a sphere into a flat surface were already analyzed in 1881 by H. Hertz (Johnson, 1985; Landau & Lifschitz, 1991)

$$F_{\text{Hertz}}(z) = \frac{4}{3}E^* \sqrt{R}(z_0 - z)^{3/2} \quad \text{for} \quad z \le z_0 \tag{2.7}$$

where the effective elastic modulus E^*

$$\frac{1}{E^*} = \frac{(1 - \mu_t^2)}{E_t} + \frac{(1 - \mu_s^2)}{E_s} \tag{2.8}$$

depends on the Young's moduli $E_{t,s}$ and the Poisson ratios $\mu_{t,s}$ of tip and surface, respectively. R is the tip radius and z_0 is the point of contact. Figure 2.3a shows two curves following this force law for a soft and hard sample, respectively.

However, this model does not include adhesion forces, which have to be considered at the nanometer scale. Two extreme cases were analyzed by Johnson et al. (1971) and Derjaguin et al. (1975). The model of Johnson, Kendall, and Roberts (JKR model) considers only the adhesion forces *inside* the contact area, whereas the model of Derjaguin, Muller, and Toporov (DMT model) includes only the adhesion *outside* the contact area. Various models analyzing the contact mechanics in the intermediate regime were suggested by other authors (see, e.g., Schwarz, 2003 for an overview).

However, in many practical cases, it is sufficient to assume that the geometrical shape of tip and sample does not change until contact has been established at $z = z_0$ and that afterwards, the tip-sample forces are given by the *DMT-M theory*, denoting Maugis' approximation to the earlier Derjaguin-Muller-Toporov model. In this approach, an offset $F_{vdW}(z_0)$ is added to the well-known Hertz model, which accounts for the adhesion force between tip and sample surface. Therefore, the DMT-M model is often also referred to as *Hertz-plus-offset model* (Schwarz, 2003). The resulting overall force law is given by

$$
F_{DMT-M}(z) = \begin{cases} -\frac{A_H R}{6z^2} & \text{for} \quad z \geq z_0 \\ \frac{4}{3} E^* \sqrt{R}(z_0 - z)^{3/2} - \frac{A_H R}{6z_0^2} & \text{for} \quad z < z_0 \end{cases} \tag{2.9}
$$

Figure 2.3b displays the resulting tip-sample forces curves for the DMT-M model for a hard and soft sample, respectively. The parameters are given in the figure caption. They represent typical values for AFM measurements under ambient conditions.

Frictional forces. During the scanning of the tip on the sample surface, there are counteracting frictional forces. These forces dissipate the kinetic energy of the moving tip-sample contact into the surface or tip material. This can be due to permanent changes in the surface itself, by scratching or indenting, or also by the excitation of lattice vibration (i.e., phonons) in the material.

Chemical binding forces. Due to the overlap of molecular orbitals there might arise specific bonding states between the tip and the surface molecules. These forces are extremely short-ranged and can be exploited to achieve atomic resolution imaging of surfaces (mostly in vacuum). Since these forces are also specific to the chemical identity of the molecules, it is conceivable to identify the chemical character of the surface atoms with AFM scans (Sugimoto et al., 2007).

Magnetic and electrostatic forces. Long-range magnetic or electrostatic force might be attractive or repulsive. They are usually measured when the tip is not in contact with the surface (i.e., "noncontact" mode). For magnetic forces, magnetic materials have to be used for tip or tip coating. Well-defined electrical potentials between tip and sample are necessary for the measurement of electrostatic forces.

More detailed information on intermolecular and surface forces relevant for AFM measurements can be found in the monographs of Israelachvili (1992), Sarid (1994), and Butt & Kappl (2010). Figure 2.4 nicely summarizes the most important ones. In principle, every type of force can be measured with an AFM.

2.2   CONTACT MODE

An AFM can be driven in different modes of operation. First, we introduce the contact mode, which is the historically the oldest one. To distinguish it from the later introduced dynamic modes, the contact mode is also sometimes referred to as static mode. However, due to its straightforwardness it can be used to easily obtain

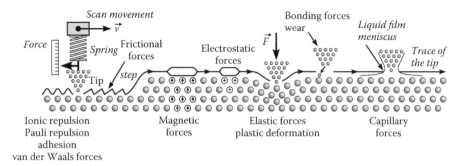

Scan movement

Force

Spring   Frictional forces

Tip   step

Ionic repulsion
Pauli repulsion
adhesion
van der Waals forces

Electrostatic forces

Magnetic forces

Bonding forces
wear

Liquid film meniscus

$\vec{F}$

Elastic forces
plastic deformation

Trace of the tip

Capillary forces

FIGURE 2.4 Summary of the forces relevant in atomic force microscopy. (Image courtesy of Udo D. Schwarz, Yale University.)

nanometer resolution images on a wide variety of surfaces. Furthermore, it has the advantage that not only the deflection but also the torsion of the cantilever can be measured. As shown by Mate et al. (1987) the lateral force can be directly correlated to the friction between tip and sample, thus extending AFM to *friction force microscopy*.

Figures 2.5a and 2.5b present typical applications of an AFM driven in contact mode. The images show a measurement of a (L-α-dipalmitoyl-phosphatidycholine [DPPC]; Fluka) film adsorbed on a mica substrate. The lateral force was simultaneously recorded with the topography and shows a contrast between the DPPC film and the substrate. This effect can be attributed to the different frictional forces on DPPC and the mica substrate and is frequently used to obtain a chemical contrast on flat surfaces (Overney et al., 1992; McKendry et al., 1998).

So far we have neglected one important issue for the operation of the AFM: The elasticity and mechanical stability of the cantilever. If the tip is slowly approached toward the surface in the static mode, the tip-sample force F_{ts} has to be counteracted by the restoring bending force of the cantilever.

$$k_{cant}(d-z) = F_{ts} \qquad (2.10)$$

This equation has a unique solution as long as the tip-sample force is continuously repulsive like the elastic Hertz force Equation 2.7. Figure 2.3c displays the resulting force you would detect with a cantilever spring constant of 5 N/m.

If you compare this graph with the original force curves in Figure 2.3a, you will immediately notice that they differ in magnitude. The measured bending curves are considerably softer and have a flatter negative slope as the original tip-sample force curves. This is due to the softness of the cantilever. If you approach toward the sample surface, the bending of the cantilever is larger than the indentation of the tip into the sample. Consequently, you measure the bending of the cantilever and not the elasticity of your sample. You might use hard cantilevers with high spring constants to avoid this effect, but then you will loose sensitivity, as a small bending of the cantilever is more difficult to detect. Another option is to rescale the x-axis to correct

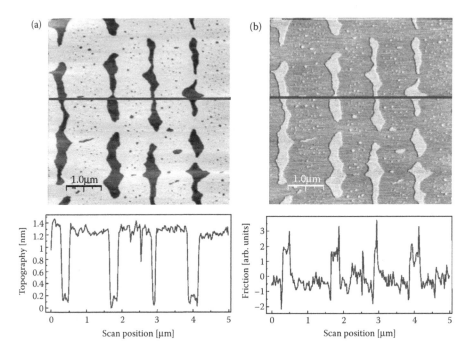

FIGURE 2.5 (a) Atomic force microscopy image obtained in contact mode of a monomolec-ular L-α-dipalmitoyl-phosphatidycholine (DPPC) film adsorbed on a mica substrate. The image is color coded, that is, dark areas represent the mica substrate and light areas the DPPC film. (b) The simultaneously recorded friction image shows lower friction on the film (dark areas) as on the substrate (light areas). The graphs at the bottom represent single scan lines obtained at the positions marked by a dark line in the images at the top. Sample and experiment by L. Chi and J.-E. Schmutz, University of Münster.

for the bending of the cantilever. This procedure, however, needs a calibration of the cantilever bending.

The situation gets even more disappointing if attractive tip-sample forces are present as shown in Figure 2.3b. In this case, the attractive forces are so large that cannot be counterbalanced by the soft cantilever anymore. Mathematically speak-ing, the gradient of the tip-sample forces is larger than the spring constant of the cantilever (Burnham & Colton, 1989). Therefore, an instability occurs if

$$k_{cant} < \frac{\partial F_{ts}(z)}{\partial z} \qquad (2.11)$$

As a result, the tip "jumps" toward the sample surface during an approach.

This effect strongly influences static mode AFM measurements in air and vac-uum where strong long-range attractive forces are present as exemplified by a typi-cal force-versus-distance curve shown in Figure 2.6. Here, the force acting on the tip recorded during an approach and retraction movement of the cantilever is depicted.

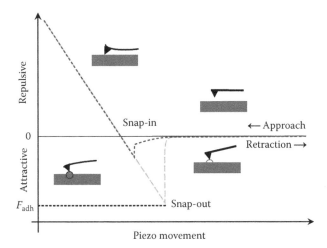

FIGURE 2.6 A schematic of a typical force-versus-distance-curve obtained in static mode. The cantilever approaches toward the sample surface. Due to strong attractive forces it "jumps" (snap-in) toward the sample surface at a specific position. During retraction, the tip is strongly attracted by the surface and the snap-out point is considerably behind the snap-in point. This results in a hysteresis between approach and retraction.

Upon approach of the cantilever toward the sample, the attractive forces acting on the tip bend the cantilever toward the sample surface. At a specific point close to the sample surface, these forces can no longer be sustained by the cantilever spring and the tip "jumps" toward the sample surface. Now tip and sample are in direct mechanical contact. A further approach toward the sample surface pushes the tip into the sample. Since the spring constant of the cantilever is usually much softer than the elasticity of the sample, the bending of the cantilever increases almost linearly.

If the cantilever is now retracted from the surface, the tip stays in contact with the sample at first because it is strongly attracted by the sample due to adhesive forces. The force F_{adh} is necessary to retract the tip from the surface. The corresponding snap-out position is always at a larger distance from the surface than the snap-in, which results in a hysteresis between approach and retraction of the cantilever. This phenomenon of mechanical instability is often referred to as the *jump-to-contact*. Unfortunately, this sudden jump can lead to undesired changes of the tip and/or sample.

The same type of measurement can be used to measure molecular binding forces or the elongation of chain-like molecules (for a review see, e.g., Janshoff et al., 2002 Hinterdorfer & Dufrene, 2006). As shown in Figure 2.7a, chain-like molecules attached to a surface might eventually pick up the tip and the resulting bending of the cantilever is detected during the retraction of the cantilever from the surface. Applications of this technique can be found in the other chapters of this book.

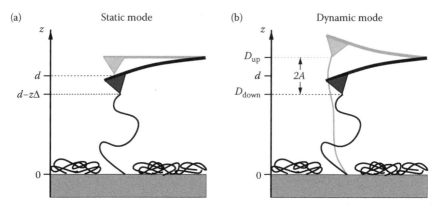

FIGURE 2.7 Two different technique for the measurement of forces acting on a chain-like molecule. (a) Typical experimental setup in the static mode where a chain-like molecule attached to a tip is stretched. Measured quantities are the bending Δz of the cantilever (i.e., the force) versus the extension of the molecule. (b) Another option is dynamic force spectroscopy where the cantilever oscillates with an amplitude A around its equilibrium position d, while it is retracted from the surface covered with chain-like molecules. If a molecule binds to the tip, the resonance frequency f of the cantilever changes in dependence of the cantilever position.

2.3   DYNAMIC MODE

Despite the success of contact mode AFM, the resolution was found to be limited in many cases (in particular for soft samples) by lateral forces acting between tip and sample. To avoid this effect, the cantilever can be oscillated in vertical direction near the sample surface. Imaging with vibrating cantilever is often denoted as *dynamic force microscopy* (DFM).

The historically oldest scheme of cantilever excitation in DFM imaging is the external driving of the cantilever at a fixed excitation frequency exactly at or very close to the cantilever's first resonance (Martin & Wickramasinghe, 1987; Martin et al., 1987; Zhong et al., 1993; Putman et al., 1994). For this driving mechanism, different detection schemes measuring either the change of the oscillation amplitude or the phase shift were proposed. Over the years, the amplitude modulation (AM) or "tapping" mode, where the oscillation amplitude is used as a measure of the tip-sample distance, has developed into the most widely used technique for imaging under ambient conditions and liquids.

Under the influence of tip-sample forces, the resonant frequency (and consequently also amplitude and phase) of the cantilever will change and serve as the measurement parameters. If the tip is approached toward the surface, the oscillation parameters amplitude and phase are influenced by the tip–surface interaction and can therefore be used as feedback channels. A certain set-point, for example, the amplitude is given, and the feedback loop will adjust the tip-sample distance such that the amplitude remains constant. The controller parameter is recorded as a function of the lateral position of the tip with respect to the sample and the scanned image essentially represents the surface topography.

A sketch of the experimental setup of a dynamic force microscope utilizing the AM technique is shown in Figure 2.8. As in the contact mode, the deflection of the cantilever is typically measured with the laser beam deflection method. During operation in conventional AM-mode, the cantilever is driven at a fixed frequency by a constant-amplitude (CA) signal originating from an external function generator, while the resulting oscillation amplitude and/or the phase shift are detected by a lock-in amplifier. The function generator supplies not only the signal for the dither piezo; its signal serves simultaneously as a reference for the lock-in amplifier in the analyzer electronics.

The tapping-mode can be operated in air and in liquids. A typical image obtained with this technique in ambient conditions is shown in Figure 2.9. For a direct comparison with the contact mode (see Figure 2.5), the sample is also DPPC adsorbed on a mica substrate. While in contact-mode the frictional forces are measured simultaneously with the topography, in dynamic mode the phase between excitation and oscillation is acquired as an additional channel. The phase image gives information

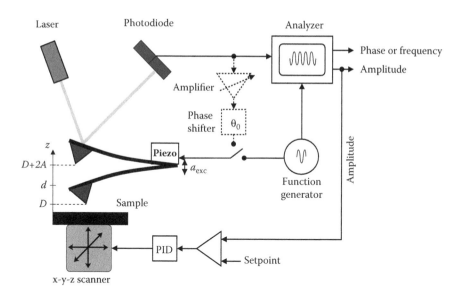

FIGURE 2.8 (**See color insert.**) Schematic drawing of the experimental setup of a dynamic force microscope where the driving of the cantilever can be switched between amplitude-modulation (AM) mode (solid lines) or frequency-modulation (FM) mode (dashed lines). While the cantilever in the AM-mode is externally driven with a frequency generator, the FM-mode exhibits a feedback loop consisting of a time ("phase") shifter and an amplifier. In both cases, we assume that the laser beam deflection method is used to measure the oscillation of the tip, which oscillates between the nearest tip-sample position D and $D + 2A$. The equilibrium position of the tip is denoted as d.

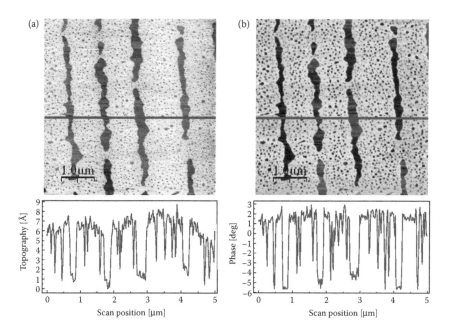

FIGURE 2.9   (a) A dynamic force microscopy image of a monomolecular L-α-dipalmitoyl-phosphatidycholine (DPPC) film adsorbed on a mica substrate in ambient conditions. (b) The phase contrast is directly related to the topography, that is, the phase is different between substrate and DPPC film. Sample and experiment by L. Chi and J.-E. Schmutz, University of Münster.

about the different material properties of DPPC and the mica substrate. It can been shown that the phase signal is closely related to the energy dissipated in the tip-sample contact (Cleveland et al., 1998; Tamayo & García, 1998; Garcia et al., 2006). A typical example for high-resolution imaging of DNA in liquid solution is shown in Figure 2.10.

However, another method to oscillate the cantilever is the *frequency-modulation* (FM) mode, which was primarily developed for application in vacuum where standard AFM cantilevers made from silicon or silicon nitride exhibit very high Q values, what makes the response of the system slow. Therefore, Albrecht et al. (1991) introduced in 1991 the FM-mode, which works well for high-Q systems and consequently developed into the dominating driving scheme for DFM experiments in UHV (Garcia & Pèrez, 2002; Morita et al., 2002; Giessibl, 2003; Hölscher & Schirmeisen, 2005). In contrast to the AM-mode, this approach features a so-called *self-driven* oscillator (Hölscher et al., 2001, 2002), which uses the cantilever deflection itself as drive signal, thus ensuring that the cantilever instantaneously adapts to changes in the resonance frequency.

Contrary to the AM technique, a dynamic force microscope driven in the FM-mode has no external driving but a feedback circuit consisting of an amplifier and a

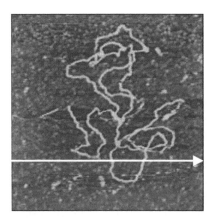 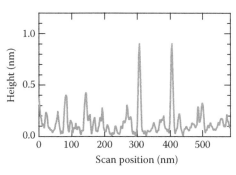

FIGURE 2.10 Topography of DNA adsorbed on mica imaged in buffer solution by tapping-mode AFM. The graph shows a single scan line obtained at the position marked by an arrow in the image on the left.

phase shifter (dashed lines in Figure 2.8). In this way, the signal of the displacement sensor is first amplified before it is phase shifted. Subsequently, it is used to excite the dither piezo driving the cantilever. Two different driving techniques have been established for use with the FM-detection scheme: The original CA mode, where the oscillation amplitude of the cantilever is held constant by the an automatic gain control (Albrecht et al., 1991), and the constant-excitation (CE) mode (Kitamura & Iwatsuki, 1995; Ueyama et al., 1998), where the excitation amplitude of the cantilever driving is kept constant. Both the CE-mode (Kobayashi et al., 2002; Okajima et al., 2003; Hölscher & Anczykowski, 2005; Ebeling et al., 2006a; Schmutz et al., 2007) and the CA-mode (Farell et al., 2005; Fukuma et al., 2005a,b; Hoogenboom et al., 2006) are frequently used in air and liquids. However, since the amplitude can be used as a feedback signal for scanning in the CE-mode, its implementation is easily possible for an existing DFM build for AM-mode applications in air and liquid. Therefore, we focus on the CE mode in the following.

2.4 THEORY OF THE DYNAMIC MODE

2.4.1 EQUATION OF MOTION

On the basis of the above description of the experimental setup, we can formulate the basic equation of motion describing the cantilever dynamics in the dynamic mode (Hölscher, 2002; Rodríguez & García, 2003):

$$m\ddot{z}(t) + \frac{2\pi f_0 m}{Q_0}\dot{z}(t) + k_{cant}(z(t) - d) = \underbrace{F_{ts}[z(t), \dot{z}(t)]}_{\text{Tip-sample force}}$$

$$+ \begin{cases} a_{exc}\,k_{cant}\cos(2\pi f_d t) & \text{for AM-mode} \\ -\frac{a_{exc}}{A}k_{cant}z(t - t_0) & \text{for FM-mode} \end{cases} \quad (2.12)$$

Here, $z(t)$ is the position of the tip at the time t; k_{cant}, m, and $f_0 = \sqrt{(k_{cant}/m)}/(2\pi)$ are the spring constant, the effective mass, and the eigenfrequency of the cantilever, respectively. Somewhat simplifying, it is assumed that the quality factor Q_0 unites the intrinsic damping of the cantilever and all influences from surrounding media such as air or liquid, if present, in a single overall value. The equilibrium position of the tip is denoted as d. The (nonlinear) tip-sample interaction force F_{ts} is introduced by the first term on the right side of the equation.

The two driving mechanisms are considered by the distinction on the right side of the equation. The external driving force of the cantilever is used for the AM-mode. Here, the driving signal is modulated with the CE amplitude a_{exc} at a fixed frequency f_d. The self-excitation of the cantilever used in the FM-mode is described by the retarded amplification of the displacement signal, that is, the tip position z is measured at the retarded time $t - t_0$. Nonetheless, a consideration of the time shift by a phase difference θ_0 is also possible, giving equivalent results. Therefore, we use "time shift" and "phase shift" as synonyms throughout this review and notice that both parameters are scaled by $\theta_0 = 2\pi f_d t_0$.

Before finishing this section, we would like to add some words of caution regarding the validity of the equation of motion Equation 2.12, as it disregards two effects that might become of importance under specific circumstances. First, we describe the cantilever by a spring-mass-model and neglect in this way higher modes of the cantilever. This is justified in most cases as the first eigenfrequency is by far dominant in typical AM-AFM experiments (see, e.g., Fig. 1 in Cleveland et al., 1998). Thus, a mathematical treatment that ignores higher modes is still able to describe and explain all major general features experimentally observed in standard DFM imaging, which is the limited goal of this review. Comparison with studies that include higher harmonics by numerical means (Stark & Heckl, 2000; Rodríguez & García, 2002; Lee et al., 2002; Stark et al., 2004) confirms this statement. It might, however, not apply if advanced signal analysis in certain DFM spectroscopy modes is intended.

Second, we assume in our model equation of motion that the dither piezo applies a sinusoidal force to the spring, but do not consider that the movement of the dither piezo simultaneously also changes the effective position of the tip at the cantilever end by the current value of the excitation $a_{exc} \cos(2\pi f_d t)$ (Lee et al., 2002; Legleiter & Kowalewski, 2005; Legleiter et al., 2006). This effect becomes important when a_{exc} is in the range of the cantilever oscillation amplitude. Fortunately, for conditions characterized by sufficiently high-quality factors, this effect can be neglected. This is usually safely the case for measurements in air, where oscillation amplitudes typically exceed excitation amplitudes by several hundred times. During operation in liquids, however, the Q factor is low, and the oscillation amplitudes might be comparable with a_{exc} (Legleiter et al., 2006).

Finally, to avoid confusion with other literature, we would like to mention some words regarding the terminology used throughout this review. Due to the frequently occurring intermitted contact between tip and sample at the lowest point of the oscillation, the AM mode introduced above has often been denoted as *tapping mode* (Zhong et al., 1993). Over the years, use of the term "tapping mode" has then evolved into a synonym for the AM-mode in many publications, disregarding whether the tip is actually making intermittent contact or not. On the other hand, if it is the operator's firm believe that no contact is established during the oscillations, the AM-mode is sometimes also referred to as "noncontact" mode. Please note, however, that the term "noncontact atomic force microscopy (NC-AFM)" is often employed in connection with the FM mode, which is mostly applied in UHV (Morita et al., 2002). This example shows already how different driving modes might be mixed up if we use the assumed type of tip-sample interaction to define a DFM technique. Therefore, to avoid confusion in this review, we will use the suitable expressions for the driving technique (AM- or FM-mode) to describe the applied AFM mode.

2.4.2 DRIVEN AND SELF-DRIVEN CANTILEVERS

To analyze the specific features of the AM- and FM-mode, it is instructive to first examine the difference of both driving terms. For simplicity, we assume in these preparatory considerations that the cantilever vibrates far away from the sample surface. Consequently, we can neglect tip-sample forces ($F_{ts} \equiv 0$), resulting in a greatly simplified equation of motion Equation 2.12. This restriction will be abandoned later.

First, we consider the situation where the DFM is driven in the AM-mode. Under these circumstances, the equation motion reduces to the well-known case of a driven and damped harmonic oscillator:

$$m\ddot{z}(t) + \frac{2\pi f_0 m}{Q_0} \dot{z}(t) + k_{cant}(z(t) - d) = a_{exc} k_{cant} \cos(2\pi f_d t) \qquad (2.13)$$

The external driving forces the cantilever to oscillate exactly at the driving frequency f_d. Therefore, the steady-state solution is given by the ansatz

$$z(t \gg 0) = d + A\cos(2\pi f_d t + \phi) \qquad (2.14)$$

where ϕ is the phase difference between the excitation and the oscillation of the cantilever. Introducing this ansatz, we obtain two functions for the amplitude and phase curves:

$$A = \frac{a_{exc}}{\sqrt{\left(1 - \frac{f_d^2}{f_0^2}\right)^2 + \left(\frac{1}{Q_0}\frac{f_d}{f_0}\right)^2}} \qquad (2.15a)$$

$$\tan\phi = \frac{1}{Q_0}\frac{f_d/f_0}{1 - f_d^2/f_0^2} \qquad (2.15b)$$

The features of such an oscillator are well known from introductory physics courses, and we will thus skip their further discussion at this point.

In contrast, the case where the cantilever is entirely self-driven is much less discussed in the literature. Here, the corresponding equation of motion reduces to

$$m\ddot{z}(t) + \frac{2\pi f_0 m}{Q_0}\dot{z}(t) + k_{\text{cant}}(z(t) - d) = -\frac{a_{\text{exc}}}{A}k_{\text{cant}}\,z(t - t_0) \qquad (2.16)$$

As the cantilever is not excited with a specific externally set frequency, the cantilever itself serves as the frequency determining element. Therefore, we make the ansatz (Hölscher et al., 2001; 2003)

$$z(t \gg 0) = d + A\cos(2\pi f t) \qquad (2.17)$$

and introduce it into Equation 2.17 into Equation 2.16. As a result we obtain a set of two coupled trigonometric equations:

$$a_{\text{exc}}\cos(2\pi f t_0) = \frac{f_0^2 - f^2}{f_0^2} \qquad (2.18a)$$

$$\frac{a_{\text{exc}}}{A}\sin(2\pi f t_0) = \frac{1}{Q_0}\frac{f}{f_0} \qquad (2.18b)$$

The two equations can be decoupled with the assumption that the time shift t_0 is set to a value corresponding to $t_0 = 1/(4f_0)$ $(= 90°)$, which simultaneously corresponds to the by far most common choice for t_0. For this value, the solution of Equation 2.18 is given by

$$f = f_0 \qquad (2.19a)$$

$$A = a_{\text{exc}}Q_0 \qquad (2.19b)$$

This simple calculation demonstrates the very specific behavior of a self-driven oscillator if the phase (or time) shift is set to 90°. In this case, the cantilever oscillates exactly with its eigenfrequency f_0. Due to this specific feature revealed by Equation 2.19a, we define that the cantilever is in *resonance* if this condition is fulfilled. The linear relationship between the oscillation and excitation amplitude is described by Equation 2.19b.

2.4.3 THEORY OF THE AM-MODE

In the first step, we assumed that the cantilever vibrates far away from the sample surface. Therefore, we neglected tip-sample forces in Equation 2.13 and finally got the well-known theory of a driven-damped harmonic oscillator.

However, if the cantilever is brought closer toward the sample surface, the tip senses the tip-sample interaction force F_{ts}, which changes the oscillation behavior of the cantilever. However, since the mathematical form of realistic tip-sample forces is highly nonlinear (see Figure 2.3b) this fact complicates the analytical solution of the equation of motion Equation 2.12. For the analysis of DFM experiments, we need to focus on steady-state solutions of the equation of motion with sinusoidal cantilever oscillation. Therefore, it is advantageous to expand the tip-sample force into a Fourier series

$$
F_{ts}[z(t),\dot{z}(t)] \approx f_{\mathrm{d}} \int_0^{1/f_{\mathrm{d}}} F_{ts}[z(t),\dot{z}(t)]\mathrm{d}t
$$

$$
+2f_{\mathrm{d}} \int_0^{1/f_{\mathrm{d}}} F_{ts}[z(t),\dot{z}(t)]\cos(2\pi f_{\mathrm{d}}t + \phi)\mathrm{d}t \times \cos(2\pi f_{\mathrm{d}}t + \phi)
$$

$$
+2f_{\mathrm{d}} \int_0^{1/f_{\mathrm{d}}} F_{ts}[z(t),\dot{z}(t)]\sin(2\pi f_{\mathrm{d}}t + \phi)\mathrm{d}t \times \sin(2\pi f_{\mathrm{d}}t + \phi)
$$

$$
+\ldots
$$

(2.20)

where $z(t)$ is given by Equation 2.14.

The first term in the Fourier series reflects the averaged tip-sample force over one full oscillation cycle, which shifts the equilibrium point of the oscillation by a small offset Δd from d to d_0. Actual values for Δd, however, are very small. For typical amplitudes used in AM-AFM in air (some nanometers to some tens of nanometers), the averaged tip-sample force is in the range of some piconewtons. The resulting offset Δd is less than 1 pm for typical sets of parameters (Hölscher et al., 2006). Since this is well beyond the resolution limit of an AM-AFM experiment in air, we neglect this effect in the following and assume $d \approx d_0$ and $D = d - A$.

For further analysis, we now insert the first harmonics of the Fourier series Equation 2.20 into the equation of motion Equation 2.12, obtaining two coupled equations (Hölscher et al., 2006)

$$
\frac{f_0^2 - f_{\mathrm{d}}^2}{f_0^2} = I_+(d,A) + \frac{a_{\mathrm{d}}}{A}\cos\phi
$$

(2.21a)

$$
-\frac{1}{Q_0}\frac{f_{\mathrm{d}}}{f_0} = I_-(d,A) + \frac{a_{\mathrm{d}}}{A}\sin\phi
$$

(2.21b)

where the following integrals have been defined:

$$
I_+(d,A) = \frac{2f_{\mathrm{d}}}{k_{\mathrm{cant}}A} \int_0^{1/f_{\mathrm{d}}} F_{ts}[z(t),\dot{z}(t)]\cos(2\pi f_{\mathrm{d}}t + \phi)\mathrm{d}t
$$

$$
= \frac{1}{\pi k_{\mathrm{cant}}A^2} \int_{d-A}^{d+A} (F_{\downarrow} + F_{\uparrow})\frac{z-d}{\sqrt{A^2 - (z-d)^2}}\mathrm{d}z
$$

(2.22a)

$$
I_-(d,A) = \frac{2f_{\mathrm{d}}}{k_{\mathrm{cant}}A} \int_0^{1/f_{\mathrm{d}}} F_{ts}[z(t),\dot{z}(t)]\sin(2\pi f_{\mathrm{d}}t + \phi)\mathrm{d}t
$$

$$
= \frac{1}{\pi k_{\mathrm{cant}}A^2} \int_{d-A}^{d+A} (F_{\downarrow} - F_{\uparrow})\mathrm{d}z
$$

$$
= \frac{1}{\pi k_{\mathrm{cant}}A^2}\Delta E(d,A)
$$

(2.22b)

Both integrals are functions of the actual oscillation amplitude A and cantilever-sample distance d. Furthermore, they depend on the sum and the difference of the tip-sample forces during approach ($F_{\downarrow}$) and retraction ($F_{\uparrow}$) manifested by the labels

"+" and "−" for easy distinction. The integral I_+ is a weighted average of the tip-sample forces $(F_\downarrow + F_\uparrow)$. On the other hand, the integral I_- is directly connected to ΔE, which reflects the energy dissipated during an individual oscillation cycle. Consequently, this integral vanishes for purely conservative tip-sample forces, where $F_\downarrow$ and $F_\uparrow$ are identical. A detailed discussion of these integrals was already given by Dürig, 2000b and Sader et al., 2005.

By now combining Equations 2.15b and 2.22b, we get a direct correlation between the phase and the energy dissipation.[*]

$$\sin\phi = -\left(\frac{A}{A_0} \frac{f_d}{f_0} + \frac{Q_0 \Delta E}{\pi k_{\text{cant}} A_0 A} \right) \tag{2.23}$$

This relationship can also be obtained from the conservation of energy principle (Cleveland et al., 1998; Tamayo & García, 1998; Garcia et al., 2006) and demonstrates that the phase signal in tapping mode is directly related to the energy dissipation caused by the tip-sample interaction.

Equation 2.21 can be used to calculate the resonance curves of a dynamic force microscope including tip-sample forces. The results are

$$A = \frac{a_{\text{d}}}{\sqrt{\left(1 - \frac{f_d^2}{f_0^2} - I_+(d,A)\right)^2 + \left(\frac{1}{Q_0} \frac{f_d}{f_0} + I_-(d,A)\right)^2}} \tag{2.24a}$$

$$\tan\phi = \frac{\frac{1}{Q_0} \frac{f_d}{f_0} + I_-(d,A)}{1 - \frac{f_d^2}{f_0^2} - I_+(d,A)} \tag{2.24b}$$

Equation 2.24a describes the shape of the resonance curve, but it is an implicit function of the oscillation amplitude A and cannot be plotted directly.

Figure 2.11a contrasts the solution of this equation (solid lines) with numerical solution (symbols). As pointed out by various authors (Gleyzes et al., 1991; Kühle et al., 1998; Wang, 1998; Aimé et al., 1999; Sasaki & Tsukada, 1999; Nony et al., 2001; Lee et al., 2002; San Paulo & García, 2002), the amplitude versus frequency curves are multivalued within certain parameter ranges. Moreover, as the gradient of the analytical curve increases to infinity at specific positions, some branches are unstable. The resulting instabilities are reflected by the "jumps" in the simulated curves (marked by arrows in Figure 2.11), where only stable oscillation states are obtained. Obviously, they are different for increasing and decreasing driving frequency. This is a well-known effect frequently observed in nonlinear oscillators that leads also to a hysteresis in amplitude versus distance curves (Figure 2.11b). As a rule of thumb, it can be said that the tip-sample forces in AM-mode are in the attractive range before the jump and repulsive after the jump (Hölscher et al., 2006). Therefore, it is highly advantageous to scan delicate surfaces with a high amplitude set-point before the jump. The increase in resolution caused by the reduction of the tip-sample forces by this procedure has been nicely demonstrated by San Paulo and García (2000).

---

[*]The "−" sign on the right side of this equation is due to our definition of the phase ϕ in Equation 2.14.

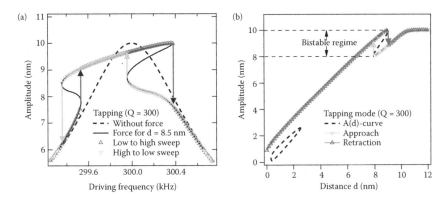

FIGURE 2.11 **(See color insert.)** (a) Resonance curve for AM-mode operation if the cantilever oscillates near the sample surface with $d = 8.5$ nm and $A_0 = 10$ nm, thereby experiencing the model force field given by Equation 2.9. The solid lines represent the analytical result of Equation 2.24a, while the symbols are obtained from the numerical solution of the equation of motion Equation 2.12. The dashed lines reflect the resonance curves without tip-sample force and are shown purely for comparison. The resonance curve exhibits instabilities ("jumps") during a frequency sweep. These jumps take place at different positions (marked by arrows) depending on whether the driving frequency is increased or decreased. (b) A hysteresis is also observed for amplitude versus distance curves. The dashed line shows the analytical result, and the symbols show the numerical solutions for approach and retraction using a driving frequency of 300 kHz and the same parameters as in (a).

2.4.4 FORCE SPECTROSCOPY USING THE AM-MODE

In the above subsections, we have outlined the influence of the tip-sample interaction on the cantilever bending and oscillation based on the assumption of a specific model force. However, in practical imaging, the tip-sample interaction is not *a priori* known. In contrast, the ability to measure the continuous tip-sample interaction force as a function of the tip-sample distance would add a tool of great value to the force microscopist's toolbox. Since the cantilever reacts to the interaction between tip and sample in a highly nonlinear way, one might wonder how that could be done.

Surprisingly, despite the long time that the AM-mode is already used, it was only recently that solutions to this inversion problem have been found (Hölscher, 2006; Lee & Jhe, 2006; Hu & Raman, 2008). We start our analysis by applying the transformation $D = d - A$ to the integral I_+ Equation 2.22a, where D corresponds to the nearest tip-sample distance as defined in Figure 2.8. Next, we note that due to the cantilever oscillation, the current method intrinsically recovers the values of the force that the tip experiences at its lower turning point, where $F_\downarrow$ necessarily equals $F_\uparrow$. We thus define $F_{ts} = (F_\downarrow + F_\uparrow)/2$, and Equation 2.22a subsequently reads as

$$I_+ = \frac{2}{\pi k_{cant} A^2} \int_D^{D+2A} F_{ts} \frac{z - D - A}{\sqrt{A^2 - (z - D - A)^2}} dz \qquad (2.25)$$

The amplitudes commonly used in AM mode are considerably larger than the interaction range of the tip-sample force. Consequently, tip-sample forces in the integration range between $D+A$ and $D+2A$ are practicably insignificant. For this so-called "large amplitude approximation" (Giessibl, 1997; Dürig, 1999), the last term can be expanded at $z \to D$ to $(z-D-A)/\sqrt{A^2-(z-D-A)^2} \approx -\sqrt{A/2(z-D)}$, resulting in

$$I_+ \approx -\frac{\sqrt{2}}{\pi k_{cant} A^{3/2}} \int\limits_{D}^{D+2A} \frac{F_{ts}}{\sqrt{z-D}} dz \qquad (2.26)$$

Introducing this equation into Equation 2.21a, we obtain the following integral equation:

$$\underbrace{\frac{k_{cant} A^{3/2}}{\sqrt{2}} \left[\frac{a_d \cos(\phi)}{A} - \frac{f_0^2 - f_d^2}{f_0^2} \right]}_{\kappa(D)} = \frac{1}{\pi} \int\limits_{D}^{D+2A} \frac{F_{ts}}{\sqrt{z-D}} dz \qquad (2.27)$$

The left-hand side of this equation contains only experimentally accessible data, and we denote this term as $\kappa(D)$, which is inherently a function of the nearest tip-sample distance D. The benefit of these transformations is that the integral equation can be inverted (Dürig, 1999) and, as a final result, we find

$$F_{ts}(D) = -\frac{\partial}{\partial D} \int\limits_{D}^{D+2A} \frac{\kappa(z)}{\sqrt{z-D}} dz \qquad (2.28)$$

It is now straightforward to recover the tip-sample force using Equation 2.28 from a "spectroscopy experiment," that is, an experiment where the amplitude and the phase are continuously measured as a function of the actual tip-sample distance $D = d - A$ at a fixed location. With this input, one first calculates κ as a function of D. In a second step, the tip-sample force is computed, solving the integral in Equation 2.28 numerically.

Additional information about the tip-sample interaction can be obtained noticing that the integral I_- is directly connected to the energy dissipation ΔE. By simply combining Equations 2.21b and 2.22b, we get

$$\Delta E = \left(\frac{1}{Q_0} \frac{f_d}{f_0} + \frac{a_{exc}}{A} \sin\phi \right) \pi k_{cant} A^2 \qquad (2.29)$$

The same result has been found earlier by Cleveland et al. (1998) using the conservation of energy principle. However, exceeding Cleveland's work, we suggest to plot the energy dissipation as a function of the nearest tip-sample distance $D = d - A$ to have the same scaling as for the tip-sample force.

A verification of the algorithm is shown in Figure 2.12, where we present computer simulations of the method by calculating numerical solutions of the equation of motion with a fourth-order Runge–Kutta method (Press et al., 1992). To be able to check both Equation 2.28 and Equation 2.29, we need to add a dissipative

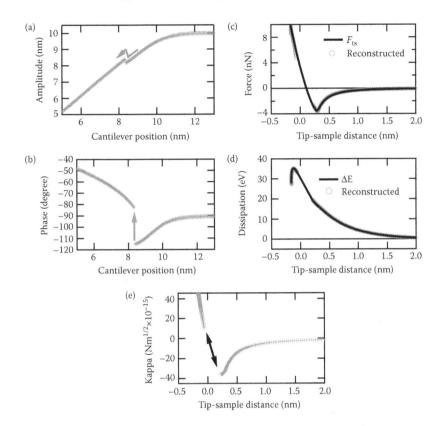

FIGURE 2.12 A numerical verification of the proposed algorithm. On the basis of the equation of motion Equation 2.12, the amplitude (a) and phase (b) versus distance curves during the approach toward the sample surface have been numerically calculated. Both curves show the instability that is typical for AM-DFM operation in ambient conditions. As described in the text, the data is used for the reconstruction of the tip-sample force (c) and the energy dissipation (d). The assumed tip-sample model interactions according to Equations 2.9c and 2.30d are plotted by solid lines. Finally, (e) reflects the $\kappa(D)$ values that can be computed from the amplitude and phase values given in (a) and (b).

component to our original model interaction force F_{DMT-M} Equation 2.9. Instead of exploring elaborate energy dissipation mechanisms, it is sufficient for principle demonstration to simply add an additional dissipative force term F_{diss}, that is, $F_{ts} = F_{air} + F_{diss}$. To characterize F_{diss}, we chose viscous damping with an exponential distance-dependence: $F_{diss} = F_0 \exp(-z/\zeta_0)\dot{z}$. The energy dissipation caused by this type of dissipation is given by (Gotsmann et al., 1999)

$$\Delta E = 4\pi^2 f_d A F_0 \zeta_0 \exp\left(-\frac{D+A}{\zeta_0}\right) I_1\left(\frac{A}{\zeta_0}\right) \quad (2.30)$$

where I_1 is the modified Bessel function of the first kind.

Figure 2.12 displays the resulting amplitude and phase versus distance curves during approach, respectively. The assumed parameters and the conservative force are the same as in Figure 2.13b while the following parameters have been used for the dissipative force: $F_0 = 10^{-6}$ Ns/m and $\zeta_0 = 0.5$ nm. Again the amplitude curve shows the previously discussed discontinuity caused by an instability.

The subsequent reconstruction of F_{ts} and ΔE based on the data provided by the amplitude and phase versus distance curves is presented in Figures 2.12c and 2.12d. The assumed tip-sample force and energy dissipation are plotted by solid lines, while the reconstructed data is indicated by symbols; the excellent agreement demonstrates the reliability of the method. Nonetheless, it is important to recognize that the often observed instability in amplitude and phase versus distance curves affects the reconstruction of the tip-sample force. If such an instability occurs, experimentally accessible $\kappa(D)$ values will feature a "gap" at a specific range of tip-sample distances D. This issue is illustrated in Figure 2.12e, where the gap is indicated by an arrow. As a consequence, one might be tempted to interpolate the missing κ-values in the gap. This is a workable solution if, as in our example, the accessible κ-values appear smooth and, in particular, the lower turning point of the $\kappa(D)$ values is clearly visible. In most realistic cases, however, the $\kappa(D)$ values will not look so smooth as in our simulation and/or the lower turning point might not be reached, and uttermost caution is advised when applying any inter- or extrapolation.

Finally, let us note two more issues: (a) The reconstruction of the energy dissipation does not require the continuous knowledge of κ-values. Thus, it is not influenced by the instability and gives reliable values also after the jump (Figure 2.12d). (b) The "large amplitude approximation" is not a prerequisite for the inversion of the tip-sample forces from the amplitude and phase data. The application of other numerical methods (Lee & Jhe, 2006; Hu & Raman, 2008; Hölscher, 2008) where the amplitude is not restricted to large values is also possible.

2.4.5 THEORY OF FM-MODE

After the analysis of the AM-mode, we now give the solution of the equation of motion Equation 2.12 for the FM-mode. As before we assume that the tip-sample force is so small and the Q-factor so high that, as a consequence, higher harmonics can be neglected. Inserting again the ansatz Equation 2.17 leads now to a set of two coupled trigonometric equations

$$\frac{a_{exc}}{A}\cos(2\pi f t_0) = \frac{f^2 - f_0^2}{f_0^2} - I_+(d,A) \tag{2.31a}$$

$$\frac{a_{exc}}{A}\sin(2\pi f t_0) = \frac{1}{Q}\frac{f}{f_0} - I_-(d,A) \tag{2.31b}$$

where we again defined the two integrals I_+ Equation 2.22a and I_- Equation 2.22b.

Both equations can be simplified for the conditions typically found in DFM experiments, where the FM-mode is applied. First, we assume that the frequency shift $\Delta f := f - f_0$ caused by the tip-sample interaction and the damping is small compared with the resonance frequency of the free cantilever ($\Rightarrow f/f_0 \approx 1$ and $f^2 - f_0^2 \approx -2\Delta f f_0$). Second, we consider that the phase shift is typically set to $90°$ in the FM-mode. In this case, the terms on the left side are given by $\cos(2\pi f t_0) = 0$ and $\sin(2\pi f t_0) = 1$. Due to these simplifications, the frequency shift and the driving amplitude can be calculated from

$$\Delta f(d,A) = -\frac{f_0}{2} I_+ = -\frac{f_0}{\pi k_{\text{cant}} A^2} \int_{d-A}^{d+A} F_{\text{ts}} \frac{z-d}{\sqrt{A^2-(z-d)^2}} dz \qquad (2.32a)$$

$$a_{\text{exc}} = \frac{A}{Q} + \frac{\Delta E(d,A)}{\pi k_{\text{cant}} A} \qquad (2.32b)$$

These two equations are valid for every type of interaction as long as the resulting cantilever oscillation is nearly sinusoidal. For completeness we note that Equation 2.32a coincides with the result for the FM-mode with constant-oscillation amplitude mostly used in vacuum (Giessibl, 1997; Dürig, 1999).

2.4.6 Force Spectroscopy Using the FM-Mode

In the above subsection, we have calculated the influence of the tip-sample interaction on the cantilever oscillation with the assumption that the tip-sample force is known. In practice, however, it is just the other way around: The tip-sample interaction is unknown. As in the AM-mode, however, it is of the great interest for the experimentalist to measure the tip-sample interaction force as a function of the tip-sample distance.

To solve this problem, we again simplify the integral Equation 2.22a to the form given in Equation 2.25. Introducing now this equation into Equation 2.32a, we obtain the following integral equation:

$$\Delta f(D,A) \approx \frac{f_0}{\sqrt{2}\pi k_{\text{cant}} A^{3/2}} \int_{D}^{D+2A} \frac{F_{\text{ts}}(z)}{\sqrt{z-D}} dz \qquad (2.33)$$

The inversion of this equation now leads to the following formula for the tip-sample interaction potential (Dürig, 1999; Hölscher et al., 2000)

$$V_{\text{ts}}(D) = \sqrt{2} \int_{D}^{D+2A} \frac{k_{\text{cant}} A(z')^{3/2}}{f_0} \frac{\Delta f(z')}{\sqrt{z'-D}} dz' \qquad (2.34)$$

Consequently, the tip-sample force is given by

$$F_{\text{ts}}(D) = -\sqrt{2}\frac{\partial}{\partial D} \int_{D}^{D+2A} \frac{\Delta \gamma(z')}{\sqrt{z'-D}} dz' \qquad (2.35)$$

where we defined the so-called *normalized frequency shift* (Giessibl, 1997)

$$\gamma(D) := \frac{k_{cant}A^{3/2}}{f_0}\Delta f(D) \tag{2.36}$$

which is independent of the oscillation amplitude, but a function of the nearest tip-sample distance D (Hölscher et al., 2000). To recover the tip-sample interaction force from a spectroscopy experiment, we have to measure the frequency shift and the driving amplitude as a function of the tip-sample distance before we calculate the normalized frequency shift. After that, we introduce this data into Equation 2.35.

For completeness we would like to mention that the "large amplitude approximation" used for the deviation of Equation 2.25 is not a prerequisite for the reconstruction of the tip-sample force. Inversion methods suitable for small amplitudes have been presented by several researchers (Dürig, 2000a; Giessibl, 2001; Pfeiffer, 2004; Sader & Jarvis, 2004). The method of Sader & Jarvis (2004), for example, can also be applied to our case and adds two additional terms to our previous formula

$$F_{ts}^{SJ}(D) = -\sqrt{2}\frac{\partial}{\partial D}\int_D^\infty \frac{k_{cant}A^{3/2}}{f_0}\frac{\Delta f(z)}{\sqrt{z-D}}$$
$$+\sqrt{2}k_{cant}\frac{\Delta f}{f_0}\left((z-D)+\frac{A^{1/2}}{4}\sqrt{\frac{z-D}{\pi}}\right)dz \tag{2.37}$$

The calculation of the energy dissipation ΔE is straightforward using Equation 2.32b

$$\Delta E(D,A) \cong \pi k_{cant}\left(A(D)\,a_{exc} - \frac{A(D)^2}{Q}\right) \tag{2.38}$$

It might be interesting to note that this equation follows also from the conservation of energy (Cleveland et al., 1998; Gotsmann et al., 1999).

Since the quantitative value of excitation amplitude a_{exc} is typically unknown in an experiment, it is convenient to determine a_{exc} from the freely oscillating cantilever with the help of Equation 2.19b. As a result, we get

$$\Delta E(D) = \frac{\pi k_{cant}A(D)}{Q}(A_0 - A(D)) \tag{2.39}$$

where A_0 is the oscillation amplitude of the freely oscillating cantilever (i.e., with negligible tip-sample force).

2.5   EXPERIMENTAL COMPARISON OF AM- AND FM-MODE

The setup presented in the previous subsections can be easily realized by combining a commercial dynamic force microscope (MultiMode AFM with NanoScope IIIa Controller, Veeco Instruments Inc.) with an additional electronics for the CE mode. In this way, it is also possible to switch between AM- and FM-mode using the same cantilever and sample. The spring constants k_{cant} of the cantilever can be determined via the resonant frequency f_0 of the freely oscillating cantilever (Sader et al., 1995), while their quality factors Q can be obtained from resonance curves (Ebeling et al., 2006b).

To illustrate the main differences between the "conventional" AM-mode and the presently much less used FM-mode in air, we present two spectroscopy experiments in Figure 2.13, where the oscillating cantilever was approached to and retracted from a mica surface in both modes. The corresponding spectroscopy curves are presented in Figures 2.13a and 2.13b. The measured quantities in the AM-mode are amplitude and phase, whereas amplitude and frequency shift are recorded in FM-mode.

As already discussed by others (Anczykowski et al., 1996; San Paulo & Garcìa, 2002; Hölscher & Schwarz, 2007), the amplitude and phase shift curves recorded in AM-mode sometimes show a significant hysteresis during approach and retraction. At specific positions (marked by arrows in Figure 2.13a), the oscillation becomes

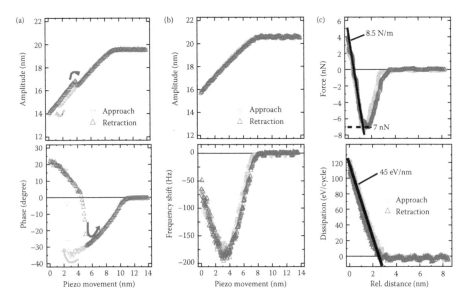

FIGURE 2.13 Examples of "spectroscopy measurements" obtained on mica in ambient conditions. (a) Amplitude and phase versus distance curves in the tapping mode. The instabilities during approach and retraction cause a hysteresis. (b) Such a behavior is not observed in the constant-excitation mode where the approach and retraction curves of the amplitude and frequency shift are identical within the noise limit. The parameters of the cantilever were $f_0 = 167\,224$ Hz, $k_{cant} = 37.5$ N/m, and $Q = 465$. (c) Using the algorithm described in the text, we reconstruct the tip-sample potential and force from the data sets shown in (b). The interaction force decreases until it reaches a minimum of -7 nN and increases again with a slope of 8.5 N/m. The origin of all x-axes has been arbitrarily set to the left of the graphs.

unstable and the cantilever jumps into another stable oscillation state. However, such a hysteresis is not present in the spectroscopy curves measured in the FM-mode due to the specific self-oscillation technique (Hölscher et al., 2002). As shown in Figure 2.13b, the particular amplitude and frequency shift curves are identical within the noise limit and piezo hysteresis for approach and retraction. The amplitude is constant until the tip senses the interaction with the sample surface and decreases continuously during further approach. The frequency shift curves show a decrease and increase of the resonant frequency with a distinct minimum.

As already pointed out, the continuous approach and retraction curves obtained in the FM mode allow the reconstruction of the tip-sample interaction by an inversion algorithm. An application of this procedure to the spectroscopy data is plotted in Figure 2.13c and reveals the tip-sample force and the energy dissipation per oscillation cycle.

The tip-sample force shows a distinct minimum of $-7\,nN$. This is the minimal force needed to retract the tip from the sample surface. Therefore, we denote it as *adhesion force* in the following. During further approach, the tip-sample force increases with a slope of 8.5 N/m as shown by a linear fit (solid line). This linear increase in the tip-sample force is caused by the contact of tip and sample. Justified by the almost linear increase, we use the term *contact stiffness* for the slope obtained by this linear fitting procedure. Interestingly, the energy dissipation curve begins to rise with the onset of the attractive tip-sample force and shows also a linear increase with a slope of 45 eV/nm. The physical origin of the energy dissipation, however, is still under discussion (Kantorovich and Trevethan, 2004; Schirmeisen and Hölscher, 2005; Hoffmann et al., 2007).

2.6 MAPPING OF THE TIP-SAMPLE INTERACTIONS ON DPPC MONOLAYERS

In the previous section, we showed that the FM technique is well suited to measure the tip-sample force at an arbitrary position. However, it is also possible to record the tip-sample interaction in a more defined way as contour maps perpendicular to the sample surface. To examine the possible resolution of this approach under ambient conditions, we recorded sets of spectroscopy curves along predefined scan lines on DPPC, which frequently serves as a model for membranes (Sackmann, 1996). The tip-sample interaction was subsequently calculated from the measured amplitude and frequency shift versus distance curves with respect to the actual scan position. The obtained curves were then plotted in a color-coded contour map showing the potential of the tip-sample interaction.

Monolayers of DPPC were prepared with the Langmuir–Blodgett technique. As shown in the topography image in Figure 2.14, the monolayers have a lateral structure of alternating stripes and channels. This specific pattern is obtained by rapidly withdrawing the mica substrate at a low monolayer surface pressure and constant temperature as described by Gleiche et al. (2000). The stripes consist of DPPC in a liquid condensed phase (LC-phase), whereas the channels between the stripes are filled with DPPC in the liquid expanded phase (LE-phase) (Chen et al., 2007). The lateral periodicity of stripes and channels depends on the parameters used during the preparation of the sample.

We imaged the sample using the oscillation amplitude as a feedback signal in the FM-mode using CE before we recorded 50 spectroscopy curves along a predefined direction marked in Figure 2.14a. All data sets were than transformed into tip-sample potential curves using the mathematical method described in Section 2.4.5. Finally, we computed the corresponding contour map as shown in Figure 2.14. The complete procedure was done by a computer script using IGOR Pro software (Wavemetrics Inc.).

The resulting color-coded image reveals the different tip-sample interaction on the stripes (LC-phase) and in the channels (LE-phase). The potential is significantly larger above the stripes ($\approx -100\,\text{eV}$) compared with the channels ($\approx -150\,\text{eV}$), as it can be seen by the color coding in Figure 2.14.

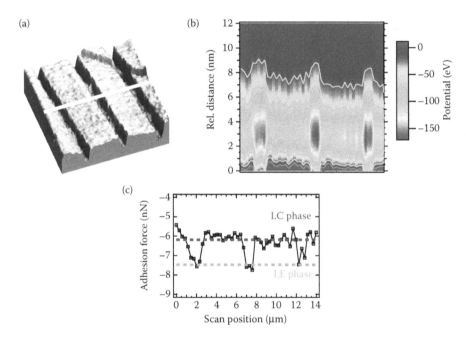

FIGURE 2.14   (See color insert.) (a) Surface plot (scan size: $14 \times 14\,\mu\text{m}^2$) of the topography of the L-α-dipalmitoyl-phosphatidycholine (DPPC) film prepared by the Langmuir–Blogdett technique. The monolayer shows alternating stripes and channels that consists of DPPC adsorbed in the liquid condensed and liquid expanded phase, respectively. The white line marks the position where we recorded the frequency shift and amplitude versus distance curves for the construction of the contour map of the tip-sample interaction potential shown in (b). The graph in (c) displays the corresponding adhesion force obtained from the data shown in (b). The parameters of the cantilever were $f_0 = 170,460\,\text{Hz}$, $k_{\text{cant}} = 39.6\,\text{N/m}$, and $Q = 492$.

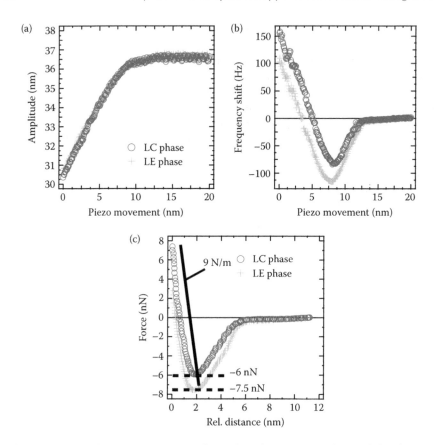

FIGURE 2.15 (a) Spectroscopy curves obtained on the stripes (circles) and the channels (crosses). The tip–sample interaction is calculated from this amplitude and frequency shift versus distance curves. (b) Using the numerical procedure described in the text, we calculated the corresponding tip–sample interaction force and potential. A significant difference between the curves is observed for distances between 2 and 6 nm.

The local stiffness and adhesion force can be determined from the force curves plotted in Figure 2.15. Here, we show amplitude and frequency shift versus distance curves and the resulting tip–sample force measured on the LC- and LE-phase. The force curves reveal an adhesion force of –6 and –7.5 nN on the stripes and channels, respectively. However, the local stiffness is about 9 N/m for both positions, and we could not determine a significant difference in the stiffness between the two phases of the DPPC monolayers. This outcome can be attributed to the fact that the repulsive interaction forces for a thin film depends strongly on its thickness (Domke & Radmacher, 1998). Therefore, the contact stiffness measured on the two phases of the DPPC monolayer is dominated by the mica substrate.

2.7 DYNAMIC FORCE SPECTROSCOPY OF SINGLE DEXTRAN MONOMERS

As already mentioned, AFM is frequently used for the precise measurement of the forces acting on chain-like molecules during their stretching. In this section, we show how the forces acting on a chain-like molecule in liquid can also be measured in the CE mode of the FM technique. The difference between the conventional static mode and our dynamic approach is depicted in Figure 2.7. In the static mode (Figure 2.7a), the cantilever is approached and retracted from the surface covered with chain-like molecules (dextran monomers in our case). If one of these molecules has bound to the tip, it is stretched during the retraction of the cantilever from the surface. A measurement of the cantilever bending Δz during the retraction is used to measure the force versus extension curve of the molecule. In the dynamic mode, however, the cantilever oscillates between the positions D_{down} and D_{up}, while it is approached and retracted from the surface (Figure 2.7b). Since the chain-like molecule exerts an attractive force to the cantilever, its fundamental features like the oscillation amplitude A and resonance frequency f will be modified. In the following, we use this effect to record the force versus extension curve together with the energy dissipation.

The dynamic force spectroscopy experiments were measured with the same set-up already described in the previous sections. For the sample preparation, the polysaccharide dextran (Dextran 500, Sigma Aldrich) was solved in pure water (≈ 0.2 g/ml). Afterwards a small amount (100 µl) of this solution was dropped onto a cleaned glass substrate. After drying, the substrate was rinsed several times with water to remove the loose molecule chains. The measurements were performed in a droplet (40–50 µl) of buffer solution (PBS Dulbecco, Biochrom AG) in an open liquid cell without an o-ring. All data sets shown here were obtained with a rectangular silicon cantilever (MikroMasch NSC36). Its spring constant in liquid was determined using the method given by by Sader et al. (1999). The used cantilever had a spring constant of $k_{cant} = 24.6$ N/m, an eigenfrequency of $f_0 = 28.3$ kHz, and a quality factor of $Q = 64$.

Figure 2.16 shows a typical data set recorded with this experimental set-up. The oscillation amplitude (Figure 2.16a) and the shift of the resonance frequency (Figure 2.16b) are displayed as a function of the cantilever-sample distance d. Both the amplitude A and the frequency shift $\Delta f = f - f_0$ remain almost constant during the approach. Only when the tip comes very close to the surface, the repulsive tip-sample interaction leads to a strong decrease (amplitude) and increase (frequency shift) of these two signals. During the retraction, however, a dextran strand adhered to the tip, resulting into a very characteristic pattern in the amplitude and in the frequency shift data. A minimum, a maximum, and a sudden jump appear as prominent features in the two curves. We observed that this pattern is a clear sign for a dextran molecule stretched during the retraction.

In the previous sections, we demonstrated how the tip-sample force can be reconstructed from the frequency shift and amplitude data via an integral equation. This approach, however, was based on the assumption that the tip-sample force decreases to zero if the tip is far away from the sample surface. In our case, however, it is

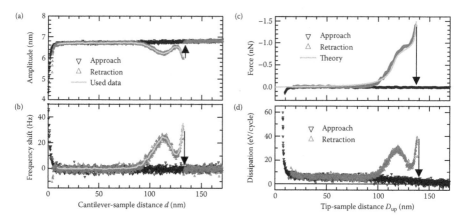

FIGURE 2.16 (See color insert.) An application of the introduced dynamic force spectroscopy technique. (a) Amplitude (A) and (b) frequency shift (Δf) curves measured during approach to and retraction from the surface covered with dextran molecules. During the retraction, one dextran molecule bound to the tip as revealed by the change of the frequency and amplitude signal. At a position of about 135 nm (see arrows), the maximum binding force was exceeded and the cantilever oscillated freely again. Only the data before this jump is used for the subsequent analysis. (c) Using Equation 2.40, the force acting on the dextran molecule (symbols) is reconstructed as a function of the actual tip position D_{up}. The experimental result is well described by a "single-click" model using only the number of molecules ($N = 266$) as fitting parameter (solid line). (d) The energy dissipated per oscillation cycle can be calculated from Equation 2.39 for approach and retraction. The zero of the x-axes is arbitrarily set to the left side of the graphs.

just the other way around. The attractive force caused by the dextran molecule is practically zero at the surface and increases during the retraction from the sample surface. Consequently, we have to switch the integration limits of the integral equation used to calculate the extension force in dynamic force spectroscopy. Using the approach of Sader and Jarvis (2004), we obtain

$$F(D_{up}) = -\frac{\partial}{\partial D_{up}} 2k_{cant} \int_{0}^{D_{up}} \frac{\Delta f(z)}{f_0} \tag{2.40}$$

$$\times \left[(D_{up} - z) + \sqrt{\frac{A(z)}{16\pi}} \sqrt{D_{up} - z} + \frac{A(z)^{3/2}}{\sqrt{2(D_{up} - z)}} \right] dz \tag{2.41}$$

where the upper turning point D_{up} is easily calculated from the transformation $D_{up} = d + A$ (see Figure 2.7b). The energy dissipated during one oscillation cycle can be directly calculated using Equation 2.39.

Due to the jump in the frequency shift and amplitude curve, we can apply these formulas only for the data on the left side of the jumps. (After the jump, the molecule is ruptured from the tip and the tip-sample interaction is practically zero.) The solid lines on the left side of the arrows in Figures 2.16a and 2.16b mark this subset. Introducing this data into Equation 2.40, we obtain the force versus distance curve displayed in Figure 2.16c.

As it is well known from previous studies of dextran (Rief et al., 1997, 1998; Haverkamp et al., 2007) the force curve reveals a kink in the range of about 700 pN. It originates from a conformational transition within each dextran monomer. Due to the external applied force, each sugar ring of the dextran monomer flips into a new conformation (Rief et al., 1998), resulting in an additional elongation (about 10%) of the monomer. This is a thermodynamic process that can be described with a "single-click" model (Haverkamp et al., 2007). The solid line in Figure 2.16c shows a fit of this model to the experimental data. The agreement is remarkable because the only fit parameter is the number of monomers in the dextran molecule. All other parameters were fixed and chosen as given by Rief et al. (1998) and Haverkamp et al. (2007).

This agreement was also observed for other stretching events. Figure 2.17 summarizes the measured force curves of three dextran strands of different length. Again, we could fit the "single-click" model to this data. Since this model is not time or velocity dependent and is based on the assumption that the number of folded and unfolded molecules are in a thermodynamic equilibrium, this outcome proves that we indeed measured the equilibrium force of the dextran molecule even with our dynamic approach.

As mentioned previously, it is also possible to calculate the dissipated energy per oscillation cycle via Equation 2.39 in dynamic force spectroscopy. The resulting dissipation curve is shown in Figure 2.16d as a function of the tip-sample distance and shows a maximum at a position corresponding to the kink in the force curve.

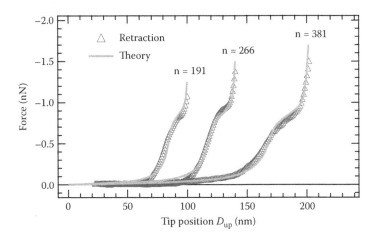

FIGURE 2.17 Force versus extension curves of three different dextran strands measured with the proposed dynamic approach. The symbols represent experimental data that is well reproduced by the theoretical force curves, where only the number of dextran monomers were used as a fit parameter ($n = 191$, 266, and 381, respectively).

We always observed this coincidence in the conservative and dissipative interaction curves. Since the kink appears at the position where most of the dextran monomers unfold, it is very likely that a large amount of energy dissipation is caused by this unfolding processes. However, the dissipation increases also shortly before the rupture of the dextran strand from the tip. At this position, the majority of the dextran monomers is already unfolded. Therefore, we conclude that other effects like hydrodynamic damping and/or the rupture of the molecule from the tip will contribute to the energy dissipation.

2.8   CONCLUSION AND PERSPECTIVES

In summary, an overview over the basic principles of AFM driven in the contact as well as in dynamic modes is presented. While the contact mode is very easy to apply and the standard technique to measure force versus distance curves, the application of the tapping-mode in air and liquids is an everyday tool in nanotechnology, enabling the imaging of sample surfaces with very high resolution. In addition, it allows the quantitative interpretation of the tip-sample interactions.

AFM is a nice example for the often observed incident that scientific progress is frequently triggered by the development of new experimental techniques. The friction force microscope (Mate et al., 1987) is an instance since it opened a complete new field of science: the analysis of friction and wear at the atomic-scale (Gnecco & Meyer, 2006; Hölscher et al., 2008). Many other recent achievements in the field of nanotechnology are unthinkable without the help of AFM and other scanning probe methods. However, despite these success stories, there is still room for improvements. From my viewpoint, the following recent developments might be of interest for the AFM enthusiast.

- Scanning a sample surface with an AFM enables the experimentalist to scan the surface topography with high resolution but additional effort has to be spent to map other physical quantities. As described in this chapter, it is possible to determine the tip-sample forces with high resolution at arbitrary sample positions. This, however, has to be done after imaging the topography. Of course, it is possible to do this on a fine grid, reconstructing again the surface topography but this is quite time-consuming. A possible solution to this problem was given by Sahin et al., (2007) who used torsional harmonic cantilevers to measure the time varying forces between tip and substrate. In this way it is possible to determine the indentation forces on soft samples like polymers during scanning.
- It is often criticized that the AFM features atomic-scale resolution but no chemical identification. Of course, it is often possible to distinguish between different materials through friction force (Figure 2.5) or phase images (Figure 2.9). Even the chirality of molecules can be detected in this way (McKendry et al., 1998). Also, on the atomic scale, it is possible to distinguish different atomic elements as nicely shown by Sugimoto et al. (2007). Previous knowledge about the sample, however, is necessary in these cases.

A very promising way to determine the chemical nature of a sample without any knowledge is the combination of AFM with infrared spectroscopy. Dazzi et al. (2005, 2007) have shown how this can be done.

This selection, however, is personal colored and one should always keep in mind the quote of Niels Bohr (1885–1962) that *"prediction is very difficult, especially about the future."* Consequently, the next breakthrough might come from a completely different direction.

REFERENCES

Aimé, J. P., Boisgard, R., Nony, L., & Couturier, G. (1999). Nonlinear dynamic behavior of an oscillating tip-microlever system and contrast at the atomic scale. *Phys. Rev. Lett.*, 82, 3388–3391.

Albrecht, T. R., Grütter, P., Horne, D., & Rugar, D. (1991). Frequency modulation detection using high-Q cantilevers for enhanced force microscope sensitivity. *J. Appl. Phys.*, 69, 668–673.

Alexander, S., Hellemans, L., Marti, O., Schneir, J., Elings, V., & Hansma, P. K. (1988). An atomic-resolution atomic-force microscope implemented using an optical lever. *J. Appl. Phys.*, 65, 164–167.

Allers, W., Schwarz, A., Schwarz, U. D., & Wiesendanger, R. (1998). A scanning force microscope with atomic resolution in ultrahigh vacuum and at low temperatures. *Rev. Sci. Instrum.*, 69, 221–225.

Anczykowski, B., Krüger, D., & Fuchs, H. (1996). Cantilever dynamics in quasinoncontact force microscopy: Spectroscopic aspects. *Phys. Rev. B*, 53, 15485–15488.

Bhushan, B. & Marti, O. (2005). Scanning probe microscopy–principle of operation, instrumentation, and probes. In B. Bhushan (Ed.), *Nanotribology and Nanomechanics – An Introduction* (pp. 41–115). Berlin, Heidelberg: Springer-Verlag.

Bilas, P., Romana, L., Kraus, B., Bercion, Y., & Mansot, J. L. (2004). Quantitative characterization of friction coefficient using lateral force microscope in the wearless regime. *Rev. Sci. Instrum.*, 75, 415–421.

Binnig, G., Quate, C. F., & Gerber, C. (1986). Atomic force microscopy. *Phys. Rev. Lett.*, 56, 930–933.

Binnig, G., Gerber, C., Stoll, E., Albrecht, T. R., & Quate, C. F. (1987). Atomic resolution with atomic force microscope. *Europhys. Lett.*, 3, 1281.

Binnig, G., Quate, C. F., & Gerber, C. (1986). Atomic force microscopy. *Phys. Rev. Lett.*, 56, 930–933.

Burnham, N. A. & Colton, R. J. (1989). Measuring the nanomechanical properties and surface forces of materials using an atomic force microscopy. *J. Vac. Sci. Technol. A*, 7, 2906.

Butt, H.-J. & Jaschke, M. (1995). Calculation of thermal noise in atomic force microscopy. *Nanotechnology*, 6, 1.

Butt, H.-J. & Kappl, M. (2010). *Surface and Interfacial Forces*. Weinheim: Wiley-VCH.

Chen, X., Lenhert, S., Hirtz, M., Lu, N., Fuchs, H., & Chi, L. (2007). Langmuir-blodgett pattering: A bottom-up way to build mesostructures over large areas. *Acc. Chem. Res.*, 40, 393–401.

Cleveland, J. P., Anczykowski, B., Schmid, A. E., & Elings, V. B. (1998). Energy dissipation in tapping-mode atomic force microscopy. *Appl. Phys. Lett.*, 72, 2613.

Cook, S. M., Schäffer, T. E., Chynoweth, K. M., Wigton, M., Simmonds, R. W., & Lang, K. M. (2006). Practical implementation of dynamic methods for measuring atomic force microscope cantilever spring constants. *Nanotechnology*, 17, 2135–2145.

Dazzi, A., Prazeres, R., Glotin, F., & Ortega, J. M. (2005). Local infrared microspectroscopy with subwavelength spatial resolution with an atomic force microscope tip used as a photothermal sensor. *Opt. Lett.*, 30(18), 2388–2390.

Dazzi, A., Prazeres, R., Glotin, F., & Ortega, J. (2007). Analysis of nano-chemical mapping performed by an afm-based acousto-optic technique. *Ultramicroscopy*, 107(12), 1194–1200.

Derjaguin, B. V., Muller, V. M., & Toporov, Y. P. (1975). Effect of contact deformations on the adhesion of particles. *J. Colloid Interface Sci.*, 53, 314–326.

Domke, J. & Radmacher, M. (1998). Measuring the elastic properties of thin polymer films with the atomic force microscope. *Langmuir*, 14, 3320–325.

Dürig, U. (1999). Relations between interaction force and frequency shift in large-amplitude dynamic force microscopy. *Appl. Phys. Lett.*, 75, 433–435.

Dürig, U. (2000a). Extracting interaction forces and complementary observables in dynamic probe microscopy. *Appl. Phys. Lett.*, 76, 1203–1205.

Dürig, U. (2000b). Interaction sensing in dynamic force microscopy. *N. J. of Phys.*, 2, 5.1–5.12.

Ebeling, D., Hölscher, H., & Anczykowski, B. (2006a). Increasing the Q-factor in the constant-excitation mode of frequency-modulation atomic force microscopy in liquid. *Appl. Phys. Lett.*, 89, 203511.

Ebeling, D., Hölscher, H., Fuchs, H., Anczykowski, B., & Schwarz, U. D. (2006b). Imaging of biomaterials in liquids: A comparison between conventional and Q-controlled amplitude modulation ("tapping mode") atomic force microscopy. *Nanotechnology*, 17, S221–S226.

Farell, A. A., Fukuma, T., Uchihashi, T., Kay, E. R., Bottari, G., Leigh, D. A., Yamada, H., & Jarvis, S. P. (2005). Conservative and dissipative forces imaging of switchable rotaxanes with frequency modulation atomic force microscopy. *Phys. Rev. B*, 72, 125430.

Fukuma, T., Ichii, T., Kobayashi, K., Yamadaa, H., & Matsushige, K. (2005a). True-molecular resolution imaging by frequency modulation atomic force microscopy in various environments. *Appl. Phys. Lett.*, 86, 034103.

Fukuma, T., Kobayashi, K., Matsushige, K., & Yamada, H. (2005b). True molecular resolution in liquid by frequency-modulation atomic force microscopy. *Appl. Phys. Lett.*, 86, 193108.

Garcia, R. & Pérez, R. (2002). Dynamic atomic force microscopy methods. *Surf. Sci. Rep.*, 47, 197–301.

Garcia, R., Gómez, C. J., Martinez, N. F., Patil, S., Dietz, C., & Magerle, R. (2006). Identification of nanoscale dissipation processes by dynamic atomic force microscopy. *Phys. Rev. Lett.*, 97, 016103.

Giessibl, F. J. (1997). Forces and frequency shifts in atomic-resolution dynamic-force microscopy. *Phys. Rev. B*, 56(24), 16010–16015.

Giessibl, F.-J. (2001). A direct method to calculate tip-sample forces from frequency shifts in frequency-modulation atomic force microscopy. *Appl. Phys. Lett.*, 78, 123–125.

Giessibl, F.-J. (2003). Advances in atomic force microscopy. *Rev. Mod. Phys.*, 75, 949–983.

Gleiche, M., Chi, L. F., & Fuchs, H. (2000). Nanoscopic channel lattices with controlled anisotropic wetting. *Nature*, 403, 173–175.

Gleyzes, P., Kuo, P. K., & Boccara, A. C. (1991). Bistable behavior of a vibrating tip near a solid surface. *Appl. Phys. Lett.*, 58, 2989–2991.

Gnecco, E. & Meyer, E., Eds. (2006). *Fundamentals of Friction and Wear on the Nanoscale.* Berlin, Germany: Springer-Verlag.

Gotsmann, B., Seidel, C., Anczykowski, B., & Fuchs, H. (1999). Conservative and dissipative tip-sample interaction forces probed with dynamic afm. *Phys. Rev. B*, 60(15), 11051–11061.

Green, C. P., Lioe, H., Cleveland, J. P., Proksch, R., Mulvaney, P., & Sader, J. E. (2004). Normal and torsional spring constants of atomic force microscope cantilevers. *Rev. Sc. Instrum.*, 75(6), 1988–1996.

Haverkamp, R. G., Marshall, A. T., & Williams, M. A. K. (2007). Model for stretching elastic biopolymers whcih exhibit conformational transformations. *Phys. Rev. E*, 75, 021907.

Hinterdorfer, P. & Dufrene, Y. F. (2006). Detection and localization of single molecular recognition events using atomic force microscopy. *Nature Methods*, 3, 347–355.

Hoffmann, R., Baratoff, A., Hug, H.-J., Hidber, H. R., v. Löhneysen, H., & Güntherodt, H.-J. (2007). Mechanical manifestations of rare atomic jumps in dynamic force microscopy. *Nanotechnology*, 18, 395503.

Hölscher, H. (2002). Q-controlled dynamic force spectroscopy. *Surf. Sci.*, 515, 517–522.

Hölscher, H. (2006). Quantitative measurement of tip-sample interactions in amplitude modulation atomic force microscopy. *Appl. Phys. Lett.*, 89, 123109.

Hölscher, H. (2008). Theory of phase-modulation atomic force microscopy with constant-oscillation amplitude. *J. Appl. Phys.*, 103, 064317.

Hölscher, H. & Anczykowski, B. (2005). Quantitative measurement of tip-sample forces by dynamic force spectroscopy in ambient conditions. *Surf. Sci.*, 579, 21–26.

Hölscher, H. & Schirmeisen, A. (2005). Dynamic force microscopy and spectroscopy. In P. W. Hawkes (Ed.), *Advances in Imaging and Electron Physics* (pp. 41–101). London: Academic Press Ltd.

Hölscher, H. & Schwarz, U. D. (2007). Theory of amplitude modulation atomic force microscopy with and without Q-control. *Int. J. Nonlinear Mech.*, 42, 608–625.

Hölscher, H., Gotsmann, B., & Schirmeisen, A. (2003). On dynamic force spectroscopy using the frequency modulation technique with constant excitation. *Phys. Rev. B*, 68, 153401.

Hölscher, H., Schirmeisen, A., & Schwarz, U. D. (2008). Principles of atomic friction: From sticking atoms to superlubric sliding. *Phil. Trans. R. Soc.* A, 366, 1869.

Hölscher, H., Ebeling, D., & Schwarz, U. D. (2006). Theory of Q-controlled dynamic force microscopy in air. *J. Appl. Phys.*, 99, 084311.

Hölscher, H., Gotsmann, B., Allers, W., Schwarz, U. D., Fuchs, H., & Wiesendanger, R. (2001). Measurement of conservative and dissipative tip-sample interaction forces with a dynamic force microscope using the frequency modulation technique. *Phys. Rev. B*, 64, 075402.

Hölscher, H., Gotsmann, B., Allers, W., Schwarz, U. D., Fuchs, H., & Wiesendanger, R. (2002). Comment on "damping mechanism in dynamic force microscopy." *Phys. Rev. Lett.*, 88, 019601.

Hölscher, H., Schwarz, A., Allers, W., Schwarz, U. D., & Wiesendanger, R. (2000). Quantitative analysis of dynamic force spectroscopy data on graphite(0001) in the contact and non-contact regime. *Phys. Rev. B*, 61, 12678–12681.

Hoogenboom, B. W., Hug, H. J., Pellmont, Y., Martin, S., Frederix, P. L. T. M., Fotiadis, D., & Engel, A. (2006). Quantitative dynamic-mode scanning force microscopy in liquid. *Appl. Phys. Lett.*, 88, 193109.

Hu, S. & Raman, A. (2008). Inverting amplitude and phase to reconstruct tip-sample interaction forces in tapping mode atomic force microscopy. *Nanotechnology*, 19, 375704.

Hutter, J. L. & Bechhofer, J. (1993). Calibration of atomic-force microscope tips. *Rev. Sci. Instrum.*, 64, 1868.

Israelachvili, J. N. (1992). *Intermolecular and Surface Forces*. London: Academic Press Ltd.

Janshoff, A., Neitzert, M., Oberdörfer, Y., & Fuchs, H. (2002). Force spectroscopy of molecular systems - single molecule spectroscopy of polymers and biomolecules. *Angew. Chem. Int. Ed.*, 39, 3212–3237.

Johnson, K. L. (1985). *Contact Mechanics*. Cambridge, UK: Cambridge University Press.

Johnson, K. L., Kendall, K., & Roberts, A. D. (1971). Surface energy and contact of elastic solids. *Proc. R. Soc. London Ser. A*, 324(1558), 301.

Kantorovich, L. N. & Trevethan, T. (2004). General theory of microscopic dynamical response in surface probe microscopy:from imaging to dissipation. *Phys. Rev. Lett.*, 93, 236102.

Kawakatsu, H., Kawai, S., Saya, D., Nagashio, M., Kobayashi, D., Toshiyoshi, H., & Fujita, H. (2002). Towards atomic force microscopy up to 100 mhz. *Rev. Sc. Instrum.*, 73(6), 2317–2320.

Kitamura, S. & Iwatsuki, M. (1995). Observation of 7×7 reconstructed structure on the silicon (111) surface using ultrahigh vacuum noncontact atomic force microscopy. *Jpn. J. Appl. Phys.*, 34, L145–L148.

Kobayashi, K., Yamada, H., & Matsushige, K. (2002). Dynamic force microscopy using fm detection in various environments. *Appl. Surf. Sci.*, 188, 430.

Kühle, A., Sorensen, A., & Bohr, J. (1998). Role of attractive forces in tapping tip force microscopy. *J. Appl. Phys.*, 81, 6562–6569.

Landau, L. D. & Lifschitz, E. M. (1991). *Lehrbuch der theoretischen Physik VII: Elastizitätstheorie*. Berlin: Akademie-Verlag.

Lee, M. & Jhe, W. (2006). General solution of amplitude-modulation atomic force microscopy. *Phys. Rev. Lett.*, 97, 036104.

Lee, S. I., Howell, S. W., Raman, A., & Reifenberger, R. (2002). Nonlinear dynamics of microcantilevers in tapping mode atomic force microscopy: A comparison between theory and experiment. *Phys. Rev. B*, 66, 115409.

Legleiter, J. & Kowalewski, T. (2005). Insight into fluid tapping-mode atomic force microscopy provided by numerical simulations. *Appl. Phys. Lett.*, 87, 163120.

Legleiter, J., Park, M., Cusick, B., & Kowalewski, T. (2006). Scanning probe acceleration microscopy (SPAM) in fluids: Mapping mechanical properties of surfaces at the nanoscale. *Proc. Natl. Acad. Sci. U.S.A.*, 103, 4813–4818.

Linnemann, R., Gotszalk, T., Rangelow, I. W., Dumania, P., & Oesterschulze, E. (1996). Atomic force microscopy and lateral force microscopy using piezoresistive cantilevers. *J. Vac. Sci. Technol. B*, 14(2), 856–860.

Lüthi, R., Meyer, E., Haefke, H., Howald, L., Gutmannsbauer, W., Guggisberg, M., Bammerlin, M., & Güntherodt, H.-J. (1995). Nanotribology: An UHV-SFM study on thin films of C_{60} and AgBr. *Surf. Sci.*, 338, 247–260.

Martin, Y. & Wickramasinghe, H. K. (1987). Magnetic imaging by "force microscopy" with 1000 Å resolution. *Appl. Phys. Lett.*, 50, 1455–1457.

Martin, Y., Williams, C. C., & Wickramasinghe, H. K. (1987). Atomic force microscope - force mapping and profiling on a sub 100-Å scale. *J. Appl. Phys.*, 61, 4723–4729.

Mate, C. M., McClelland, G. M., Erlandsson, R., & Chiang, S. (1987). Atomic-scale friction of a tungsten tip on a graphite surface. *Phys. Rev. Lett.*, 59(17), 1942–1945.

McKendry, R., Theoclitou, M.-E., Rayment, T., & Abell, C. (1998). Chiral discrimination by chemical force microscopy. *Nature*, 391, 566–568.

Meyer, G. & Amer, N. M. (1988). Novel optical approach to atomic force microscopy. *Appl. Phys. Lett.*, 53, 1045–1047.

Meyer, E., Hug, H.-J., & Bennewitz, R. (2004). *Scanning Probe Microscopy – The Lab on a Tip.* Berlin: Springer-Verlag.

Morita, S., Wiesendanger, R., & Meyer, E. (Eds.) (2002). *Noncontact Atomic Force Microscopy.* Berlin: Springer-Verlag.

Moser, A., Hug, H.-J., Jung, T., Schwarz, U. D., & Güntherodt, H.-J. (1993). A miniature fibre optic force microscope scan head. *Meas. Sci. Technol.*, 4, 769–775.

Munday, J. N., Capasso, F., & Parsegian, V. A. (2009). Measured long-range repulsive casimir-lifshitz forces. *Nature*, 457.

Neumeister, J. M. & Ducker, W. A. (1994). Lateral, normal, and longitudinal spring constants of atomic force microscopy cantilevers. *Rev. of Sci. Instrum.*, 65(8), 2527–2531.

Nony, L., Boisgard, R., & Aimé, J.-P. (2001). Stability criterions of an oscillating tip-cantilever system in dynamic force microscopy. *Eur. Phys. J. B*, 24, 221–229.

Okajima, T., Sekiguchi, H., Arakawa, H., & Ikai, A. (2003). Self-oscillation technique for afm in liquids. *Appl. Surf. Sci.*, 210, 68.

Overney, R. M., Meyer, E., Frommer, J., Brodbeck, D., Lüthi, R., Howald, L., Güntherodt, H.-J., Fujihira, M., Takano, H., & Gotoh, Y. (1992). Friction measurements on phase-separated thin films with a modified atomic force microscope. *Nature*, 359, 133–135.

Parsegian, V. A. (2006). *Van der Waals Forces.* Cambrigde: Cambridge University Press.

Pfeiffer, O. (2004). *Quantitative dynamische Kraft- und Dissipationsmikroskopie auf molekularer Skala.* PhD thesis, Universität Basel.

Press, W. H., Tekolsky, S. A., Vetterling, W. T., & Flannery, B. P. (1992). *Numerical Recipes in C.* Cambridge: Cambridge University Press.

Putman, C. A. J., Vanderwerf, K. O., Degrooth, B. G., Vanhulst, N. F., & Greve, J. (1994). Tapping mode atomic force microscopy in liquid. *Appl. Phys. Lett.*, 64, 2454–2456.

Rief, M., Fernandez, J. M., & Gaub, H. E. (1998). Elastically coupled two-level system a a model for biopolymer extensibility. *Phys. Rev. Lett.*, 81, 4764–4767.

Rief, M., Oesterhelt, F., Heymann, B., & Gaub, H. E. (1997). Single molecule force spectroscopy on polysaccharides by atomic force microscopy. *Science*, 275, 1295–1297.

Rodríguez, T. R. & García, R. (2002). Tip motion in amplitude modulation (tapping-mode) atomic-force microscopy: Comparison between continuous and point-mass models. *Appl. Phys. Lett*, 80, 1646–1648.

Rodríguez, T. R. & García, R. (2003). Theory of Q-control in atomic force microscopy. *Appl. Phys. Lett.*, 82, 4821–4823.

Rugar, D., Mamin, H. J., & Guethner, P. (1989). Improved fiber-optic interferometer for atomic force microscopy. *Appl. Phys. Lett.*, 55(25), 2588–2590.

Sackmann, E. (1996). Supported membranes: scientific and practical applications. *Science*, 271, 43–48.

Sader, J. E. (1995). Parallel beam approximation for v-shaped atomic force microscope cantilevers. *Rev. Sc. Instrum.*, 66(9), 4583–4587.

Sader, J. E. & Jarvis, S. P. (2004). Accurate formulas for interaction force and energy in frequency modulation force spectroscopy. *Appl. Phys. Lett.*, 84, 1801–1803.

Sader, J. E., Chon, J. W. M., & Mulvaney, P. (1999). Calibration of rectangular atomic force microscope cantilevers. *Rev. Sci. Instrum.*, 70(10), 3967–3969.

Sader, J. E., Larson, I., Mulvaney, P., & White, L. R. (1995). Method for the calibration of atomic force microscope cantilevers. *Rev. Sci. Instrum.*, 66, 3789–3798.

Sader, J. E., Uchihashi, T., Higgins, M. J., Farrell, A., Nakayama, Y., & Jarvis, S. (2005). Quantitative force measurements using frequency modulation atomic force microscopy—theoretical foundations. *Nanotechnology*, 16, S94–S101.

Sahin, O., Maganov, S., Su, C., Quate, C., & Solgaard, O. (2007). An atomic force microscope tip designed to measure time-varying nanomechanical forces. *Nature Nanotechnology*, 2, 507–514.

San Paulo, A. & García, R. (2000). High-resolution imaging of antibodies by tapping-mode atomic force microscopy: Attractive and repulsive tip-sample interaction regimes. *Biophys. J.*, 78, 1599–1605.

San Paulo, A. & García, R. (2002). Unifying theory of tapping-mode atomic-force microscopy. *Phys. Rev. B*, 66, 041406.

Sarid, D. (1994). *Scanning Force Microscopy—With Applications to Electric, Magnetic, and Atomic Forces*. Oxford: Oxford University Press.

Sasaki, N. & Tsukada, M. (1999). Theory for the effect of the tip-surface interaction potential on atomic resolution in forced vibration system of noncontact afm. *Appl. Surf. Sci.*, 140, 339–343.

Schäffer, T. E. & Fuchs, H. (2005). Optimized detection of normal vibration modes of atomic force microscope cantilevers with the optical beam deflection method. *J. Appl. Phys.*, 97, 083524.

Schirmeisen, A. & Hölscher, H. (2005). Velocity dependence of energy dissipation in dynamic force microscopy: Hysteresis versus viscous damping. *Phys. Rev. B*, 72, 045431.

Schmutz, J.-E., Hölscher, H., Ebeling, D., Schäfer, M. M., & Anczykowski, B. (2007). Mapping the tip-sample interactions on DPPC and DNA by dynamic force spectroscopy under ambient conditions. *Ultramicroscopy*, 107, 875–881.

Schwarz, U. D. (2003). A generalized analytical model for the elastic deformation of an adhesive contact between a sphere and a flat surface. *J. Coll. Interf. Sci.*, 261, 99–106.

Schwarz, U. D., Köster, P., & Wiesendanger, R. (1996). Quantitative analysis of lateral force microscopy experiments. *Rev. Sci. Instrum.*, 67, 2560–2567.

Stahl, U., Yuan, C. W., de Lozanne, A. L., & Tortonese, M. (1994). Atomic force microscope using piezoresistive cantilevers and combined with a scanning electron microscope. *App. Phy. Lett.*, 65(22), 2878–2880.

Stark, R. W. & Heckl, W. (2000). Fourier transformed atomic force microscopy: Tapping mode atomic force microscopy beyond the hookian approximation. *Surf. Sci.*, 457, 219–228.

Stark, R. W., Schitter, G., Stark, M., Guckenheimer, R., & Stemmer, A. (2004). State-space model of freely vibrating and surface-coupled cantilever dynamics in atomic force microscopy. *Phys. Rev. B*, 69, 085412.

Stifter, T., Marti, O., & Bhushan, B. (2000). Theoretical investigation of the distance dependence of capillary and van der waals forces in scanning force microscopy. *Phys. Rev. B*, 62, 13667–13673.

Sugimoto, Y., Pou, P., Abe, M., Jelinek, P., Perez, R., Morita, S., & Custance, O. (2007). Chemical identification of individual surface atoms by atomic force microscopy. *Nature*, 446, 64–67.

Tamayo, J. & García, R. (1998). Relationship between phase shift and energy dissipation in tapping-mode scanning force microscopy. *Appl. Phys. Lett.*, 73, 2926–2928.

Tortonese, M., Barrett, R. C., & Quate, C. F. (1993). Atomic resolution with an atomic force microscope using piezoresistive detection. *App. Phy. Lett.*, 62(8), 834–836.

Ueyama, H., Sugawara, Y., & Morita, S. (1998). Stable operation mode for dynamic noncontact atomic force microscopy. *Appl. Phys. A*, 66, S295–S297.

Varenberg, M., Etsion, I., & Halperin, G. (2003). An improved wedge calibration method for the lateral force in atomic force microscopy. *Rev. Sci. Instrum.*, 74, 3362.

Wang, L. (1998). Analytical descriptions of the tapping-mode atomic force microscopy responce. *Appl. Phys. Lett.*, 73, 3781–3783.

Wolter, O., Bayer, T., & Greschner, J. (1991). Micromaschined silicon sensors for scanning force microscoy. *J. Vac. Sci. Technol. B*, 9, 1353.

Yuan, C. W., Batalla, E., Zacher, M., de Lozanne, A. L., Kirk, M. D., & Tortonese, M. (1994). Low temperature magnetic force microscope utilizing a piezoresistive cantilever. *App. Phy. Lett.*, 65(10), 1308–1310.

Zhong, Q. D., Inniss, D., Kjoller, K., & Elings, V. B. (1993). Fractured polymer/silica fiber surface studied by tapping mode atomic force microscopy. *Surf. Sci. Lett.*, 290, L688–L692.

3 Theoretical Models in Force Spectroscopy

Raymond W. Friddle

CONTENTS

3.1 INTRODUCTION

The theory of dynamic force spectroscopy derives from the general concept of thermal escape over an energy barrier under the influence of an external field. This practice is not new and has several analogous counterparts in a wide variety of systems. What one often finds when considering the fundamental description of physical systems is that the same general dynamic model holds across a wide variety of processes. This is the case in forced bond rupture, where the description simplifies to a system in contact with its surroundings while an external forcing shifts the equilibrium from an initial state to some final state. The aim here is to formulate models that allow us to learn something useful about the system under study, preferably extrapolated to the equilibrium, unperturbed case.

The seminal works of Evan Evans (Evans & Ritchie, 1997; Evans, 2001; Evans and Williams, 2001) brought the theory and technique of DFS to an audience of modern scientists in the biophysical and materials sciences who were eager to make use of the burgeoning advancement in nanoscale technologies to study fundamental processes in physical chemistry, ultimately with single-molecule resolution. But the history of the theory of this stochastic process reaches much further back into the works of Tobolsky and Eyring (1943), Zhurkov (1965), Bell (1978) and many others who investigated the effects of force on the lifetime of materials under stress. Since then, a number of modifications have been proposed that aim to improve upon the basic theory of forced escape over a barrier. While it is not possible to consider every approach here, an attempt is made to present a general analysis of the typical scenarios encountered in the laboratory.

3.1.1 UNIMOLECULAR VERSUS BIMOLECULAR SYSTEM

The molecular systems that are commonly studied by single-molecule manipulation techniques fall under two main categories: (1) Unimolecular systems, characterized by force-induced conformational changes, which disrupt intramolecular bonds; (2) Bimolecular systems, which describe the intermolecular binding between two separate molecules (Figure 3.1). In bulk, these two systems are governed by the following equilibrium constants.

$$A \rightleftharpoons B \, , \; K_{eq} = \frac{[A]}{[B]} \qquad\qquad \text{unimolecular}$$

$$AB \rightleftharpoons A + B \, , \; K_{eq} = \frac{[AB]}{[A][B]} \quad \text{bimolecular}$$

where $[\,i\,]$ is the concentration of species i. In the case of the unimolecular system, A and B define *different conformations of the same molecule*. The probability of finding unimolecular systems in one conformation or another is governed by the equilibrium free energy between states within the energy landscape of the individual molecule. Therefore, equilibria between the states of unimolecular systems are independent of the concentration of molecules in the system. Examples include the folded/unfolded conformations of proteins and ribonucleic acid (RNA) or the isomers of a moiety. On the contrary, bimolecular equilibria are controlled by the concentration of both molecules involved in the reaction. This can be inferred from the units in the equilibrium constant of inverse concentration (nominally M^{-1}). For example, if molecule

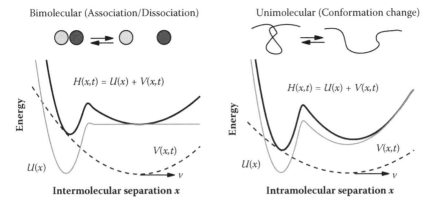

FIGURE 3.1 The two primary systems encountered in force spectroscopy of biomolecules involve intermolecular bonding and intramolecular conformational changes. While the intermolecular potential $U(x)$ of bimolecular bonds are typically characterized by a single well defining the bound state, the addition of a spring-like external potential $V(x)$ can lead to a double-well, bistable system. In a unimolecular system containing two states, the intramolecular potential is inherently bistable and thus the external field acts to tilt the balance between the two conformations.

A is in excess, while molecule B is present at low levels, then the fraction of AB pairs found relative to the total number of molecules available will be small. As the concentration of B is increased, the proportion of molecules reacted to form AB will increase as well. While more complicated theoretically, this property places bimolecular interactions in an exciting arena of their own. Investigating the strength of bimolecular systems by force and finding clear connections between the single-molecule measurement and ensemble bimolecular equilibria is still a relatively new area of study.

3.1.2 Probing Interactions by Force

When the molecule of interest is linked to a force transducer, we will assume that the primary degree of displacement occurs along the direction of the applied force. Thus, the analysis in this chapter, unless otherwise noted, will be along one-dimensional space x. When an inter- or intramolecular potential $U(x)$ is perturbed by an external potential $V(x,t)$, the total energy of the system at every position is their sum,

$$H(x,t) = U(x) + V(x,t) \tag{3.1}$$

where $H(x,t)$ is the Hamiltonian of the system. Typically, the force transducer is spring-like, and therefore applies a parabolic potential $V(x,t) = \frac{1}{2}k_{cant}(x - vt)^2$ to the system, where k_{cant} is the force constant of the spring. We use the symbol k_{cant} to refer to the commonly used cantilever in atomic force microscopes, however, the treatment from here on is general to any force transducer, which is approximately linear in force with extension. The Hamiltonian becomes a time-dependent potential, which is warped and tilted as the external field is swept along x. In general, the superposed probe potential $V(x,t)$ has a *global* effect on the energy landscape of the molecular system and does more than simply lowering transition barriers. Following the procedure of Shapiro & Qian (1997); Shapiro and Qian (1998), Figure 3.2 shows the points of stability (i.e., minima and maxima of the Hamiltonian) for a Lennard-

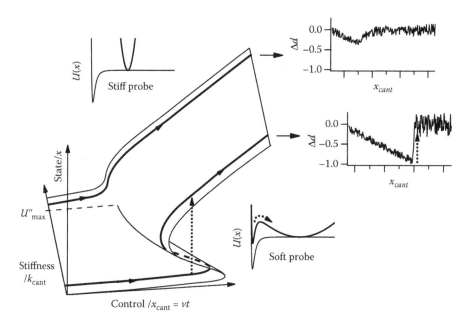

FIGURE 3.2   When the intermolecular potential of a bond, such as the Lennard-Jones potential depicted here, is under the influence of a spring-like probing potential, the combined energy landscape will be either monostable, or bistable, depending on the probing potential's parameters. The critical points of the combined bond/probe potential, or minima and maxima, define the macroscopic state x of the system (such as bound, unbound, or at the transition state). The probe potential squeezes the underlying bond potential in proportion to the spring constant k_{cant}. When the spring constant is larger than the maximum stiffness of the intermolecular potential, the combined landscape is monostable (upper thick curve). When the spring constant is smaller than this critical value, the combined landscape evolves into a bistable system consisting of two minima separated by a barrier. Moving the probe potential minimum $x_{cant} = vt$ will, at stiff k_{cant}, pull the bond apart in a continuous fashion, or, at soft k_{cant}, the bond will discontinuously snap apart due to thermal activation over the energy barrier (dashed arrow). Therefore, the critical points x form a surface when plotted as a function of the two parameters k_{cant} and x_{cant}. Simulated deflection signals (Δd) reflect the dynamic trajectories encountered in the laboratory for two examples of the probe stiffness, where Δd is the observed deflection of the force transducer away from its equilibrium (zero force) deflection.

Jones 6-12 potential under the influence of a parabolic pulling potential. The figure shows that the addition of a spring potential to a bond with a single metastable minimum can result in a monostable or bistable system depending on two important control parameters: the location of the spring minimum $x_{cant} = vt$ and the stiffness of the spring k_{cant}. The resulting surface of stability points contains a fold when the spring constant is less than the largest gradient of the bond potential $k_{cant} < U''_{max}$. The fold designates the emergence of a barrier (unstable maximum) separating two metastable minima. When translating the probe minimum x_{cant} from left to right (i.e., away from the bond minimum), the system passes through the region of bistability and can discontinuously jump from one metastable minimum to another. Such is the case when a bond ruptures. The same is true when moving right to left (toward the bond), but the jump to the bound state will typically occur at a smaller displacement from the bond minimum than the jumping off location. Hence, a hysteresis is observed when comparing the approach and retract jump positions during a nonequilibrium measurement. On the other hand, when the spring constant is greater than this critical value $k_{cant} > U''_{max}$, the fold vanishes, and only a single monostable minimum exists for all translations of the probe minimum. In most force spectroscopy experiments, the probe stiffness is much weaker than the bond, leading to the discontinuous snapping-on and snapping-off of the probe to and from the surface. Hence, we will consider the use of soft probes to be a given assumption throughout this chapter.

3.1.3 DIFFUSION AND ESCAPE OVER A BARRIER

The current theory of molecular bond rupture is a culmination of over 100 years of discoveries that have provided experimental and theoretical insight into the physics of molecules in condensed phases. It is, therefore, useful to begin by discussing the principles underlying the microscopic motion of molecules. A particle in solution undergoes rapid collisions with the surrounding solvent. Each impulse from a neighboring molecule imparts some kinetic energy to the particle. But shortly thereafter ($\sim 10^{-12}$ s), the particle experiences another kick of random strength and direction. These collisions have two primary effects. The first is to induce irregular motion of the particle. The second is to slow down the particle through friction. Since these random and frictional forces arise from the same source, they are closely related through Einstein's *fluctuation–dissipation theorem*. The motion observed is random and known as diffusive (or Brownian) motion and is fundamental to particles that are both free in solution, or under the influence of externally applied force fields. Here, we will derive the basic laws describing the motion of particles, beginning with Einstein's derivation of diffusion.

Assume that during an interval of time τ, a particle in solution makes a movement Δ, which is independent of the movements of any other particle. As each movement takes place, we also assume that it is independent of the previous movement. In other words, the particle has no memory of the last step it took. Each step distance is assumed to occur with some probability $P(\Delta)$. Note also that the step distance Δ is the *magnitude* away from the current position, we are not concerned with which direction it moves away. Consider a distribution of the concentration of many such particles $\rho(x, t)$ at time t. The above assumptions constitute a Markov process that is governed by

$$\rho(x,t+\tau) = \int P(\Delta)\rho(x+\Delta,t)d\Delta \tag{3.2}$$

Einstein's approach to solving this equation begins by expanding both sides (first order in t, and second order in Δ) and integrating over all step sizes,

$$\rho(x,t) + \tau\frac{\partial\rho(x,t)}{\partial t} =$$

$$\rho(x,t)\underbrace{\int P(\Delta)d\Delta}_{= 1} + \frac{\partial\rho(x,t)}{\partial x}\underbrace{\int \Delta P(\Delta)d\Delta}_{= 0} + \frac{\partial^2\rho(x,t)}{\partial x^2}\underbrace{\int \frac{\Delta^2}{2!}P(\Delta)d\Delta}_{= \frac{\langle\Delta^2\rangle}{2}}$$

The second term is an even valued function, $P(\Delta) = P(-\Delta)$, multiplied by an odd function, Δ, so the integral becomes odd and sums to zero. We, therefore, arrive at the *diffusion equation* (Einstein, 1905),

$$\frac{\partial\rho(x,t)}{\partial t} = D\frac{\partial^2\rho(x,t)}{\partial x^2} \tag{3.3}$$

The constant $D = \frac{\langle\Delta^2\rangle}{2\tau}$ is the diffusion coefficient, which governs how far a particle will diffuse with time. Solutions to this differential equation can be found given that we know the boundary conditions for the system of interest. Equation 3.3 is equivalent to Fick's second law of diffusion but with the local concentration $c(x,t)$ replaced by the probability density $\rho(x,t)$. Fick's first law in one dimension states that the flux of particles through a point is directly proportional to, and in opposite direction of, the concentration gradient at that point:

$$J = -D\frac{\partial\rho(x,t)}{\partial x} \tag{3.4}$$

Equation 3.4 conveys the diffusive driving force for mass transport, while Equation 3.3 is the equation of motion for an ensemble of particles diffusing free in solution.

A more general formulism arises from describing how a particle would behave while diffusing in the midst of a force field. That is, the potential energy $U(x)$ that the particle experiences is no longer flat, but has varying gradients, or forces, with x that define the forces on the particle. Consider that, in addition to concentration gradients, there is an additional component to the flux arising from forces that drive the particles with some velocity dx/dt:

$$J = -D\frac{\partial\rho(x,t)}{\partial x} + \rho(x,t)\frac{dx}{dt} \tag{3.5}$$

To approximate the velocity of the particles at x, we consider the *Langevin equation* of motion of a particle of mass m in a force field $-\partial U(x)/\partial x$:

$$m\frac{d^2x}{dt^2} = -\eta\frac{dx}{dt} - \frac{\partial U(x)}{\partial x} + \xi(t) \tag{3.6}$$

where $\xi(t)$ is a randomly fluctuating force, and η is the damping coefficient of the system. Under strong damping (large η), we view each particle rapidly gaining and losing velocity and therefore we assume negligible inertial acceleration ($\frac{m}{\eta}d^2x/dt^2 \cong 0$) and random fluctuations ($\xi(t)/\eta \cong 0$) compared with the deterministic force field $\frac{1}{\eta}\frac{-\partial U(x)}{\partial x}$. Thus, solving for dx/dt and inserting Equation 3.5 in to Fick's second law of conservation of mass, $\frac{\partial \rho(x,t)}{\partial t} = -\frac{\partial}{\partial x}J$ (see Equation 3.3), we have the *Smoluchowski Equation*,

$$\frac{\partial \rho(x,t)}{\partial t} = \frac{\partial}{\partial x}\left(\underbrace{\frac{k_B T}{\eta}\frac{\partial \rho(x,t)}{\partial x}}_{\text{Diffusion}} + \underbrace{\frac{1}{\eta}\rho(x,t)\frac{dU(x)}{dx}}_{\text{Drift}} \right) \tag{3.7}$$

where we have used the Einstein relation $D = k_B T/\eta$, and identified the diffusion due to local changes in concentration, and the drift component due to the force field $-\partial U(x)/\partial x$. Notice that in the absence of a field ($\partial_x U(x) = 0$), we recover the diffusion Equation 3.3. The Smoluchowski equation describes the probability $\rho(x,t)$ of finding a particle at x after a time t has passed since the start of the initial conditions. Over long times, the distribution of particles reaches equilibrium such that $\partial \rho(x,t)/\partial t = 0$ and we find

$$\frac{\partial}{\partial x}\left(\frac{\partial \rho(x,t)}{\partial x}\right) = \frac{\partial}{\partial x}\left(-\frac{\rho(x,t)}{k_B T}\frac{\partial U(x)}{\partial x}\right)$$
$$\frac{\partial \rho(x,t)}{\rho(x,t)} = -\frac{1}{k_B T}\frac{\partial U(x)}{\partial x}\partial x$$
$$\rho(x,t) = \rho_0 e^{-U(x)/k_B T} \tag{3.8}$$

which is the *Boltzmann distribution* of states at thermal equilibrium.

The Smoluchowski equation is the equation of motion for the distribution $\rho(x,t)$ of an ensemble of particles in a field of force. For a pair of molecules, the field of force is the intermolecular potential that attracts and repels the two. To understand how the Smoluchowski equation describes bond rupture, we require a more generalized picture of a particle diffusing out of a bound state, an area of study formally known as transition-state-theory (TST). As we show below, the solution of Smoluchowski diffusion over an energy barrier was ultimately shown in an elegant form by Kramers (1940).

Recall from Equation 3.7, the flux of particles through a point x, amidst an intermolecular field $U(x)$ is

$$J(x,t) = -\frac{k_B T}{\eta}\frac{\partial \rho(x,t)}{\partial x} - \frac{1}{\eta}\rho(x,t)\frac{dU(x)}{dx} \tag{3.9}$$

Let us consider a smooth potential $U(x)$ that contains a minimum at x_0 and maximum at x_β, such as in Figure 3.3. We envision a steady state in which a source of particles

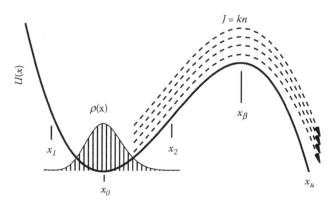

FIGURE 3.3 The prescription for deriving Kramers rate of escape, k. The distribution of particles in the well at x_o provide a source, while the location outside the barrier x_u acts as a sink. Therefore, a steady flow of particles, or flux J, is passing over the barrier. The flux is defined as the number of particles available n times rate of escape. Thus, estimating the flux and integrating over the distribution of particles in the well provides the transition rate (see discussion in text).

in the well provide a constant flux of particles over the barrier (steady-state conditions). To the far left, the steeply rising energy is essentially a reflecting boundary. To the right, after particles pass over the barrier and reach x_u, they are completely unbound and carried away from the potential indefinitely. Therefore, we define x_u as an absorbing boundary with $\rho(x_u) \equiv 0$. To calculate the flux over the barrier, we look to integrate over x_o to x_u. Multiplying both sides of Equation 3.9 by an integrating factor $\exp(U(x)/k_BT)$, we have

$$\frac{\partial \rho(x)}{\partial x} e^{U(x)/k_BT} + \frac{\rho(x)U'(x)}{k_BT} e^{U(x)/k_BT} = -\frac{\eta J}{k_BT} e^{U(x)/k_BT} \qquad (3.10)$$

$$\frac{d}{dx}\left(\rho(x)e^{U(x)/k_BT}\right) = -\frac{\eta J}{k_BT} e^{U(x)/k_BT} \qquad (3.11)$$

Where we have exploited the fact that the left side of Equation 3.10 is just the chain rule derivative of $\rho(x)\exp(U(x)/k_BT)$. Upon integration, and noting that $\rho(x_u) = 0$, we have

$$J = \frac{k_BT}{\eta} \frac{\rho(x_0)e^{U(x_0)/k_BT}}{\int_{x_0}^{x_u} e^{U(x)/k_BT}dx} \qquad (3.12)$$

Now the flux J is the steady flow of particles moving from a source, the well, over the barrier and into a sink (or carried away). To determine the actual rate k of particles escaping per unit time, we recognize that the flux is the product of the rate with the number of particles available to escape, $J = kn$. Therefore, we need to find how the particle concentration changes along x. Define n as the number of particles in the well. Near the bottom of the well, around x_0, the flux is nearly zero and we can approximate the number of particles by the quasi-equilibrium Boltzmann distribution, $\rho(x) = \rho(x_0)e^{(U(x_0)-U(x))/k_BT}$. Thus, summing over an appropriate range around the minimum, say x_1 to x_2, we have the population of particles available to cross the barrier

$$n = \rho(x_o) \int_{x_1}^{x_2} e^{(U(x_o)-U(x))/k_BT} dx \tag{3.13}$$

Finally, the thermally activated rate of escape is given by the flux over the population, $k = J/n$

$$k = \frac{k_BT}{\eta} \left(\underbrace{\int_{x_0}^{x_u} e^{U(x)/k_BT} dx}_{\substack{\text{Barrier} \\ \text{region}}} \underbrace{\int_{x_1}^{x_2} e^{-U(x)/k_BT} dx}_{\substack{\text{Well} \\ \text{region}}} \right)^{-1} \tag{3.14}$$

Equation 3.14 is the celebrated *Kramers' escape rate*. Because the well region is concave up, the number of occupied states peak at the minimum and fall off with distance due to the negative exponent in energy, $e^{-U(x)/k_BT}$. Likewise, since the barrier region is concave down the term $e^{U(x)/k_BT}$ peaks at the maximum and falls off with distance due to the positive exponent in energy. Thus, both integrals converge and are related to the tendency of particles to reside near the barrier and well.

A widely used solution to Equation 3.14 for an overdamped system can be found from a simplification of the energy landscape as shown in Figure 3.3. Here, the well around x_0 and the barrier around x_β are approximated to first order by parabolic potentials of the form $U(x) = \kappa_0(x-x_0)^2/2$ and $U(x) = \Delta U - \kappa_b(x-x_\beta)^2/2$, respectively, where ΔU is the relative energy barrier height. Using these potentials for the integrals in Equation 3.14 from x_1 to x_2 and x_0 to x_u, respectively, and carrying integration out to $\pm\infty$, we have the Kramers (1940) rate of escape for an overdamped particle:

$$k = \frac{\sqrt{\kappa_0\kappa_b}}{2\pi\eta} e^{-\Delta U/k_BT} \tag{3.15}$$

Equation 3.15 retains the Arrhenius form of activated rate processes $Ae^{-\Delta U/k_BT}$ and additionally includes properties related to the shape of the energy landscape, namely the curvatures of the bound state $\kappa_0 = U''(x_0)$ and barrier $\kappa_b = |U''(x_\beta)|$.

3.2 TRANSITION RATE UNDER FORCE

As detailed above, the rate of passage over an energy barrier k, or equivalently the mean waiting time for passage $\tau = k^{-1}$, is a function of the energy pathway between the meta-stable bound state minimum at x_0 ($U'(x_0) = 0, U''(x_0) > 0$) and the highest unstable barrier, or transition state at x_β ($U'(x_\beta) = 0, U''(x_\beta) < 0$). For simple, smooth potentials, $U(x)$, which are described by a steep pathway connecting the bound and transition states, the effect of a small to moderate external force on the potential will lead to a linear perturbation of the potential. Take, for example, the phenomenological Arrhenius law of activated processes, $k = Ae^{-\Delta U/k_B T}$, where A is a frequency prefactor and $\Delta U = U(x_\beta) - U(x_0) > 0$ is the relative barrier height. Owing to the exponential, the perturbations to the energy barrier produce the most significant effects on escape. Thus, taking an expansion of the logarithm of the rate under an applied force F, we have

$$\ln k(F) = \ln k(0) + F\frac{\partial \ln k(0)}{\partial F} + \frac{1}{2}F^2\frac{\partial^2 \ln k(0)}{\partial F^2} + \cdots \tag{3.16}$$

and keeping the first two terms one recovers the transition rate as a function of applied force:

$$k(F) \cong k^0 e^{F\cdot\frac{\partial \ln k(0)}{\partial F}} \tag{3.17}$$

In Equation 3.17, $k^0 \equiv k(0)$ is the unperturbed transition rate. We see immediately that for small to intermediate loads, a unique property of a system can be determined through application of force. That is, the change in the log-transition rate with force $\partial \ln k(0)/\partial F$ is inaccessible by any other means and provides a property specific to the internal energy of the system under study. The basic form of Equation 3.17 was used by Tobolsky and Eyring (1943) and later by Zhurkov (1965) in modeling the timescales of material failure under tensile load, where the constant $\partial \ln k(0)/\partial F$ was assumed to be a material-specific parameter. Later, Bell (1978) formed a clearer definition for this parameter. Bell rationalized that a large enough force will cancel out the barrier to passage, and if the total energy is truly linear with force, then the force gradient must be proportional to the distance between the minimum and maximum of the potential barrier, or $\partial \ln k(0)/\partial F \equiv (x_\beta - x_0)/k_B T$.

3.2.1 The Bell Model

To derive the transition rate under force in Equation 3.17 within a more physical context, consider a simple bond under the influence of a pulling potential. The pulling potential provides the external force that acts to drive the bond apart. In most cases, or as an approximation, a parabolic potential, or spring, is used to generate the pulling force:

$$V(x, x_{\text{cant}}) = \frac{1}{2}k_{\text{cant}}(x - x_{\text{cant}})^2 \tag{3.18}$$

where k_{cant} is the force constant of the spring and x_{cant} is the distance the external potential minimum is from the bond minimum. The bound state of the bond is located at $x_0 = 0$ and is contained by a sharp, steep barrier inhibiting any significant displacement of the bound state due to the modest applied force. Therefore, because we assume the bound state to be located at $x_0 = 0$, the pulling potential can be rewritten in terms of the force F acting on the bound state minimum ($F = k_{\text{cant}}x_{\text{cant}}$):

$$V(x, F) = \frac{1}{2}k_{\text{cant}}\left(x - \frac{F}{k_{\text{cant}}}\right)^2 \tag{3.19}$$

Although the pulling potential spans a large region of x space, we are only concerned with the *local* behavior of the pulling potential in the region between $x_0 = 0$

and x_β. Under our assumptions of a sharp barrier, only a small tangential portion of the pulling potential spans the distance between the bound state and the barrier. Expanding $V(x,F)$, we find

$$V(x,F) = \frac{F^2}{2k_{cant}} - Fx + \frac{1}{2}k_{cant}x^2 \tag{3.20}$$

Equation 3.20 contains two primary effects that a pulling potential will have on the bond potential (see Figure 3.4): The first term $F^2/2k_{cant}$ is the energy that the bound state minimum is raised relative to the pulling potential minimum. This parameter has no bearing on the kinetic transition rate, but as shown later, it is key in determining thermodynamic equilibrium between states, as well as the work done on the bond by the pulling potential. The second term, $-Fx$, is the primary potential gradient $-F$ along the intermolecular separation x. The last term, $\frac{1}{2}k_{cant}x^2$ is a small nonlinear contribution that is typically significant at small forces. The total barrier height under the influence of $V(x,F)$ becomes

$$\Delta U(F) = U(x_\beta) + V(x_\beta,F) - [U(0) + V(0,F)] \tag{3.21}$$

$$= \Delta U_0 - Fx_\beta + \frac{1}{2}k_{cant}x_\beta^2 \tag{3.22}$$

Inserting the linear term into the Arrhenius escape rate with $k(F) = Ae^{-\Delta U(F)/k_BT}$, we find the transition rate under force:

$$k(F) = k^0 \exp\left[\left(Fx_\beta - \frac{1}{2}k_{cant}x_\beta^2\right)/k_BT\right] \tag{3.23}$$

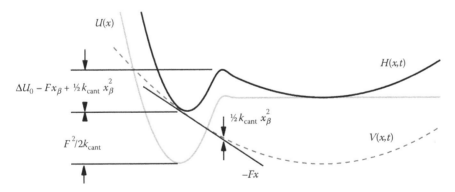

FIGURE 3.4 Schematic of the effects of pulling on a bond with a spring as detailed in the text. The barrier is located at x_β. The energy of the bond minimum is raised relative to the minimum of the pulling potential by $F^2/2k_{cant}$. In addition, the difference between the barrier and bond minimum is reduced by $-Fx_\beta$ with a small enhancement $+\frac{1}{2}k_{cant}x_\beta^2$ due to the finite curvature of the pulling potential. Therefore, as the spring potential is pulled further away, the bound state of the bond becomes both energetically unfavorable and kinetically accelerated toward unbinding.

For soft springs and large forces, the effect of the quadratic term, $\frac{1}{2}k_{cant}x_\beta^2$ is negligible, although it is more complete to include this term when considering slow pulling rates and small forces. Dropping this term and keeping the linear contribution gives the transition rate commonly attributed to Bell (1978):

$$k(F) = k^0 \exp\left[F x_\beta / k_B T\right] \qquad (3.24)$$

3.2.2 The Rebinding Rate

When a bond is loaded by a Hookean spring potential (i.e., cantilever, bead in an optical trap, etc.), the system becomes bistable when the spring minimum is pulled beyond the barrier to unbinding (Shapiro & Qian, 1997; Evans, 2001; Friddle et al., 2008). Thus, the system evolves into a two-state process for all rates of applied force, beyond the force that leads to the bifurcation into two metastable minima. This need not only apply to purely Hookean pulling potentials alone. For example, polymer linkers will also possess a minimum in their potential energy with extension, and in the case of a random flight (or Gaussian) chain, the polymer behaves like a Hookean spring in the entropic regime of small end-to-end extensions (Flory, 1969). On the basis of the simple Arrhenius law of activated transitions, the rebinding rate from a parabolic spring potential will be most strongly affected by the energy barrier located at the transition state x_β. This barrier is changing as the force applied to the bond increases. According to Equation 3.19, this energy barrier is

$$\Delta U_{reb} = V(x_\beta, F) = \frac{1}{2}k_{cant}\left(x_\beta - \frac{F}{k_{cant}}\right)^2 \qquad (3.25)$$

Under the simple model of a constant attempt frequency k_{on}^0, the rebinding rate $k_{on} = k_{on}^0 e^{-\Delta U_{reb}/k_B T}$ follows as

$$k_{on}(F) = k_{on}^0 \exp\left[-\frac{1}{2}k_{cant}\left(x_\beta - \frac{F}{k_{cant}}\right)^2 / k_B T\right] \qquad (3.26)$$

3.2.3 Corrections for Finite Barriers and Spring Pprobes

The Bell model of the transition rate (Equation 3.24) has found ubiquitous success in modeling the statistics of mechanically forced molecular systems. The Bell rate is incredibly simple. The attempt frequency and the distance between the minimum to the barrier are both assumed to be independent of the applied force. In addition, the effects that a finite force constant of the pulling spring might have on the underlying bond potential are not explicit in the Bell model. Resolving the latter issue is fairly straightforward. Recalling Equation 3.23, the complete unbinding rate due to a spring of stiffness k_{cant} is given by

$$k(F) = k^0 \exp\left[\left(F x_\beta - \frac{1}{2}k_{cant}x_\beta^2\right) / k_B T\right] \qquad (3.27)$$

which implies that without accounting for the spring constant the apparent force-free unbinding rate is reduced by a factor $\exp(-\frac{1}{2}k_{cant}x_\beta^2/k_BT)$ (Walton et al. 2008):

$$k_{eff}^0 = k^0\exp(-\frac{1}{2}k_{cant}x_\beta^2/k_BT) \tag{3.28}$$

The reason the Bell model has been so effective at modeling forced rupture may stem from the temperatures commonly explored in the laboratory. At large T thermally activated escape will readily occur before the barrier height is significantly close to vanishing under force. This will enable passage over energy barriers before appreciable perturbation of the location of x_β occurs. Applying large enough forces to perturb the underlying bond potential will require fast loading rates to compete with spontaneous unbinding. Such fast pulling speeds will incur additional hydrodynamic forces that are not accounted for in the models presented here, which assume a conservative force field acts on the bond potential. Nevertheless, small probes can be used that may diminish hydrodynamic effects at large pulling speeds, and low temperatures may be explored that will enable the observation of perturbation of the underlying potential by force. In the context of force spectroscopy, Evans explored this topic for a variety of model potentials (Evans & Ritchie, 1997). The approach taken by Garg (1995), which was later adopted to model force spectroscopy data (Dudko et al. 2003), is to utilize a cubic potential $U(x)$ to approximate the bond potential, which under a force field $-Fx$ can be expressed as (Dudko et al. 2006):

$$U(x) - Fx = \frac{3}{2}\Delta U_0\frac{x}{x_\beta} - 2\Delta U_0\left(\frac{x}{x_\beta}\right)^3 - Fx \tag{3.29}$$

where ΔU_0 and x_β are the barrier height and minimum-to-barrier distance of the unperturbed bond. The force-dependent barrier and force constants can be derived through the first and second derivatives of Equation 3.29 at the extrema $\pm\frac{x_\beta}{2}[1 - 2Fx_\beta/(3\Delta U)]$ (Garg, 1995):

$$\Delta U(F) = \Delta U_0(1 - F/F_c)^{3/2} \tag{3.30}$$

$$\kappa_0(F) = \kappa_0(1 - F/F_c)^{1/2} \tag{3.31}$$

$$\kappa_b(F) = \kappa_b(1 - F/F_c)^{1/2} \tag{3.32}$$

where $F_c = \frac{3}{2}\Delta U_0/x_\beta$ is the critical force at which the barrier vanishes due to the applied force, which coincides with the maximum gradient of the force-free potential $U'(x = 0)$. These functions are then entered into the overdamped limit of Kramers escape rate (Equation 3.5) in which the minimum and maximum of the potential are approximated as parabolic:

$$k(F) = \frac{\sqrt{\kappa_0(F)\kappa_b(F)}}{2\pi\eta}e^{-\Delta U(F)/k_BT} \tag{3.33}$$

$$= \frac{\sqrt{\kappa_0\kappa_b}}{2\pi\eta}(1 - F/F_c)^{1/2}e^{-\Delta U_0(1-F/F_c)^{3/2}/k_BT} \tag{3.34}$$

3.3   BOND LIFETIME UNDER CONSTANT FORCE

A common approach to determining the lifetime τ of a bond, or the force-dependence of the lifetime $\tau(F)$, is through applying a constant force and waiting until the bond ruptures. The waiting time probability $P_i(t|s)$ is the probability the system remains in state i uninterrupted over the time interval [s,t] (Talkner, 2003):

$$P_i(t|s) = \exp\left[-\int_s^t k_i dt\right] \qquad (3.35)$$

where k_i is the rate of escaping state i and can be constant or time dependent. The probability density of transition times $\rho(t)$ is the negative derivative of the waiting time probability. In terms of the lifetime of the bond τ the density of lifetimes under constant k_i is given by

$$\rho(\tau) = -\frac{P_i(\tau|0)}{d\tau} = k_i P_i(\tau|0) \qquad (3.36)$$

Say that at $t = 0$, we instantaneously switch on a constant pulling force F on the bond. If the unbinding rate under force is $k_{\text{off}}(F)$, the corresponding distribution of lifetimes under constant force is

$$\rho(\tau) = k_{\text{off}}(F)\exp\left[-k_{\text{off}}(F)\tau\right] \qquad (3.37)$$

The mean lifetime of the bond as a function of constant force is as expected:

$$\langle\tau\rangle(F) = \int_0^\infty \tau\rho(\tau)d\tau = \frac{1}{k_{\text{off}}(F)} \qquad (3.38)$$

$$= \frac{1}{k_{\text{off}}^0}\exp\left[-Fx_\beta/k_B T\right] \qquad (3.39)$$

where $k_{\text{off}}(F)$ is expressed as the Bell rate in Equation 3.24.

3.4   FORCE RAMP BOND RUPTURE

The intuitive notion of bond strength comes from simply pulling on a bond until it breaks. However, dissociation of weak, noncovalent bonds occurs in the absence of force. When a force is applied that increases with time, the rupture force becomes a function of the force-dependent kinetics of dissociation and the rate of increasing force. Modeling the statistics of this process allows access to the physical properties of the bond through proper analysis of the rupture data. In the strictest sense of a real bond rupture experiment, the process is defined by *at least* two states— the *initial state*, which may be the formed bond, or the folded protein, and the *final state*, which may be the dissociated bond, or the unfolded protein. Several alternative metastable states may also exist that can increase the complexity of the problem. However, the ubiquitous appearance of cooperativity in the dynamics of soft-matter usually reduces the effective problem to a two-state switching process. The inclusion of initial and final states is also required to define changes of important thermodynamic functions, such as the free energy. As a starting point, we model the system as a Markovian two-state process with time-dependent rates, which is described by the following coupled equations:

$$\frac{d}{dt}p_{on}(t) = -k_{off}(t)p_{on}(t) + k_{on}(t)p_{off}(t) \tag{3.40}$$

$$\frac{d}{dt}p_{off}(t) = k_{off}(t)p_{on}(t) - k_{on}(t)p_{off}(t) \tag{3.41}$$

where $k_{off}(t)$ and $k_{on}(t)$ are the unbinding and binding transition rates, while $p_{on}(t)$ and $p_{off}(t)$ are, respectively, the probabilities of finding the system in the bound and unbound states at an observation time t. These probabilities have initial conditions

$$p_{on}(0) = 1, \quad p_{off}(0) = 0 \tag{3.42}$$

and $p_{on}(t) + p_{off}(t) = 1$.

3.4.1 FIRST-PASSAGE APPROXIMATION

When the system is driven quickly, the rebinding rate of the system can be neglected. This is due to the fact that the waiting time for the system to rebind becomes longer with increasing force, and the speed with which large forces are reached is fast. In this case, the master equation in Equation 3.40 reduces to the first-order rate process (Evans & Ritchie, 1997),

$$\frac{d}{dt}p_{on}(t) = -k_{off}(t)p_{on}(t) \tag{3.43}$$

Beginning with the entire population of states bound $p_{on}(0) = 1$ and making the substitution of force for time $\frac{1}{r_f}dF = dt$, we must solve for the probability of remaining bound up to a force F:

$$\int_{1}^{p_{on}} \frac{dp'_{on}}{p'_{on}} = -\frac{1}{r_f}\int_{0}^{F} k_{off}(F')dF' \tag{3.44}$$

Carrying out the integration of the left side of Equation 3.44 yields the probability of remaining bound for general unbinding rates:

$$p_{on}(F) = \exp\left[-\frac{1}{r_f}\int_{0}^{F} k_{off}(F')dF'\right] \tag{3.45}$$

We see that, due to neglecting reentry ($k_{on}(t) \approx 0$), the probability $p_{on}(t)$ is exactly the waiting time probability $P_{on}(t|s) = \exp[-\int_{s}^{t} k_{off}(u)du]$ of residing *uninterrupted* in the bound state over the time interval from $s = 0$ to t.

3.4.1.1 Distribution

The distribution of first-passage times, or first rupture force, is straightforward. The probability density of first-passage times $\rho_{fp}(t)$ coincides with the probability of unbinding events during the interval $[t, t + dt]$ divided by the duration of the interval dt. This density is equivalent to the negative derivative of the waiting time probability of the bound state, $\rho_{fp}(t) = -dp_{on}(t)/dt$. In terms of the force, the density is given by

$$\rho_{fp}(F) = \frac{1}{r_f} k(F) p_{on}(F) \tag{3.46}$$

$$= \frac{1}{r_f} k(F) \exp\left[-\frac{1}{r_f}\int_0^F k(F') dF'\right] \tag{3.47}$$

To understand the behavior of Equation 3.46, consider the initial moment when $F = 0$. In this case, the probability of finding the system bound is $p_{on}(0) = 1$ and the density of transition times is equivalent to the kinetic transition rate $\rho_{fp}(0) = k_{off}(0)$. As the force on the bound state increases, the rate of unbinding increases, while the probability of bound states decreases. Thus, the probability $p_{on}(F)$ acts to modulate the unbinding transition rate $k_{off}(F)$ in proportion to the fraction of the system still occupying the bound state. Under the Bell rate, Equation 3.24, the probability density of first-passage rupture forces is (Evans & Ritchie, 1997):

$$\rho_{fp}(F) = \frac{1}{r_f} k(F) \exp\left[-\frac{k_B T}{r_f x_\beta}(k(F) - k(0))\right] \tag{3.48}$$

3.4.1.2 Mean and Most Probable Rupture Force

A useful data set for analyzing the characteristics of a bond under force is the trend of the rupture force with loading rate. The unbinding transition rate (Equation 3.24) depends only on the instantaneous value of the force, not the history of the force with time. Therefore, $k(F)$ is independent of loading rate. However, from the first-passage distribution in Equation 3.46, we see that to describe the frequency of rupture events we must scale $k(F)$ by the probability $p_{on}(F)$ that the bond is still formed at force F. This probability is loading-rate-dependent because the fraction of bonds remaining at a force F will depend on the length of time F/r_f spent to reach that force. This can be inferred directly from the first-order rate process in Equation 3.43, when expressed in terms of the force, $dp_{on}(F)/dF = -\frac{1}{r_f}k_{off}(F)p_{on}(F)$, which shows that the rate of losing bound states with force is inversely proportional to the loading rate, r_f. Hence for faster loading rates, the bound state tends to persist to higher forces.

Two common, and very similar, statistics of the rupture force distribution are the mean $\langle F \rangle$ and most probable (or mode) F^*. The mean of the first-passage rupture force distribution (Equation 3.46) can be determined by any of the following relationships:

$$\langle F \rangle = \int_0^\infty F \rho_{fp}(F) \, dF \tag{3.49}$$

$$= \frac{1}{r_f} \int_0^\infty F k_{\text{off}}(F) p_{\text{on}}(F) \, dF \tag{3.50}$$

$$= \int_0^\infty p_{\text{on}}(F) \, dF \tag{3.51}$$

$$= \int_0^1 F(p_{\text{on}}) \, dp_{\text{on}} \tag{3.52}$$

where $F(p_{\text{on}})$ is the inverse of the probability of finding the system in the bound state, Equation 3.45, which can be solved analytically for simple transition rates such as Equation 3.24.

Under the Bell transition rate in Equation 3.24, the mean rupture force under the first-passage model is

$$
\begin{aligned}
\langle F \rangle &= \int_0^\infty p_{\text{on}}(F) \, dF \\
&= \int_0^\infty \exp\left[-\frac{1}{r_f} \int_0^F k_{\text{off}}(F') \, dF' \right] dF \\
&= \exp\left[\frac{k^0 k_B T}{r_f x_\beta} \right] \int_0^\infty \exp\left[-\frac{k^0 k_B T}{r_f x_\beta} e^{F x_\beta / k_B T} \right] dF \\
&= \frac{k_B T}{x_\beta} \exp\left[\frac{k^0 k_B T}{r_f x_\beta} \right] \int_{\frac{k^0 k_B T}{r_f x_\beta}}^\infty \frac{e^{-u}}{u} \, du \\
&= \frac{k_B T}{x_\beta} \exp\left[\frac{k^0 k_B T}{r_f x_\beta} \right] E_1\left[\frac{k^0 k_B T}{r_f x_\beta} \right]
\end{aligned}
\tag{3.53}
$$

where $E_1(z)$ is the exponential integral. Equation 3.53 increases from zero force linearly with loading rate r_f, then follows a nonlinear trend, and asymptotically approaches the commonly used form at large loading rate (Williams, 2003),

$$\lim_{r_f \to \infty} \langle F \rangle \cong \frac{k_B T}{x_\beta} \ln\left[\frac{r_f x_\beta e^{-\gamma}}{k_{\text{off}}^0 k_B T} \right], \tag{3.54}$$

where $\gamma = 0.577\ldots$ is the Euler constant. Note that the exponential integral follows the simple interpolation $e^z E_1(z) \cong \ln(1 + e^{-\gamma}/z)$ over all values of the argument z.

The most probable rupture force (or mode, F^*) coincides with the maximum of the rupture force distribution. Using the first-passage distribution in Equation 3.46, Evans & Ritchie (1997) derived a useful expression for the mode by taking $d\rho_{fp}(F^*)/dF = 0$ and expressing the following relationship between loading rate and unbinding rate:

$$0 = \frac{1}{r_f} \frac{dk(F)}{dF} e^{-\frac{1}{r_f} \int_0^F k(F')dF'} - \frac{1}{r_f^2} k(F)^2 e^{-\frac{1}{r_f} \int_0^F k(F')dF'}$$

$$\frac{1}{r_f} k(F) = \frac{1}{k(F)} \frac{dk(F)}{dF}$$

$$k(F)\Big|_{F^*} = r_f \frac{d\ln k(F)}{dF}\Big|_{F^*} \tag{3.55}$$

Inserting the Bell transition rate of Equation 3.24 into Equation 3.55, the most probable rupture force under the first-passage model is

$$F^* = \frac{k_B T}{x_\beta} \ln\left[\frac{r_f x_\beta}{k^0 k_B T}\right] \tag{3.56}$$

When comparing with the fast loading rate expansion of Equation 3.54, the mean is shifted lower than the mode $\langle F \rangle = F^* - \frac{k_B T}{x_\beta} \gamma$ because the distribution is skewed to low forces due to the finite probability of bond rupture down to zero force (see Figure 3.5). As the loading rate decreases, the difference between the mode and mean decreases as well. It is also important to note that the first-passage distribution does not have a peak when the argument in the logarithm of Equation 3.56 is less

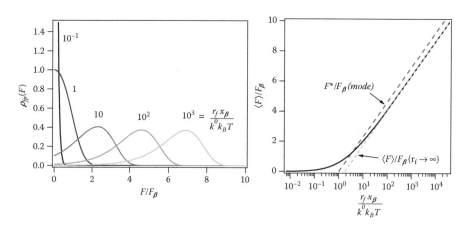

FIGURE 3.5 First-passage models of bonds rupture. The distribution of rupture forces (Equation 3.48) are shown calculated at various normalized loading rates. The mean rupture force, Equation 3.53, as a function of normalized loading rate is also shown in units of the thermal force scale $F_\beta = k_B T/x_\beta$. The mean crosses between a shallow, linearly increasing regime into a strongly nonlinear regime when the normalized loading rate is unity. Shown for comparison are the high loading rate limit, Equation 3.54, and most probable (mode), Equation 3.56, rupture force functions.

than unity. That is, the mode is only defined when $r_f \geq k^0 k_B T / x_\beta$. Caution should be taken when applying the first-passage solution for the mean or mode to force spectra in general because, in principle, the first-passage model is correct only when (1) the bond is driven at very large loading rates, which negate rebinding effects and/or (2) the linkage to the bond is very soft, such as a long polymer, which imparts a large degree of entropic freedom to the unbound state. Otherwise, as the loading rate is decreased, the force spectrum can enter into a linear-response regime whereby the rebinding rate is effective at small forces, and the simple first-passage model is no longer valid. This near-equilibrium scenario is detailed below.

3.4.2 REVERSIBLE TWO-STATE APPROXIMATION

As explained at the beginning of this section, the complete description of bond rupture includes the dynamics of both the bound state (on) and the unbound state (off):

$$r_f \frac{d}{dF} p_{on}(F) = -k_{off}(F) p_{on}(F) + k_{on}(F) p_{off}(F) \qquad (3.57)$$

where $k_{off}(F)$ and $k_{on}(F)$ are the rates of leaving the bound and unbound states, respectively, and $p_{on}(F)$ or $p_{off}(F)$ are, respectively, the probabilities of finding the system bound or unbound at a force F. These probabilities have initial conditions

$$p_{on}(0) = 1, \quad p_{off}(0) = 0 \qquad (3.58)$$

and $p_{on}(F) + p_{off}(F) = 1$. At equilibrium $dF/dt = r_f = 0$, Equation 3.57 yields the principle of detailed balance

$$k_{off}(F) p_{on}(F) = k_{on}(F) p_{off}(F) \qquad (3.59)$$

which states that the number of transitions per unit time between the pair of states are equal. Rewriting this equality in terms of the relative population of the two states, we have

$$\frac{p_{off}(F)}{p_{on}(F)} = \frac{k_{off}(F)}{k_{on}(F)} = \exp\left[-\Delta G(F)/k_B T\right] \qquad (3.60)$$

where $\Delta G(F)$ is the free energy difference between the bound and unbound states when held at an external load F. It turns out that, under the simple analytic transition rates derived above, a straightforward expression for $\Delta G(F)$ can be found. Starting from the rebinding rate in Equation 3.26, we can expand the term in parentheses to find

$$k_{on}(F) = k_{on}^0 \exp\left[\left(-\frac{F^2}{2k_{cant}} + F x_\beta - \frac{1}{2} k_{cant} x_\beta^2\right)/k_B T\right] \qquad (3.61)$$

$$= k_{off}(F) \exp\left[\left(\Delta G_0 - \frac{F^2}{2k_{cant}}\right)/k_B T\right] \qquad (3.62)$$

or,

$$\frac{k_{\text{off}}(F)}{k_{\text{on}}(F)} = \exp\left[-\left(\Delta G_0 - \frac{F^2}{2k_{\text{cant}}}\right)/k_B T\right] \qquad (3.63)$$

where $k_{\text{off}}(F)$ is the unbinding transition rate in Equation 3.23, and $\Delta G_0 = -k_B T \ln(k_{\text{off}}^0/k_{\text{on}}^0)$ is the equilibrium free energy difference between the bound and unbound states. Therefore, the free energy change is modulated by a quadratic dependence on force due to the Hookean nature of the applied force. Equation 3.63 provides a unique force value for the system that is defined when the unbinding and rebinding transition rates are equal (Evans, 2001; Friddle et al., 2008),

$$F_{\text{eq}} = \sqrt{2k_{\text{cant}}\Delta G_0} \qquad (3.64)$$

which we will call the equilibrium force of the system.

3.4.2.1 Mean

Away from equilibrium $r_f > 0$, an analytical solution for the master Equation 3.57 is not possible under transition rates changing with time. It might appear that the complexity of this complete two-state description relegates us to only consider the case of forced processes driven very far from equilibrium, such as the first-passage treatment above. However, it is actually possible to formulate an extremely accurate description of the mean rupture force for this two-state system upon closer analysis of the process. Inspection of the relative unbinding and rebinding rates in Equation 3.63, reveals behavior essential to describing the complete two-state system theoretically. For forces less than F_{eq}, the rebinding rate dominates the dynamics, acting to keep the system in the bound state. At forces beyond F_{eq}, the rebinding rate falls off rapidly due to the $-F^2/2k_{\text{cant}}$ exponent, leaving the unbinding rate $k_{\text{off}}(F)$ as the only significant contributor to the dynamics. So we see that once the applied load is greater than F_{eq}, the system enters the kinetic unbinding regime, whereas for forces below F_{eq} the bound state persists. This realization allows for a considerable simplification of the two-state master Equation 3.57 by imposing that $p_{\text{on}}(F \leq F_{\text{eq}}) = 1$ and $k_{\text{on}}(F > F_{\text{eq}}) \cong 0$:

$$\int_1^{p_{\text{on}}} \frac{dp'_{\text{on}}}{p'_{\text{on}}} = -\frac{1}{r_f}\int_{F_{\text{eq}}}^F k_{\text{off}}(F')dF' \qquad (3.65)$$

Inserting the Bell rate (Equation 3.24), and using the same procedure as in Equation 3.53, we find (Friddle et al. 2011)

$$\langle F \rangle = F_{\text{eq}} + \frac{k_B T}{x_\beta}\exp\left[\frac{k_{\text{off}}(F_{\text{eq}})k_B T}{r_f x_\beta}\right]\text{E}_1\left[\frac{k_{\text{off}}(F_{\text{eq}})k_B T}{r_f x_\beta}\right] \qquad (3.66)$$

where $k_{\text{off}}(F_{\text{eq}}) = k_{\text{off}}^0\exp[F_{\text{eq}}x_\beta/k_B T]$, and $\text{E}_1(z)$ is the exponential integral. Equation 3.66, plotted in Figure 3.6, increases from the equilibrium force, F_{eq}, linearly with loading rate r_f, then follows a nonlinear trend, and asymptotically

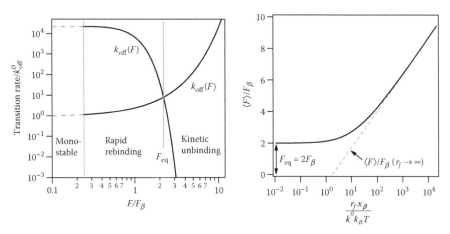

FIGURE 3.6 The force dependence of the unbinding and rebinding transition rates reveal a route to simplifying the two-state description of forced bond rupture. At and near zero force, the bound state is monostable because the probe minimum coincides with the minimum of the bond potential. After applying a small force, the system enters a bistable regime where the rebinding rate is negligibly impeded by the small reentry barrier, and thus rapid rebinding acts to prolong the stability of the bound state further. Only after pulling beyond the equilibrium force F_{eq} does the rebinding rate rapidly fall short of the unbinding rate and the system enters the kinetic regime. The resulting force spectrum, Equation 3.66, descends asymptotically to the equilibrium force for vanishing loading rate. In the fast loading rate regime, the first-passage behavior is recovered, as shown by comparison to Equation 3.54. Force plotted in units of the thermal force scale $F_\beta = k_B T / x_\beta$.

approaches the commonly used form (Equation 3.54) at large loading rate $\langle F \rangle \cong \frac{k_B T}{x_\beta} \ln \left[\frac{r_f x_\beta e^{-\gamma}}{k_{off}^0 k_B T} \right]$, where $\gamma = 0.577...$ is the Euler constant. Note that the exponential integral follows the simple interpolation $e^z E_1(z) \cong \ln(1 + e^{-\gamma}/z)$, over all values of the argument z.

3.5 MULTIPLE BONDS UNDER FORCE

In real single-molecule force spectroscopy experiments, single intermolecular bonds are difficult to achieve. To measure a true *single-molecule* system comprising of only one pair of interacting molecules, one must tune a number of parameters such as the density of molecules on the surfaces of the force apparatus, the contact time, as well as the contact force. Even after careful consideration of these factors, the number of bonds formed is typically random from measurement to measurement. Thus, there is usually no clear way to know how many molecules will take part in each force measurement *a priori*. One must rely on post measurement analysis using statistical assumptions or, as shown below, the use of flexible linkers (polymers) of known mechanical properties to deduce the number of bonds formed from each force-extension trajectory. In some cases, the effects of multiple bonds are precisely what one is interested in exploring. Multivalency pertains to the natural or synthetic creation of a construct, or cluster, bearing multiple identical ligands that enhance the binding of the construct to a molecule, surface, or cell. Multivalent interactions are important because they arise often in biological systems and can be harnessed to improve the lifetime and strength of an adhesive bond between a pair of molecules or objects. If one envisions the surfaces of the force apparatus to be part of the cluster, the *valency* between the two surfaces refers to the number of identical bonds formed.

3.5.1 PARALLEL CORRELATED BONDS

The simplest multivalent system is found when N identical bonds are rigidly coupled to one another (Figure 3.7). Because they move together in concert, the transition state to rupture is unchanged, however, the depth of the barrier and the curvatures at the barrier and minimum are scaled according to

	Single bond		N bonds
Barrier distance	x_β	$\rightarrow$	x_β
Well stiffness	κ_0	$\rightarrow$	$N\kappa_0$
Barrier stiffness	κ_b	$\rightarrow$	$N\kappa_b$
Energy barrier	ΔU_0	$\rightarrow$	$N\Delta U_0$

From Kramers model of forced escape, we see the force constants at the well and barrier contribute factors $\sqrt{N\kappa_0}$ and $\sqrt{N\kappa_b}$, however, the increased number of bonds will also contribute to an increased effective damping coefficient. Assuming this to be linear $N\eta$ (Evans & Williams, 2002), the effects due to N cancel in the frequency

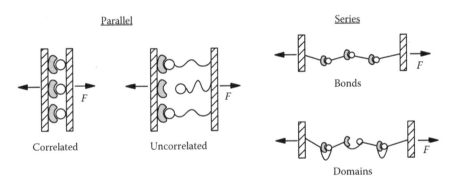

FIGURE 3.7 Four common configurations of multiply bonded systems. Parallel bonded networks may consist of close, rigidly linked subunits whose dynamics are correlated, or they may exist as separate, uncorrelated entities, each held to the cluster by a flexible tether. Examples of series bonded systems are simply polymer chains with a series of weak bonds that fail irreversibly, or polymers with folded domains that can reversibly switch between unfolded and folded states under force.

prefactor, and thus the primary contributor to the escape rate of all N bonds in concert arises from the deeper potential barrier (Evans & Williams, 2002):

$$k_{N\to0}(F) = k^0\exp\left[-\frac{(N-1)\Delta U}{k_B T} + \frac{F x_\beta}{k_B T}\right] \tag{3.67}$$

where $k^0 = \frac{\sqrt{\kappa_0 \kappa_b}}{\eta} e^{-\Delta U/k_B T}$ is the single-bond escape rate.

3.5.2 SERIES BONDS AND DOMAINS

There are two primary cases to consider when the molecule under manipulation contains multiple bonds arranged in series (Figure 3.7). The first case is simply bonds that hold the chain together. For example, a polymer such as poly(ethylene-glycol) (PEG) consists of a number of ethylene-glycol monomers connected in parallel. When under a large tensile load, the breakage of any one bond breaks the entire polymer in half irreversibly. The second case is domains within the polymer chain that change configuration under force. PEG in aqueous solution fits in this category as well because the conformation of each monomer can switch between helical or planar isomers depending on the level of force (Oesterhelt et al. 1999). Upon decreasing force, the domains can switch back to their helical form. Another common example of domains are the folded immunoglobulin domains of Titin, which unfold under force (Rief et al. 1997). The basic difference between series bonds and series domains is that domains are generally reversible upon decreasing force, whereas series bonds typically are not.

In either the series bond or series domain case, the rate of breaking or configurational switch under increasing force is given by the rate for a single sub-unit times the number of subunits available (Evans & Williams, 2002; Williams, 2003):

$$k_{N\to N-1}(F) = N k^0\exp\left[\frac{F x_\beta}{k_B T}\right] \tag{3.68}$$

where N is the number of closed bonds or domains in the chain. The mean rupture force of the first bond in the series is (see Equation 3.53)

$$\langle F \rangle = \frac{k_B T}{x_\beta}\exp\left[\frac{N k^0 k_B T}{r_f x_\beta}\right] E_1\left[\frac{N k^0 k_B T}{r_f x_\beta}\right] \tag{3.69}$$

or in the fast loading rate limit ($r_f \to \infty$)

$$\langle F \rangle \cong \frac{k_B T}{x_\beta}\left(\ln\frac{r_f x_\beta e^{-\gamma}}{k^0 k_B T} - \ln N\right) \tag{3.70}$$

Therefore, at large loading rates, the rupture force of one bond in a chain is less than one lone bond $\langle F \rangle_1$ by the log of the number of bonds in series $\langle F \rangle \cong \langle F \rangle_1 - \frac{k_B T}{x_\beta}\ln N$.

When the chain consists of domains that switch configuration upon increasing or decreasing force, one must account for both forward and reverse processes in the description. Let us assume there are N domains total in the chain, and at any point in time, N_{off} are unfolded and N_{on} are folded, such that $N = N_{on} + N_{off}$. The rates of observing unfolding and folding follow as

$$k_{N_{on} \to N_{on}-1}(F) = N_{on}k_{off}(F) = N_{on}k_{off}^0 \exp\left[\frac{Fx_\beta}{k_BT}\right] \tag{3.71}$$

$$k_{N_{off} \to N_{off}-1}(F) = N_{off}k_{on}(F) = N_{off}k_{on}^0 \exp\left[\frac{-Fx_\alpha}{k_BT}\right] \tag{3.72}$$

where k_{off} and k_{on} are, respectively, the unfolding and folding rates for individual domains. The simple form of the rates in Equations 3.71 and 3.72 imply that the total distance $(x_\beta + x_\alpha)$ between the minima of the folded and unfolded states does not change appreciably under force. This is not true in general, however, we assume that for slow to intermediate loading-rates, thermal activation over the barrier should occur before appreciable distortion of the underlying energy potential takes place. Under the simple assumption that force is linear with time $dF = r_f dt$, the equation of motion for finding N_{on} domains folded at force F is given by

$$r_f\frac{dN_{on}}{dF} = -k_{N_{on} \to N_{on}-1}(F) + k_{N_{off} \to N_{off}-1}(F) \tag{3.73}$$

$$= -N_{on}k_{off}(F) + N_{off}k_{on}(F) \tag{3.74}$$

Dividing both sides by the total number of domains N produces the rate equation for the probability $p_{on}(F)$ of finding a domain folded at force F. In this model, each domain constitutes a two-level system. Each domain is coupled to one another by the intervening polymer linkage. When both k_{on} and k_{off} are slow compared with the rate of the driving force, the domains cannot keep up with the changing force, and the force at which they fold and unfold will lag behind the force they would unfold at when driven quasi-statically. That is, each domain will be driven away from equilibrium. Rief, Fernandez, and Gaub presented a simple Monte Carlo method of analyzing this two-level system while also accounting for the coupling of the folded and unfolded states with the nonlinear polymer extension under force (Rief et al. 1998). On the other hand, when either k_{on} or k_{off} are fast enough such that the longest characteristic relaxation time $\tau_{rel}(F) = [k_{on}(F) + k_{off}(F)]^{-1}$ is small compared with the timescale of the measurement, then the system will remain close to equilibrium throughout. In this quasi-equilibrium case, a solution for the relative number of domains in the folded and unfolded states is found directly from the Boltzmann distribution:

$$\frac{N_{on}(F)}{N_{off}(F)} = \exp\left[-\Delta G(F)/k_BT\right] \tag{3.75}$$

$$= \frac{k_{on}(F)}{k_{off}(F)} \tag{3.76}$$

$$= \exp\left[-\left(\Delta G_0 + F(x_\beta + x_\alpha)\right)/k_BT\right] \tag{3.77}$$

Given the total number of domains is fixed $N = N_{on} + N_{off}$, we can use Equation 3.75 to write down the fraction of folded and unfolded states as

$$\frac{N_{on}(F)}{N} = \frac{1}{1 + \exp[\Delta G(F)/k_B T]} \tag{3.78}$$

$$\frac{N_{off}(F)}{N} = \frac{1}{1 + \exp[-\Delta G(F)/k_B T]} \tag{3.79}$$

where $\Delta G(F) = \Delta G_0 + F(x_\beta + x_\alpha)$. Convention is important here. We take the unperturbed free energy difference between the folded and unfolded states to be a negative number $\Delta G_0 < 0$ to indicate the folded state is favored at zero force. Oesterhelt et al. (1999) applied the relations in Equations 3.78 and 3.79 in a rather elegant way to model the force-extension behavior of PEG in aqueous solution. PEG monomers have been found to be stabilized in a helical, trans-trans-gauche conformation in water. In organic solvent, this stabilization is lost and PEG monomers take a planar, trans-trans-trans conformation. Oesterhelt et al. found that the water-mediated helical form of PEG could be driven into the planar form by stretching. Therefore, using the extended freely-jointed-chain model (eFJC) to describe the polymer's extension z with force, the contour lenth L_C was taken to depend on force as well:

$$z(F) = \left[\coth\left(\frac{FL_k}{k_B T}\right) - \frac{k_B T}{FL_k}\right]\left[L_C(F) + \frac{N_k F}{\kappa_s}\right] \tag{3.80}$$

where L_k is the Kuhn length, N_k is the number of segments of length L_k, and κ_s is the segment elasticity. Thus, if a folded or helical segment takes on a length l_f and an unfolded or planar segment takes a length l_u, then the contribution to the total contour length from each of the two forms are given by $l_f N_{on}/N_k$ and $l_u N_{off}/N_k$ or

$$L_C(F) = N_k\left[\frac{l_f}{1 + e^{\Delta G(F)/k_B T}} + \frac{l_u}{1 + e^{-\Delta G(F)/k_B T}}\right] \tag{3.81}$$

3.5.3 PARALLEL INDEPENDENT BONDS: KINETIC REGIME

A multivalent cluster is typically viewed as a construct of identical, noninteracting ligands joined together by flexible polymer linkers. Each ligand binds specifically to one of many receptors on a surface. Here, we assume the cluster to be linked to the end of a force transducer of stiffness k_{cant}. We will consider the case in which the linker end-to-end extension is small (small force approximation) such that the Gaussian chain model for a random flight polymer is adequate (Kratky & Porod, 1949; Flory, 1969). Thus, the polymer linker has a linear force-extension relation with force constant k_{poly}. Given the fact that typical polymers are much softer than mechanical force transducers ($k_{poly} \ll k_{cant}$), we assume the applied force on the bonds to be dominated by the force-extension relation of the polymer.

The rate of transition from N to $N-1$ bonds is enhanced by the increased number of bonds available to unbinding, but the load on each bond is divided over the whole

(Williams, 2003),

$$k_{N \to N-1}(F) = Nk^0 \exp\left[\frac{Fx_\beta}{Nk_BT}\right] \tag{3.82}$$

Because all N bonds are uncoupled, their failure is independent from one another. Hence, we describe the complete failure as a Markov process, such that the total time required for all bonds to fail is simply the sum of lifetimes of each step in the unbinding pathway.

$$\begin{array}{ccccccccc} N & \to & (N-1) & \to & (N-2) & \to & \cdots & \to & 1 & \to & 0 \\ \tau_N & & \tau_{N-1} & & \tau_{N-2} & & & & \tau_2 & \tau_1 \end{array}$$

where the lifetime of each step is given by the inverse of Equation 3.82,

$$\tau_N(F) = \frac{1}{k_{N \to N-1}(F)} \tag{3.83}$$

Thus, the rate of unbinding all N bonds is given by the inverse of the sum of each individual lifetime in the sequence

$$k_{N \to 0}(F) = \frac{1}{\sum_{n=1}^{N} \tau_n} = \left[\sum_{n=1}^{N} \frac{1}{nk^0 \exp\left[\frac{Fx_\beta}{nk_BT}\right]}\right]^{-1} \tag{3.84}$$

We will first consider when the N bonds are loaded quickly, such that the force on the system drives the bonds apart irreversibly. In general, however, the scheme above illustrating the steps in the unbinding pathway will also have arrows pointing to the left to account for rebinding. When unbinding is completely irreversible, we describe the process by first-passage statistics. As the force increases rapidly on the cluster of bonds, the failure of the first bond will result in a sudden increase in load on the remaining bonds. Therefore, it is reasonable to assume that for fast loading rates, the failure of the remaining bonds will occur soon after, if not immediately after, the rupture of the first bond. Under this assumption, we can treat the dynamics in the same manner as we would treat a single bond (Evans & Williams, 2002; Williams, 2003). That is, the failure of all N bonds will occur with a rate given by Equation 3.84, and most frequently at a force F^*. As explained in Equation 3.55, Evans & Williams (2002) showed that for first-passage processes, the most probable rupture force F^* follows from a relationship between loading rate and unbinding rate:

$$k(F)\Big|_{F^*} = r_f \frac{d\ln k(F)}{dF}\Big|_{F^*} \tag{3.85}$$

Operating the relationship in Equation 3.85, on the transition rate in Equation 3.84, a transcendental equation for the most probable force of failure F^* is found (Williams, 2003):

$$\frac{1}{r_f} = \frac{x_\beta}{k^0 k_BT} \sum_{n=1}^{N} \frac{1}{n^2} \exp\left[\frac{-F^* x_\beta}{nk_BT}\right] \tag{3.86}$$

Numerical solution to Equation 3.86 is shown in Figure 3.8. Notice that the enhancement in rupture force is modest in this nonequilibrium regime. That is, the rupture force of N bonds is less than N times the rupture force of one! The reason is because of the increased probability of detachment when multiple bonds are available. Consider the limit of vanishing force, such that the lifetime of one ligand–receptor complex is τ_1 and we assume the same Markov process for unbinding of each ligand takes place. The total mean lifetime of two ligands is

$$\tau_2 = \tau_1/2 + \tau_1 = 1.5\tau_1$$

which is not even twice the lifetime of one bond! The total lifetime of the three gives $\tau_3 \approx 1.8\tau_1$ and that of four gives only $\tau_4 \approx 2.1\tau_1$. Under the irreversible scheme we are considering here, the total lifetime of N bonds at zero force is the harmonic

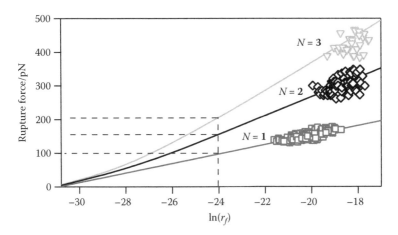

FIGURE 3.8 (See color insert.) Rupture of multiple parallel bonds for 1, 2, and 3 multivalent clusters bewteen the cancer marker Mucin-1 and its antigen fragment. Data points reproduced with permission from Sulchek, T. et al. 2005. *Proc. Nat. Acad. Sci. USA*, 102 (46), 16638–16643. Solid lines are numerical solutions of Equation 3.86. Only the single-bond, $N = 1$, curve is fit to the data to determine the kinetic unbinding rate and transition state, while the $N = 2$ and $N = 3$ curves are predictions for bivalent and trivalent bonding. The dashed lines illustrate that, for a given loading rate, including two or three bonds to the cluster does not add a significant increase in total rupture force. In fact, at the chosen loading rate, the rupture force for $N = 3$ bonds is just twice that of the single $N = 1$ case. This is due to the increased probability of rupture when multiple bonds are present. The enhancement over a single bond improves for increased loading rate, but even at extremely fast loading, the rupture force for N bonds is limited to less than N times the single-bond rupture force. As discussed in the text, binding enhancement due to multivalency is most prominent in the equilibrium regime where rebinding and entropic effects stabilize the bound state.

number H_N times the lifetime of one

$$\tau_N = \tau_1 \sum_{n=1}^{N} \frac{1}{n} = H_N \tau_1$$

For large N, the harmonic number has the asymptotic expansion

$$H_N = \ln N + \gamma + \mathcal{O}(N^{-1})$$

where $\gamma = 0.577...$ is the Euler constant. This means that, at zero load, the total lifetime of 100 bonds is just over five times the lifetime of one bond. Another way to think of this is to consider the limit of an ensemble of identical bonds $N \to \infty$. The time to wait for the first to fail is $\tau_1/N \to 0$. Under finite loading rate, the enhancement of the most-probable rupture force for N bonds F_N^* over the single bond F_1^* improves with increased loading rate. However, it can be shown analytically (Williams, 2003) that in general,

$$F_N^* < NF_1^*$$

This can be inferred from the fact that, for uncorrelated bonds, the only way the deterministic enhancement $F_N^* = NF_1^*$ can be reached in the kinetic regime, is if the loading rate is fast enough to overcome thermal fluctuations. Otherwise, the first bond will fail N times faster than the deterministic rate would require (Equation 3.84), and the remaining $N - 1$ bonds will fail immediately. This might seem to suggest that adding more ligands to a multivalent cluster does very little in terms of enhancing the lifetime of the cluster. However, this analysis has been entirely based on a one-way, irreversible Markov process. At equilibrium, or when the loading rate is sufficiently slow, each of the unbinding steps have a finite probability of rebinding as well. The effects of reversibility on the lifetime, and final rupture force, of the multivalent cluster can lead to extremely strong enhancement over the single-bond case. This problem is analyzed through a statistical mechanical perspective in the next section.

3.5.4 PARALLEL IINDEPENDENT BONDS: EQUILIBRIUM REGIME

The kinetic analysis above is specific to multibonded systems driven rapidly away from equilibrium such that unbinding/rebinding fluctuations are not important. However, for slow pulling speeds in which the timescale of the pulling process is slow compared with the fluctuations of the molecules into and out of their bound states, then the dynamics are changed considerably. The problem of determining a continuous force spectrum spanning equilibrium to kinetic regimes is haltered by the complexity of stochastic dynamics involved for more than one bond. The above analysis is evidence of this complexity given that modeling the kinetic regime alone is only possible through numerical solution of Equation 3.86. In keeping with the spirit of an equilibrium force F_{eq}, which we defined for a single bond earlier in the chapter (Equation 3.64), we will derive an approximate solution for the equilibrium force of separating a multivalent cluster from a surface by way of statistical mechanics.

Consider a number M of available surface receptors to which N ligands of the cluster are bound on average. We use the term average here to account for the fact that fluctuations in the instantaneous number of bound ligands will randomly change N with time. The relationship between M and N is not trivial. M is defined by the number of receptors *available* to the cluster ligands. Thus, M is the areal density of receptors multiplied by the area of the surface accessible to the end of a polymer linker within the cluster. To tune M at constant linker length, receptor density can be changed; or at constant receptor density, the length of the linkers can be changed. The bound state partition function of a single ligand–receptor pair under the influence of a polymer in the linear force-extension approximation is given by

$$Z_{1,\mathrm{on}}(F) = \frac{1}{\lambda}\sqrt{\frac{2\pi k_B T}{\kappa_0}}e^{(\Delta U_0 - F^2/(2k_{\mathrm{poly}}))/k_B T} \tag{3.87}$$

An unbound ligand will retract back to diffuse around the polymer's potential minimum and therefore the single unbound partition function follows as

$$Z_{1,off} = \frac{1}{\lambda}\sqrt{\frac{2\pi k_B T}{k_{\mathrm{poly}}}} \tag{3.88}$$

Note that at equilibrium only the relative levels of energy minima and their curvatures (harmonic approximation) govern the populations and free energy difference between bound and unbound states. The unbound state is the minimum of the linker potential, which is arbitrarily set to zero energy. The force term appearing in Equation 3.87 is due to the linker raising the minimum of the bound state by an amount $F^2/(2k_{\mathrm{poly}})$ (see Equation 3.20). If the cluster forms N bonds with the surface receptors, the total partition function of the bound state is given in terms of Equation 3.87 as

$$Z_{N,\mathrm{on}} = \frac{M!}{N!(M-N)!}(Z_{1,\mathrm{on}}(F/N))^N \tag{3.89}$$

$$= \frac{M!}{N!(M-N)!}\left(\frac{1}{\lambda}\sqrt{\frac{2\pi k_B T}{\kappa_0}}e^{(\Delta U_0 - F^2/(N^2 2k_{\mathrm{poly}}))/k_B T}\right)^N \tag{3.90}$$

where the prefactor is the binomial coefficient, which accounts for the number of different combinations of N occupied receptor sites out of M total (see Figure 3.9). The N uncoupled single-bond partition functions are multiplied, and the force experienced at the transducer is reduced by $1/N$ at each bond. When all N ligands are unbound, we divide by $N!$ to account for each identical ligand in the construct being indistinguishable from one another:

$$Z_{N,\mathrm{off}} = \frac{1}{N!}\left(\frac{1}{\lambda}\sqrt{\frac{2\pi k_B T}{k_{\mathrm{poly}}}}\right)^N \tag{3.91}$$

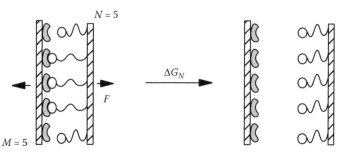

FIGURE 3.9 Equilibrium free energy of a multivalent cluster under force. N ligand–receptor interactions are considered to occur on average at equilibrium. M receptors are assumed to be within reach of any of the N receptors. This allows for "mixing" of the ligands on the surface receptors and contributes an entropic term in the free energy favoring the bound state. The illustration depicts the case of a cluster bearing five ligands, while three are bound to the surface receptors on average. The absolute free energy of the bound state alone depends only on the number of *combinations* of N ligands on M receptors, which in this example is 10. However, the free energy change between the N-bonded cluster and the free cluster depends on the number of *permutations* of N ligands with M receptors that is 60. This large degeneracy enhances the free energy cost to forcibly driving the system to the completely unbound state near equilibrium.

The free energy difference between the N-bonded cluster and the free cluster is given by

$$\Delta G_N(F) = -k_B T \ln \frac{Z_{N,\text{off}}}{Z_{N,\text{on}}}, \qquad (3.92)$$

which when bound and unbound states are balanced by the applied force $\Delta G_N(F_{N,\text{eq}}) = 0$, and we find the equilibrium force

$$F_{N,\text{eq}} = \sqrt{N^2 2 k_{\text{poly}} \Delta G_1 + N 2 k_{\text{poly}} k_B T \ln\left(\frac{M!}{(M-N)!}\right)} \qquad (3.93)$$

where

$$\Delta G_1 = \Delta U_0 - k_B T \ln \sqrt{\frac{\kappa_0}{k_{\text{poly}}}} \qquad (3.94)$$

is the free energy of detaching a single ligand–receptor pair by transferring the applied force with the polymer linker of force constant k_{poly}. In Equation 3.93, the second term under the root accounts for the entropy imparted by total number of configurations, or permutations $\frac{M!}{(M-N)!}$ of associating N ligands with N receptors out of the M available. This can be a considerably large number when $M > N$. Even when $M = N$, an entropic enhancement is found due to the different configurations that can occur between N ligands on N receptors (i.e., $N! = N(N-1)(N-2)\cdots 1$). All such configurations are identical in energy, which is important near equilibrium when spontaneous unbinding and rebinding events occur faster than the timescale of the measurement. Hence, increasing the degeneracy of the bound state enhances the stability of the cluster-surface interaction, and thus increases the equilibrium rupture force $F_{N,eq}$.

Equation 3.93 reveals a detail about tethered single bonds not considered in the previous sections on single-bond systems. If only a single ligand is linked to the force probe by a flexible linker, then the average bond valency is obviously $N = 1$. But if the density of the receptors is high enough such that more than one is available $M > 1$, then Equation 3.93 does not reduce to the equilibrium force for a single-ligand single-receptor bond $F_{eq} = \sqrt{2k_{poly}\Delta G_1}$, but more generally, it retains the entropic term and reduces to

$$F_{1,eq} = \sqrt{2k_{poly}\left(\Delta G_1 + k_B T \ln M\right)} \tag{3.95}$$

which suggests an approach to determining the surface density of receptor sites using mono-valent force spectroscopy, when the single-ligand/single-receptor free energy ΔG_1 is known.

3.5.5 COUNTING BOND VALENCY IN FORCE MEASUREMENTS

When multiple molecules are present on the contacting surfaces, intentionally or otherwise, it is ultimately desirable to know how many participate in the measurement. For molecules linked rigidly to the contacting surfaces, their rupture is most likely correlated such that their failure is highly cooperative. The appearance of peaks in the rupture force histogram may indicate the force quanta for each valency, however, this is only well-resolved if the individual distributions for each valency are small compared with the mean. Otherwise, the resulting distribution is broadened by the presence of multiple bonds with no direct distinction of each valency in the histogram. Below is two approaches toward dealing with the bond valency problem in force measurements. The first is based on a statistical approach, which is appropriate for molecules whose discrete rupture events are indistinguishable. This is usually the case for molecules linked closely to the probe whereby rupture of the bonds is correlated. The second is based on the linear scaling of stiffness with the number of parallel polymer tethers, which can be used to deduce bond valency when the molecules of interest are linked via flexible tethers.

3.5.5.1 Poisson Analysis of Rupture Events

Evans & Williams (2002) showed how the fraction of occurrences of adhesion events within a total sample of attempted force measurements can be related to the Poisson distribution to estimate the mean number of bonds formed in the sample, and the fraction of those which are single bonds. If one prepares the experiment such that the probability of bond formation between the contacting surfaces is low, then one can assume Poisson statistics govern the number of bonds formed on each contact. To make proper use of such an analysis, the same protocol should be used for each force measurement, such as contact time and force. For a large sample size, the discrete probability of finding n bonds formed when the sample mean is $\overline{N}$ is given by the Poisson distribution:

$$P(n;\overline{N}) = e^{-\overline{N}}\frac{\overline{N}^n}{n!} \tag{3.96}$$

The sum over all probabilities for which an adhesion event occurs ($n \neq 0$) gives the fraction of all possible adhesion events for any bond valency. Therefore, the fraction of occurrences of adhesion events N_e within the total sample of attempted force measurements N_{tot} is equal to the sum of probabilities with $n \geq 1$:

$$\sum_{n=1}^{\infty} e^{-\overline{N}}\frac{\overline{N}^n}{n!} = 1 - e^{-\overline{N}} = \frac{N_e}{N_{tot}} \tag{3.97}$$

which gives the mean number of bonds formed in terms of the fraction of adhesion events:

$$\overline{N} = \ln\left(\frac{1}{1 - N_e/N_{tot}}\right) \tag{3.98}$$

With this solution for $\overline{N}$, one can solve for $P(n = 1;\overline{N})$ in Equation 3.96 to find the fraction of single-bond events out of the total number of samples:

$$\frac{N_1}{N_{tot}} = \left(1 - \frac{N_e}{N_{tot}}\right)\ln\left(\frac{1}{1 - N_e/N_{tot}}\right) \tag{3.99}$$

Likewise, the fraction of single bonds out of the number of successful adhesion events follows from $P(n = 1;\overline{N})/(N_e/N_{tot})$ as

$$\frac{N_1}{N_e} = \left(\left(\frac{N_e}{N_{tot}}\right)^{-1} - 1\right)\ln\left(\frac{1}{1 - N_e/N_{tot}}\right) \tag{3.100}$$

Rewriting Equation 3.98 in terms of $\overline{N}$, we can solve for the frequency of successful adhesion events that give single bonds on average:

$$\left(\frac{N_e}{N_{tot}}\right)_{\overline{N}=1} \cong 0.63 \tag{3.101}$$

However, inserting this frequency in Equation 3.100, we find that only 53% of the successful events will contain single bonds, while the remaining 47% will consist of a mixture of doubles, triples, and so on. Therefore, aiming for a target of single bonds on average will still leave the measurements containing a significant portion (nearly half!) of higher bond numbers. Alternatively, one can look to filter out the higher bond numbers by decreasing the frequency of interactions further. For example, to achieve adhesion events containing 90% single bonds, one would aim for an adhesion event frequency of about 18%. This translates to 16% of single bonds out of the total number of attempted measurements. To put this in perspective, to measure 100 single-bond events, one would need to make more than 600 measurements, resulting in an excess of 500 "empty" measurements.

3.5.5.2 Valency from Polymer Tethers

The use of polymer linkers provides a direct approach to determining the true number of bonds involved in a multivalent force measurements. The force-extension behavior of a variety of polymers have been well-characterized ill date. In most cases, polymers under tension in solution are described by the worm-like-chain, the freely-jointed-chain (FJC), or variations on these models. The primary disadvantage of polymer tethers is the nonlinear force with extension that does not directly lend the data to simple dynamic models where linearly applied forces $F(t) = r_f t$ are assumed. However, accounting for the nonlinearity can be done numerically, or one can assume the force in the region around the unbinding event is linear and use the compliance of the tether in this region for the effective loading rate (Friedsam et al. 2003).

Once the parameters of the single polymer force-extension model are determined, a model to estimate the number of multiple bonds is straightforward. Assuming the polymer dynamics remain at equilibrium throughout the pulling process (i.e., the F vs. z curve is independent of pulling speed), then the force due to N_p identical parallel polymers, at any extension z, is just N_p times the force of one polymer. The model that describes the single polymer tether is simply recast using F_{tot}/N_p in place of the force F, where F_{tot} is the total force applied to the multivalent cluster (see Figure 3.10). For example, referring to the discussion above for the commonly used PEG polymer tethers (see Equations 3.75 through 3.81), the extension of N_p parallel PEG tethers is

$$z(F_{tot}/N_p) = \left[\coth\left(\frac{F_{tot}L_k}{N_p k_B T}\right) - \frac{N_p k_B T}{F_{tot}L_k}\right]\left[L_C(F_{tot}/N_p) + \frac{N_k F_{tot}}{N_p \kappa_s}\right] \qquad (3.102)$$

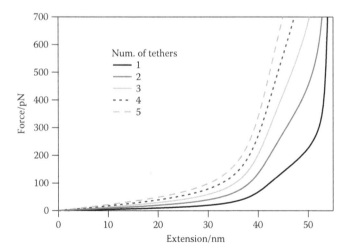

FIGURE 3.10 (See color insert.) Theoretical force-extension curves of multiple, parallel PEG polymers from numerical solution of Equation 3.102. (Adapted from Sulchek, T. et al. 2006. *Biophys. J.*, 90 (12), 4686–4691.)

with force-dependent contour length

$$L_C(F_{tot}/N_p) = N_k \left[\frac{l_f}{1 + e^{\Delta G(F_{tot}/N_p)/k_B T}} + \frac{l_u}{1 + e^{-\Delta G(F_{tot}/N_p)/k_B T}} \right] \quad (3.103)$$

Assuming the contour length may vary slightly between measurements, the only free parameters in the model are the number of Kuhn segments N_k and number of tethers N_p, which are completely uncoupled parameters. This approach for determining bond valency was first employed by Sulchek et al. (2005, 2006) to determine bond valency in PEG-tethered Mucin-1/antibody binding, which led to direct verification of the multivalent model of Williams (2003) for parallel independent bonds (see Equation 3.86 and Figure 3.8).

3.6 THERMODYNAMICS, WORK, AND FLUCTUATION THEOREMS

We've seen in the preceding sections that, under increasing load, the force at which a bond breaks is not a unique property of the bond alone, but it also depends on how rapidly we increase the load. Thus, the observed rupture force is acquired under nonequilibrium conditions. However, we are typically interested in probing systems to determine their characteristic properties when the system is in thermodynamic equilibrium. Of fundamental importance is the equilibrium free energy of the system, which defines the work exchanged by the system with its surroundings when the work done on or by the system is completely reversible. There are varying definitions of free energy, the most prominent being Helmholtz A (common to physicists) and Gibbs G (common to chemists). However, in the case of single-molecule events, the difference between the two is subtle:

$$G = A + pV \quad (3.104)$$

where p is the constant pressure and V the final volume of the system. For single-molecule transformations, we assume the volume change will be negligible and therefore Gibbs and Helmholtz free energies can be used interchangeably:

$$G \cong A = -k_B T \ln Z \quad (3.105)$$

where Z is the partition function of the system. Throughout this chapter, we chose to use G to define the free energy of a system and ΔG to designate the change in free energy over a reversible transformation.

3.6.1 FLUCTUATION THEOREMS

The second law of thermodynamics states that the ensemble average work $\langle W \rangle$ done on a system cannot be less than the equilibrium free energy:

$$\langle W \rangle \geq \Delta G \quad (3.106)$$

However, for small systems, such as individual molecules, fluctuations can significantly broaden the distribution of work $\rho(W)$ for a given process. Although the mean work will always exceed the free energy for a driven process, a fraction of trajectories within the distribution $\rho(W)$ will have work values less than ΔG. If the system parameters are changed rapidly, the system is driven far from equilibrium and the distribution $\rho(W)$ will be broad. On the other hand, if the system parameters are changed quasi-statically, the distribution $\rho(W)$ approaches a delta function centered at the free energy ΔG of the transformation. Consider now a system that is initially at thermal equilibrium. An increasing force is applied that drives the system to a new state. Then, the system reaches a new equilibrium and is driven backward in time (decreasing force) at the same rate to the original state. Accumulating many realizations of this process produces probability distributions $\rho_{for}(W)$ and $\rho_{rev}(W)$ of the work done on the system along the forward and reverse processes. Crooks (1999) showed that the relative probability of a work value $W = w$ during the forward process to the probability of a work value $W = -w$ during the reverse process is given by the following detailed fluctuation theorem as

$$\frac{\rho_{for}(W)}{\rho_{rev}(-W)} = \exp\left[(W - \Delta G)/k_B T\right] \tag{3.107}$$

This relationship indicates that the unique work value at which $\rho_{for}(W') = \rho_{rev}(-W')$ is equal to the free energy, $W' = \Delta G$. Noting that the dissipated work is $W_d = W - \Delta G$, we can also express the Crooks (1999) theorem as

$$\frac{\rho_{for}(W_d)}{\rho_{rev}(-W_d)} = \exp\left[W_d/k_B T\right] \tag{3.108}$$

Rearranging Equation 3.107 and integrating W over all work values $[-\infty, \infty]$ yields the following integral fluctuation theorem:

$$\int \rho_{for}(W)\exp\left[-W/k_B T\right] dW = \int \rho_{rev}(-W)\exp\left[-\Delta G/k_B T\right] dW$$

$$\langle\exp\left[-W/k_B T\right]\rangle = \exp\left[-\Delta G/k_B T\right] \tag{3.109}$$

Equation 3.109 is Jarzynski's nonequilibrium work relation Jarzynski (1997). This relation states that the ensemble average over the Boltzmann-weighted work is equivalent to the Boltzmann-weighted free energy. Application of these theorems is contingent on a proper definition of the work done on the system of interest. We will briefly address this topic for two commonly encountered cases in the next section.

3.6.2 WORK AND FREE ENERGY IN FORCED TRANSITIONS

The *thermodynamic work* W performed on a system follows as the integral over the time rate of change of the Hamiltonian covering an observation time from 0 to t_f:

$$W = \int_0^{t_f} \frac{\partial H(x(t),t)}{\partial t} dt \tag{3.110}$$

where in the case of bond rupture, the Hamiltonian is the sum of the time-invariant intermolecular potential of the bond $U(x)$ and the time-dependent pulling potential applied to the bond $V(x,t)$:

$$H(x(t),t) = U(x(t)) + V(x(t),t) \tag{3.111}$$

As illustrated in Figure 3.1, two common scenarios for $U(x)$ arise: (1) a unimolecular system, such as a protein or RNA, described by a bistable potential; or (2) a bimolecular system, such as a ligand–receptor bond, which is described by a single metastable state within the potential. In the second case, the total Hamiltonian becomes bistable due to the pulling potential $V(x,t)$. However, the resulting solutions for the work done on these two systems differs significantly.

Unimolecular system with bistable U(x). We will assume the bistable potential has two minima separated by a constant distance Δx, with the initial state at $x(0) = 0$. We will also take the simple assumption that a single transition from state 1 to state 2 occurs in the observation window. While under the influence of a pulling potential $V(x,t) = \frac{1}{2}k_{\text{cant}}(x - vt)^2$, the system will switch states from $x(t) = 0$ to $x(t) = \Delta x$ at an arbitrary time t_s (or switching force $F_s = k_{\text{cant}}vt_s$). Referring to Figure 3.11, the work then follows as

$$W = \int_0^{t_s} k_{\text{cant}}v^2t\,dt + \int_{t_s}^{t_f} k_{\text{cant}}v(vt - \Delta x)dt \tag{3.112}$$

$$= k_{\text{cant}}vt_s\Delta x - \frac{1}{2}k_{\text{cant}}\Delta x^2 + \frac{1}{2}k_{\text{cant}}(vt_f - \Delta x)^2 \tag{3.113}$$

$$= \underbrace{F_s\Delta x - \frac{1}{2}k_{\text{cant}}\Delta x^2}_{W_{\text{molecule}}} + \underbrace{\frac{1}{2}k_{\text{cant}}(F_f/k_{\text{cant}} - \Delta x)^2}_{W_{\text{spring}}(t_f)} \tag{3.114}$$

where the work is separated into energy used to switch the molecule between states and the energy finally stored in the spring $W_{\text{spring}}(t_f) = V(\Delta x, t_f)$ after stretching it up to the final time t_f, or equivalently, final force $F_f = k_{\text{cant}}vt_f$. Equation 3.114 states that, although the driven transition occurs at force F_s, the total work done on the system depends on how long you pull on it. This is true because the thermodynamic work accounts for the entire system, including the pulling device (Jarzynski, 2006, 2007).

Note, however, that we assume the pulling potential to move slowly enough through the solution to contribute a conservative force to the system. Therefore, $W_{\text{spring}}(t_f)$ is the reversible work done on the spring, independent of pulling speed. The total reversible work, or free energy, for the bistable system will then be a combination of the minimal work to switch states of the molecule ΔG_0 plus the reversible work to stretch the spring:

$$\Delta G = \Delta G_0 + W_{\text{spring}}(t_f) \tag{3.115}$$

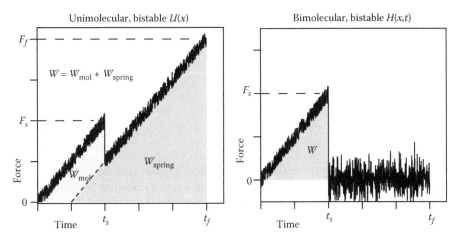

FIGURE 3.11 Idealized schematic of the work done on (a) a unimolecular system such as a protein or RNA molecule and (b) a bimolecular system such as a ligand–receptor pair. Symbols correspond to those found in the text. In (a), the molecular potential $U(x)$ is already bistable, which leads to a total work done on the combined molecule/probe system W that depends on the final pulling time t_f. The two contributions to the total work, the work done on the molecule W_{mol} and the work done on stretching the spring W_{spring} are separable, and illustrated by different shading. In (b), only the total Hamiltonian $H(x,t)$ of the combined intermolecular and probe potentials is bistable, which means the probing spring is an integral component of the two-state system. Once the bond ruptures, the particle resides around the minimum of the pulling spring for the remainder of the pulling process, and no further work is done on the spring. Thus, the work done on the system is the area under the force-distance trajectory, but it is not dependent on the final observation time t_f.

Thus, the dissipated energy W_d will be independent of the final work done on the spring:

$$W_d = W - \Delta G \tag{3.116}$$
$$= W_{\text{molecule}} + W_{\text{spring}}(t_f) - (\Delta G_0 + W_{\text{spring}}(t_f)) \tag{3.117}$$
$$= W_{\text{molecule}} - \Delta G_0 \tag{3.118}$$

which is typically the quantity of interest when probing molecular systems.

Bimolecular system with single metastable U(x). Here, we consider a bimolecular bond with a single minimum in $U(x)$ located at $x = 0$. The total Hamiltonian evolves into a bistable system when the minimum of the pulling potential $V(x,t) = \frac{1}{2}k_{\text{cant}}(x - vt)^2$ is pulled sufficiently far past the barrier to rupture. The minimum of the pulling potential defines the second state and is located at vt. Therefore, assuming negligible displacement of the bound state minimum, the system will switch from $x = 0$ to $x = vt$ upon unbinding. Again, assuming unbinding occurs at an arbitrary time t_s (or unbinding force $F_s = k_{cant}vt_s$). The thermodynamic work is

$$W = \int_0^{t_s} k_{cant}v^2 t\,dt + \int_{t_s}^{t_f} k_{cant}v(vt - vt)\,dt \qquad (3.119)$$

$$= \frac{1}{2}k_{cant}v^2 t_s^2 \qquad (3.120)$$

$$= \frac{F_s^2}{2k_{cant}} \qquad (3.121)$$

Unlike the unimolecular system above, the thermodynamic work done on a bimolecular system is independent of the final observation time t_f because no work is done on the molecule when it breaks away and resides around the minimum of the pulling potential (see Figure 3.11). While this is a sufficient approximation, it is not necessarily true in general, as hydrodynamic effects will contribute further dissipation as the probe is dragged through the fluid. Therefore, Equation 3.121 is valid at slow to moderate pulling speeds when viscous damping is negligible.

The dissipated energy in this case is given by

$$W_d = W - \Delta G \qquad (3.122)$$

$$= \frac{F_s^2}{2k_{cant}} - \Delta G^0 \qquad (3.123)$$

Therefore, the work done on bond rupture (dissociation of a bimolecular complex) can be derived from the integral over the force-extension trajectory or, in the case of a bond loaded by a spring of stiffness k_{cant}, the work is simply one-half the square of the rupture force over the spring constant.

3.7 SUMMARY

Single-molecule manipulation techniques are arguably the only method available to study the fundamental energy landscapes that govern inter- and intramolecular bonding and mechanics. Initial developments provided access to a previously inaccessible parameter of a bond—the transition state. Early work has also been concerned primarily with kinetics, under the assumption that driven molecular systems are too far from equilibrium to permit exploration of equilibrium properties. But recent developments in nonequilibrium thermodynamics have provided amazing tools, such as fluctuation theorems, which link nonequilibrium measurements to their equilibrium roots. In addition, recent findings have also shown that bond rupture experiments can be performed very close to equilibrium conditions, allowing estimates of bond-free energies directly. Therefore, force spectroscopy should no longer be thought of as a purely irreversible, kinetic regime technique. The dynamics can be reversible under appropriate conditions and in principle, the full breadth of information ranging from equilibrium free energies to activation barriers to kinetics should be accessible. Some exciting future developments of the technique will most likely be found in relating bond rupture of single bonds to their bulk, ensemble equilibrium counterparts. In addition, the theory and experimental use of stiff probes, as opposed to the commonly used soft probes, may provide deeper insights into interaction potentials through continuously probing the energy landscape without discontinuous instabilities.

REFERENCES

Bell, G.I., 1978. Models for the specific adhesion of cells to cells, *Science*, 200, 618–627.

Crooks, G., 1999. Entropy production fluctuation theorem and the nonequilibrium work relation for free energy differences, *Phys. Rev. E*, 60 (3), 2721–2726.

Dudko, O.K., Filippov, A.E., Klafter, J., and Urbakh, M., 2003. Beyond the conventional description of dynamic force spectroscopy of adhesion bonds, *Proc. Natl. Acad. Sci. USA*, 100 (20), 11378–11381.

Dudko, O.K., Hummer, G., and Szabo, A., 2006. Intrinsic rates and activation free energies from single-molecule pulling experiments, *Phys. Rev. Lett.*, 96 (10), 108101-1-108101-4.

Einstein, A., 1905. Über die von der molekularkinetischen theorie der wärme geforderte bewegung von in ruhenden flüssigkeiten suspendierten teilchen, *Ann. Phys.*, 322, 549–560.

Evans, E., 2001. Probing the relation between force — Lifetime — and chemistry in single molecular bonds, *Annu. Rev. Biophys. Biomolec. Struct.*, 30, 105–128.

Evans, E. and Ritchie, K., 1997. Dynamic strength of molecular adhesion bonds, *Biophys. J.*, 72 (4), 1541–1555.

Evans, E. and Williams, P.M., 2002. *Physics of Bio-Molecules and Cells*, vol. 75, Springer and EDP Sciences, Heidelberg.

Flory, P.J., 1969. *Statistical Mechanics of Chain Molecules*, New York, Interscience Publishers, reprinted by Hanser publishers, 1989.

Friddle, R.W., Noy, A., and De Yoreo, J.J., 2011. Rethinking the dynamics of bond-breaking, *submitted*.

Friddle, R.W., Podsiadlo, P., Artyukhin, A.B., and Noy, A., 2008. Near-equilibrium chemical force microscopy, *J. Phys. Chem. C*, 112 (13), 4986–4990.

Friedsam, C., Wehle, A.K., Kuhner, F., and Gaub, H.E., 2003. Dynamic single-molecule force spectroscopy: bond rupture analysis with variable spacer length, *J. Phy. Condensed Matt*, 15 (18), S1709–S1723.

Garg, A., 1995. Escape-field distribution for escape from a metastable potential well subject to a steadily increasing bias field, *Phys. Rev. B*, 51 (21), 15592–15595.

Jarzynski, C., 1997. Nonequilibrium equality for free energy differences, *Phys. Rev. Lett.*, 78 (14), 2690–2693.

Jarzynski, C., 2006. Work fluctuation theorems and single-molecule biophyics, *Prog. Theor. Phys. (suppl.)*, 165 (1), 1–17.

Jarzynski, C., 2007. Comparison of far-from-equilibrium work relations, *C.R. Phys.*, 8, 495–506.

Kramers, H.A., 1940. Brownian motion in a field of force and diffusion model of chemical reactions, *Physica*, 7, 284–304.

Kratky, O. and Porod, G., 1949. Röntgenuntersuchung gelöster fadenmoleküle, *Rec. Trav. Chim. Pays-Bas.*, 68, 1106–1123.

Oesterhelt, F., Rief, M., and Gaub, H.E., 1999. Single molecule force spectroscopy by AFM indicates helical structure of poly(ethylene-glycol) in water, *New J. Phy.*, 1, 6.1–6.11.

Rief, M., Fernandez, J., and Gaub, H., 1998. Elastically coupled two-level systems as a model for biopolymer extensibility, *Phys. Rev. Lett.*, 81 (21), 4764–4767.

Rief, M., Gautel, M., Oesterhelt, F., Fernandez, J., and Gaub, H., 1997. Reversible unfolding of individual titin immunoglobulin domains by AFM, *Science*, 276 (5315), 1109–1112.

Shapiro, B. and Qian, H., 1998. Hysteresis in force probe measurements: A dynamical systems perspective, *J. Theor. Biol.*, 194 (4), 551–559.

Shapiro, B.E. and Qian, H., 1997. A quantitative analysis of single protein-ligand complex separation with the atomic force microscope, *Biophys. Chem.*, 67, 211–219.

Sulchek, T., Friddle, R., Langry, K., Lau, E., Albrecht, H., Ratto, T., DeNardo, S., Colvin, M., and Noy, A., 2005. Dynamic force spectroscopy of parallel individual Mucin1-antibody bonds, *Proc. Nat. Acad. Sci. USA*, 102 (46), 16638–16643.

Sulchek, T., Friddle, R., and Noy, A., 2006. Strength of multiple parallel biological bonds, *Biophys. J.*, 90 (12), 4686–4691.

Talkner, P., 2003. Statistics of entrance times, *Physica A*, 325 (1–2), 124–135.

Tobolsky, A. and Eyring, H., 1943. Mechanical properties of polymeric materials, *J. Chem. Phys.*, 11 (3), 125–134.

Walton, E.B., Lee, S., and Van Vliet, K.J., 2008. Extending Bell's model: how force transducer stiffness alters measured unbinding forces and kinetics of molecular complexes, *Biophys. J.*, 94 (7), 2621–2630.

Williams, P., 2003. Analytical descriptions of dynamic force spectroscopy: behaviour of multiple connections, *Anal. Chim. Acta*, 479 (1), 107–115.

Zhurkov, S.N., 1965. Kinetic concept of the strength of solids, *Int. J. Fract. Mech.*, 1, 311.

4 Immobilization and Interaction Strategies in DFS of Biomolecular Partners

Boris B. Akhremitchev

CONTENTS

4.1   INTRODUCTION

In dynamic force spectroscopy (DFS), mechanical forces are applied to intermolecular bond understudy to characterize strength of biomolecular recognition and extract kinetic parameters of bond dissociation. To perform such measurements, interacting partners should be attached to the opposing surfaces of atomic force spectroscopy (AFM) probe and the substrate. Thus, successful immobilization is vital for obtaining

133

meaningful results in DFS experiments. Different attachment strategies can be used as described in the recently published review articles (Hinterdorfer & Dufrene 2006; Bizzarri & Cannistraro, 2010). Two basic requirements for such attachments are that (1) the mechanical strength of attachment significantly exceeds the strength of interaction understudy and that (2) various effects arising from attaching biomolecules do not prohibit the molecular interpretation of the measured biorecognition events. While the first requirement is rather obvious, the second requirement demands careful consideration of various effects in attaching biomolecules. Ideally, measured interactions between surfaces that are equipped with recognition partners correspond to breaking of a single recognition bond under study. Moreover, for accurate interpretation of experimental results, it is important that conditions of measurements match assumptions of theoretical models that are used for data reduction. Number of effects that make typical DFS experiments less than ideal are related to attachment of molecules. Effects detrimental to accuracy of DFS experiments include (1) effects of surfaces on affinity of recognition between biomolecules, (2) effects of spurious (nonspecific) interactions, (3) effects of multiple recognition events that might occur during the measurements, (4) effects of polymeric linkers (which are often used to attach biomolecules to the surfaces) on extracted parameters of interaction. The latter two effects arise from effects of attachment density and geometry and from non-Hookean spring elasticity of polymeric tethers (the non-Hookean spring aspect is outside the scope of this chapter). Some of these effects might occur simultaneously, and therefore fulfilling the second requirement stated above is not a trivial task. Therefore, advantages and disadvantages of different interaction strategies are considered below.

The chapter is organized in the following way: First, immobilization strategies are described and requirements for functional attachment of biomolecules are indicated. Then, we describe physical and chemical methods of attaching biomolecules, indicating corresponding strengths and weaknesses of different approaches. In the following section, problem of separating the single rupture events from the multiple rupture events is described. The chapter concludes with a brief summary.

4.2 IMMOBILIZATION STRATEGIES

Immobilization of biological molecules at interfaces is hardly a new task (Hermanson, 1996; Wong et al. 2009). However, because of the single-molecule nature of interactions studied by DFS and because of the mechanical nature of the measurements, DFS experiments require immobilization approaches that differ in several aspects from other experimental techniques like chromatography, surface plasmon resonance, and capillary electrophoresis. In DFS, interaction between recognition partners is measured by analysis of detected rupture events. Registered rupture events might have spurious nature. For example, false ruptures might be detected if molecules are not attached strongly enough to surfaces and rupture occurs at the attachment point or when nonspecific bonds are being ruptured. In addition, if ruptures of multiple bonds occur simultaneously, then measured distribution of rupture forces might be affected. These effects should be avoided by selecting a proper immobilization approach.

Immobilization approaches used in other techniques are often valued by the surface density of attached molecules (Mallik et al. 2004). In DFS experiments, too high surface density is detrimental because it hampers analysis of interactions from a single pair of recognition partners. Another significant difference is that DFS uses a nanoscale probe that mechanically contacts the substrate. Often, area of the tip-sample contact significantly exceeds dimensions of molecules. Consequently, spurious interactions might occur if two surfaces adhere to each other. Figure 4.1 illustrates these effects.

To prevent these and other detrimental effects, polymeric linkers are often used to tether molecules to surfaces. One advantage in using relatively long linkers is to remove contribution of nonspecific interactions between the probe and the surface in analysis of recognition events by using tip-sample separation at rupture as a discriminating parameter (see Figure 4.1c). Such use of the linkers has been proposed in an early force spectroscopy work by Hinterdorfer et al. (Hinterdorfer et al. 1996) and this approach became a rather common method to separate rupture forces of different nature (e.g., specific and nonspecific interactions) (Willemsen et al. 1998; Hinterdorfer et al. 2002; Kuhner et al. 2004; Ratto et al. 2004; Bonanni et al. 2005; Kienberger et al. 2005; Ray and Akhremitchev, 2005; Averett et al. 2008; Guo et al. 2008; Guo et al. 2010b; Jiang et al. 2010; Mayyas et al. 2010; Nguyen et al. 2011). However, nonspecific interaction of tethered molecule or the linker with the substrate still might contribute to the measured rupture forces (see Figure 4.1c). When using polymeric linkers, it is important that polymers are water soluble, do not alter recognition of interacting molecules, and do not participate in immobilization reactions (except in a desired way). The most widely used polymeric linkers are poly(ethylene glycol) molecules (PEG). Surfaces coated with PEG resist protein adsorption (Ma et al. 2006). However, PEG itself adsorbs onto hydrophobic and hydrophilic surfaces

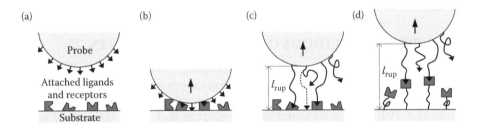

FIGURE 4.1   Schemes of attachment of recognition partners to the atomic force spectroscopy probe and substrate. Panels (a) and (b) illustrate attaching molecules without tethers. Panel (b) illustrates that in such attachment both nonspecific interactions and multiple interactions are expected. In addition, interactions might be significantly hindered by surfaces. Panel (c) illustrates attaching one of the molecules by polymeric tether. This reduces probability of nonspecific interactions and alleviates the steric hindrance. Panel (d) illustrates the double-tether approach. Here, nonspecific interactions are eliminated from analysis by considering ruptures occurring at the sum of the tether lengths; the steric hindrance is also greatly reduced. However, both the single-tether and the double-tether approaches do not prevent multiple bond ruptures from occurring.

in aqueous media (Pagac et al. 1997; Shar et al. 1999; Kim et al. 2003). Moreover, mechanical compression forces might cause strong adsorption of PEG onto silicon nitride probe; the resulting physical attachment is capable of withstanding pulling forces upto hundreds of piconewton (Oesterhelt et al. 1999). Therefore, additional test experiments are required to determine the nature of measured rupture events (Averett et al. 2008; Guo et al. 2010b).

Problem of nonspecific interaction of linkers or tethered molecules with the substrate can be solved by using a double-tether approach (illustrated in Figure 4.1d) where both interacting partners are tethered to surfaces by relatively long linkers (Kuhner et al. 2004; Ratto et al. 2004; Ray & Akhremitchev, 2005). This double-tether approach also helps to remove steric hindrance by surfaces and allows to discriminate (although not perfectly) single-molecule interactions from multiple interactions (Guo et al. 2008; Guo et al. 2010a; Mayyas et al. 2010). Multiple bond effects will be considered in Section 4.4 in more detail. Another method to alleviate nonspecific adsorption problem is to use experimental methodology where the probe does not apply compressive forces onto the sample surface (Ludwig et al. 1999; Sekiguchi et al. 2003; Guo et al. 2010b). However, this methodology is not widely used because it remains challenging to repeatedly approach the substrate without applying compressive forces to a distance that can be bridged by the linkers.

From a point of view of the weakest bond holding the immobilized biomolecules on surfaces, attachment methods can be divided in two groups: methods that use physical adsorption or molecular recognition forces and methods that use covalent attachment. Although physical bonds are generally weaker than covalent bonds, using such attachment schemes is adequate when physisorption is strong enough for force spectroscopy methods that utilize low pulling forces (e.g., optical tweezers and magnetic tweezers techniques) (Averett et al. 2008; Neuman & Nagy, 2008). Below we describe various immobilization strategies in more detail, indicating the strengths and weaknesses of different approaches.

4.3 PHYSICAL METHODS OF ATTACHING MOLECULES

Physical methods of attaching biological molecules in AFM experiments were introduced at the early stages of development of force spectroscopy technique (Lee et al. 1994; Florin et al. 1994; Moy et al. 1994; Radmacher et al. 1994; Chilkoti et al. 1995; Fritz et al. 1998; Willemsen et al. 1998; Baumgartner et al. 2000). In the past decade, this technique still has been used in AFM-based experiments (Sekiguchi et al. 2003; Zhang & Moy, 2003; Averett et al. 2008). This approach is often used to attach cells to the AFM probes (Zhang & Moy, 2003; Zhang et al. 2004; Wojcikiewicz et al. 2006) and in other force spectroscopy methods like biomembrane force probe, optical and magnetic tweezers (Evans et al. 2004; Pincet & Husson, 2005; Ferrer et al. 2008; Todd et al. 2008). In DFS experiments, this approach has been mostly replaced by chemical methods of attaching molecules that are described in the following subsection. Therefore, here we only briefly describe immobilization strategy using physical forces and point out advantages and disadvantages of this approach.

The most common physical method of attaching biomolecules utilizes strong specific bonds formed between biotin and avidin (other proteins that can be used in this approach are streptavidin and neutravidin) (Chilkoti et al. 1995). Because avidin is tetrameric, upto four biotin molecules can be bound to one protein molecule simultaneously. Therefore, avidin strongly binds to surfaces that display biotin groups and is yet capable of binding to additional biotin molecules. Consequently, this system can be used as a universal platform to attach molecules to surfaces because biotin that attaches to the surface-immobilized avidin can be chemically conjugated to other molecules of interest. Typical sample preparation consists of three steps: (1) preparing surface with attached biotin molecules; (2) coating surface with avidin; and (3) coupling biotin-conjugated molecule of interest to the avidinated substrate. The initial biotin-coated surface can be prepared by adsorption of biotinylated bovine serum albumin (BSA) (Lee et al. 1994; Moy et al. 1994; Florin et al. 1994; Chilkoti et al. 1995) or by chemical methods (Evans et al. 2001; Evans et al. 2004; Pincet & Husson, 2005).

Universality and relative simplicity of sample preparation using biotin–avidin platform are the main merits of this approach. However, there are significant drawbacks that preclude the widespread use of this approach. The most significant drawbacks are (1) the relative weakness of the biotin–avidin bond in comparison to covalent bonds and (2) difficulty in preparing low-attachment density of molecules on surfaces. As a result of the first aspect, the rate of dissociation under applied force is the sum of dissociation rates of molecules under study and the dissociation rate of biotin–avidin bond (Patel et al. 2004). Therefore, if the force-dependent dissociation rate of molecular bond under study is comparable to the dissociation rate of biotin–avidin bond, at some fraction of experiments the latter might be broken, thus distorting the distribution of rupture forces. Such distortion might result in significant errors in extracted kinetic parameters of bond dissociation. However, this aspect might be of minor importance when studying bonds that are considerably weaker than biotin–avidin bonds. The second aspect mentioned above stems from the tetrameric nature of avidin. Consequently, it is likely that more than one biotin molecule will attach to one avidin molecule and thus probability to form two molecular bonds instead of one becomes significant (Guo et al. 2010a). Moreover, if the surface density of avidin molecules is not low, even larger number of molecular bonds might be formed. This effects result in large number of rupture transitions detected in a single force curve and in wide distribution of rupture forces (Lee et al. 1994; Moy et al. 1994; Zhang & Moy, 2003). Consequently, distribution of rupture forces coming from a single molecular bond cannot be extracted reliably from the data.

Physisorption of molecules directly to the substrate is an alternative to using biotin–avidin linking described above. Typically, molecules are adsorbed on flat mica or other substrates without making a covalent bond (Radmacher et al. 1994; Fritz et al. 1998; Willemsen et al. 1998; Baumgartner et al. 2000; Averett et al. 2008). If physisorbed molecules strongly adhere to the substrate and are significantly diluted on the surface, two drawbacks that are mentioned above can be avoided. For example, it has been shown that specific knob-hole interaction between different parts of fibrin molecule can be observed for both, physisorbed and chemically immobilized fibrinogen (Averett et al. 2008). However, to make this observation, chemical immobilization is still necessary to perform comparison. Therefore, the main advantage of using physical immobilization methods—it's simplicity—is lost.

4.4 CHEMICAL METHODS OF ATTACHING BIOMOLECULES

Chemical methods of attachment of recognition partners have been employed from the early stages of development of DFS methodology (Dammer et al. 1995; Dammer et al. 1996; Gad et al. 1997; Willemsen et al. 1998; Fritz et al. 1998; Yip et al. 1998; Ros et al. 1998; Dettmann et al. 2000; Schwesinger et al. 2000). Attachment of recognition partners to the substrates occurs via covalent bonds that are considerably stronger than forces between recognition partners (Grandbois et al. 1999). Biomolecules of interest are attached to the substrates via functional groups; typically, these are primary amine and thiol groups that either occur naturally in biomolecules or can be added to facilitate the immobilization. These molecules are covalently attached to the solid surfaces (substrate and the probe), and thus these surfaces also should be chemically reactive. Covalent attachment can be made via linkers of different lengths. Below, we first describe chemical attachment of recognition partners via functional groups, then we describe chemistry at the solid substrates and discuss using polymeric linkers between molecules understudy and the substrates. Finally, we provide two examples of the sample preparation protocols.

4.4.1 CHEMICAL REACTIONS TO IMMOBILIZE BIOMOLECULES

Biological molecules are usually immobilized using amine, carboxyl, or thiol groups (Hermanson, 1996). Often there are many amine and carboxyl groups available at the surface of protein molecules and therefore immobilization is not selective. Selective immobilization can be achieved by using natural surface thiols of proteins (which are rare), introducing additional thiol groups at specific sites or by newer methods described in a recent review by Wang et al. (Wong et al. 2009). The most commonly used chemistries (Hermanson, 1996; Wong et al. 2009) are shown in the Figure 4.2.

Amine coupling shown in Figure 4.2a uses ester of N-hydroxysuccinimide (NHS) (Dammer et al. 1995; Dammer et al. 1996; Friedsam et al. 2004; Neuert et al. 2006; Odorico et al. 2007; Gilbert et al. 2007; Averett et al. 2008; Averett et al. 2009; Mayyas et al. 2010). There are many commercially available cross-linkers or polymeric tethers with NHS functional group. Also, NHS esters are prepared when carboxyl groups are activated as described below. Compounds containing NHS group usually have low solubility in water, therefore, for coupling in aqueous solutions water-soluble esters of N-hydroxysulfosuccinimide (sulfo-NHS) can be used. Coupling is efficient at slightly basic conditions at pH 7–8. However, basic conditions speed up the hydrolysis of NHS esters (at pH 8 half-life of NHS esters is approximately 1 h) (Lomant & Fairbanks, 1976; Staros et al. 1986; Hermanson, 1996). Therefore, coupling is typically performed in phosphate buffers with

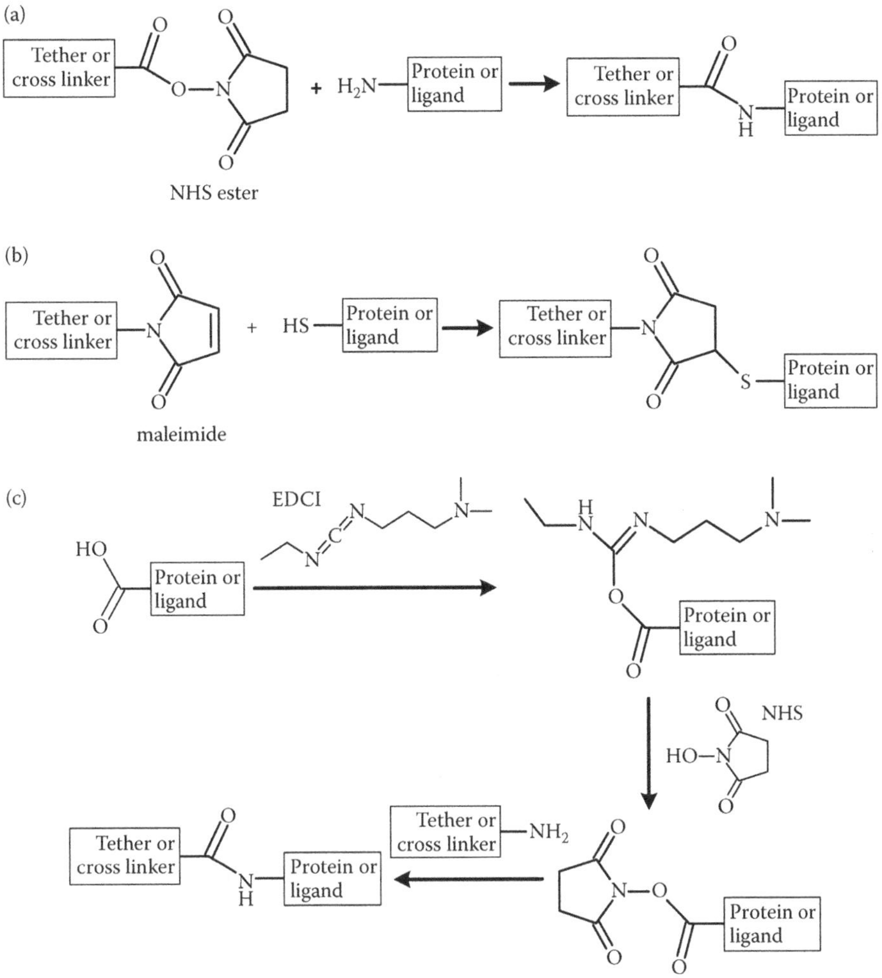

FIGURE 4.2 Typical coupling schemes to (a) amine, (b) thiol, and (c) carboxyl functional groups.

pH 7.2–7.5 (amine containing buffers like Tris should be avoided because of reactivity with NHS group). In organic solvents, few volume percent of organic base pyridine can be added to achieve efficient coupling. If hydrolysis of NHS esters presents a problem, then amines can be coupled to aldehydes by forming and subsequently reducing the Schiff base as shown in Figure 4.5. However, coupling of amines to aldehydes is slower than coupling to NHS esters (Bonanni et al. 2005).

Coupling to thiols is often achieved using reagents containing maleimide groups as shown in Figure 4.2b, resulting in stable thioether bond (Bartels et al. 2003; Hukkanen et al. 2005; Ray & Akhremitchev, 2005; Ray et al. 2006; Kim et al. 2009).

Maleimide compounds might couple to amines, but coupling to thiols is strongly favored at pH below 7.5 (Hermanson, 1996; Wong et al. 2009). Also, maleimide compounds hydrolyze in water, though hydrolysis of maleimides is slower than for NHS esters (Hermanson, 1996). Another method to immobilize molecules by their thiol group is coupling using iodoacetyl reactive groups. This coupling method is less efficient than maleimide coupling (Mallik et al. 2007).

Molecules naturally displaying thiols at their surface or containing thiols that are purposely incorporated into their structure can be coupled directly to gold substrates or probes using high affinity of gold toward organic sulfur (Ulman, 1991; Ros et al. 1998; Sekiguchi et al. 2003; Bonanni et al. 2005; Love et al. 2005; Gilbert et al. 2007; Jiang et al. 2010; Nguyen et al. 2011). This method of immobilization involves adsorption of molecules from solution onto gold surface; this deposition is straightforward to perform as it requires no prior modification of gold surface except for initial cleaning. Possible drawbacks of this approach are that interaction with gold might affect affinity of recognition and proximity of the surface might restrict motion of molecules (Bizzarri & Cannistraro, 2009). Also, high density of attachment at surfaces might promote "contamination" of single-molecule recognition events with multiple interactions.

Carboxylic groups are often coupled by using carbodiimide coupling agent (Figure 4.2c shows coupling agent 1-ethyl-3-(3-dimethylaminopropyl)carbodiimide also known as EDCI, EDC, or EDAC) and NHS to form an amino-reactive NHS ester (Hermanson, 1996). This ester then can be coupled to amine groups of other molecules (Ratto et al. 2004; Friedsam et al. 2004; Neuert et al. 2006; Odorico et al. 2007; Averett et al. 2008; Mayyas et al. 2010; Jiang et al. 2010). Hydrolysis of NHS can be avoided by performing the first two steps at pH 5–6 and then rising pH to 7.2–7.5 to perform NHS ester coupling to amines. Another popular carbodiimide that activates carboxyl groups for the formation of amide bonds with primary amines and can be used for esterification of alcohols is dicyclohexylcarbodiimide (DCC) (Hermanson, 1996; Jung et al. 2007). DCC is a known allergen and a sensitizer that causes skin rashes; therefore, precautions should be taken when working with this chemical to avoid skin contacts.

4.4.2 Chemical Modification of Substrates

In DFS experiments, the recognition partners are attached to surfaces directly or via polymeric linkers. Gold surfaces and surfaces that exhibit silanol groups are most widely used in DFS. Gold strongly binds to organic sulfur, thus eliminating the step of activating silanols. On the other hand, gold coating on the AFM probes makes probes duller, thus increasing probability of multiple interactions. Silanol chemistry is attractive because silanol groups are present on surfaces of silicon, silicon nitride, and glass after cleaning in acids (Hinterdorfer et al. 2002). Below, we first describe functionalization of gold surfaces and then functionalization of surfaces using silanol groups.

4.4.2.1   Functionalization of Gold Surfaces

Gold is a convenient substrate material because it does not oxidize in conditions of DFS experiments, gold-coated substrates, or probes are commercially available or can be prepared using standard deposition techniques. In addition, chemical immobilization of molecules on gold is relatively well researched in comparison to many other substrates (Ulman, 1991; Love et al. 2005). As an alternative, galvanic displacement can be used to deposit copper on AFM probes (Fritz, Carraro, & Maboudian, 2001). However, it remains unclear whether this technique offers significant advantages for DFS experiments.

As noted above, molecules containing thiols can be immobilized on gold surfaces directly. Molecules that do not contain thiol groups can be immobilized using cross linkers equipped with thiols and another functional group that can be used in the subsequent chemical reactions. Table 4.1 shows several chemicals that have been used in DFS experiments to functionalize gold surfaces.

Mercaptohexadecanoic acid and mercaptoundecanoic acid are the most common compound for functionalization of gold surfaces in DFS (Schmitt et al. 2000; Ratto et al. 2004; Friedsam et al. 2004; Odorico et al. 2007; Averett et al. 2008; Friddle et al. 2008; Averett et al. 2009). These molecules form stable self-assembled monolayers on gold (Ulman, 1991; Friedsam et al. 2004) that can be made amino-reactive using coupling reactions shown in Figure 4.2c. Other commercially available thiols that can be used for functionalization of gold surfaces include 1,8-octanedithiol and cysteamine (Hukkanen et al. 2005; Sulchek et al. 2005; Odorico et al. 2007).

TABLE 4.1

Compounds for Functionalization of Gold Surfaces

Name of Compound	Chemical Formula	Functional Group for Further Modification
Mercaptoundecanoic acid and mercaptohexadecanoic acid	$HS-(CH_2)_{10}-COOH$ $HS-(CH_2)_{15}-COOH$	Carboxyl
1,8-Octanedithiol	$HS-(CH_2)_8-SH$	Thiol
Cysteamine	$HS\diagup\diagdown NH_2$	Amine
11,11'-Dithio-bis-(undecanoic acid N-hydroxysuccinimide ester)		Amino-reactive NHS ester

---

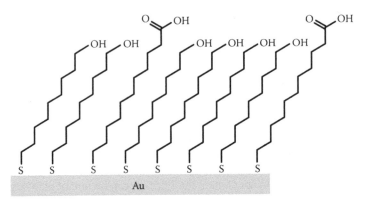

FIGURE 4.3 Formation of carboxy-functionalized self-assembled monolayer on gold. Adding shorter hydroxyl-terminated chains decreases surface density of immobilized molecules and decreases surface adhesion to the atomic force spectroscopy probe.

Following such modification, thiol and amine coupling chemistries shown in Figures 4.2a and 4.2b can be used in further coupling reactions. To decrease density of reactive groups, functional groups can be diluted with alcohols that have a slightly shorter chain as illustrated in Figure 4.3 (Friedsam et al. 2004; Hukkanen et al. 2005).

Custom chemical synthesis can employed to prepare thiol-containing molecules that facilitate further functionalization (Dammer et al. 1995; Dammer et al. 1996; Gilbert et al. 2007). Amino-reactive molecule that binds to gold is shown in Table 4.1. It is not clear whether preparation of such complex molecules has advantage over more common approaches. In addition, water-soluble heterobifunctional polymeric tethers with thiol at one terminus and either carboxyl or amine at the other terminus are commercially available. It might be expected that such tethers will find a wide spread use in DFS applications.

4.4.2.2 Functionalization of Surfaces Displaying Silanol Groups

Silicon is often used as a substrate to covalently immobilize molecules (Kaas & Kardos, 1971; Vandenberg et al. 1991; Aswal et al. 2006). Immobilization of molecules typically involves reactions with silanol surface groups ($\equiv Si - OH$). Silanols are usually present on oxidized surface of silicon and on surface of silicon nitride probes. Their concentration can be increased by cleaning in acidic or basic solutions. For example, glass can be treated for several hours in 5% HCl solution or for 30 min in hot piranha solution (25% sulfuric acid and 15% hydrogen peroxide) or for 1 h in 10% sodium hydroxide solution (Hermanson, 1996). Several compounds that can be used to chemically functionalize silanols are listed in Table 4.2.

The most common technique to functionalize silanols is silanization with organic silanes that carry amine functional group. Surfaces functionalized with amines can be used in further immobilization steps. Chemical compound APTES shown in

TABLE 4.2
Examples of Compounds for Functionalization of Silanols

Name of Compound	Chemical Formula	Generated Functional Group
3-Aminopropyl triethoxysilane (APTES, APTS, APS)		Amine
3-Aminopropyl silatrane		Amine
Ethanolamine hydrochloride		Amine
3-Mercaptomethyl diethylethoxysilane		Thiol
3-Glycidoxypropyl methyldiethoxysilane		Epoxy

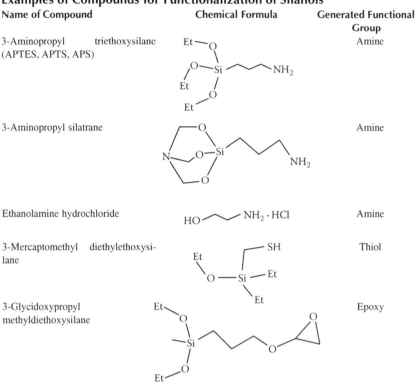

Table 4.2 is often used to aminate surfaces in DFS experiments (Ros et al. 1998; Yip et al. 1998; Bartels et al. 2003; Ratto et al. 2004; Chtcheglova et al. 2008; Ahmad et al. 2011). Silanes that have other alkoxy- or chloro-groups also can be used in silanization, the latter compounds are considerably more moisture sensitive. Other inorganic surfaces that exhibit hydroxyls can be used as substrates for silanization reaction, for example, aluminum oxide (Vandenberg et al. 1991; Hermanson, 1996).

The main disadvantage of functionalization with APTES or other di- or trivalent silanes is that formation of monolayer is competing with hydrolysis and polymerization of APTES molecules and subsequent formation of covalent APTES aggregates on surfaces (Vandenberg et al. 1991; Ulman, 1991; Fadeev & McCarthy, 2000; Howarter & Youngblood, 2006). Figure 4.4a shows both the "ideal" surface functionalization with APTES as well as foreseeable deposition of APTES in the polymeric state. Such aggregation presents a problem because chains of polymerized silane can

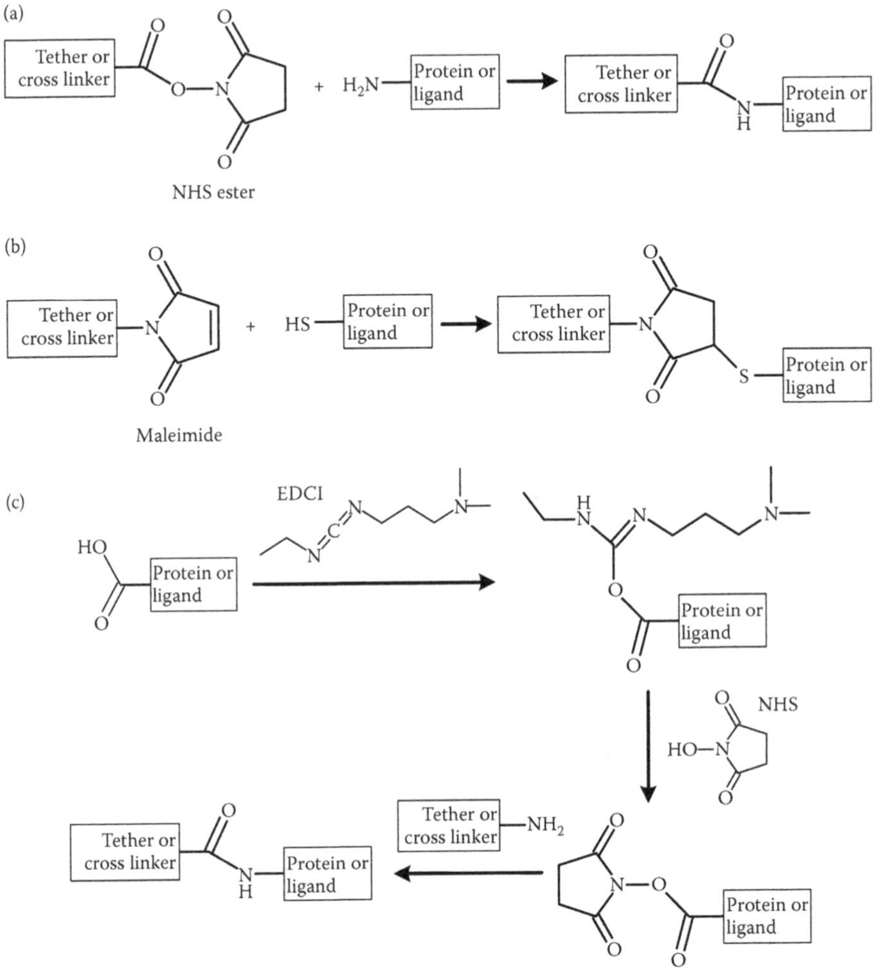

FIGURE 4.4 Functionalization of surface silanols with (a) 3-aminopropyltriethoxysilane and (b) with ethanolamine. In part (a) possible reaction outcomes are shown: (i) the desired outcome; (ii) possible side reactions.

interact with the AFM probe and produce spurious unbinding events that might be mistaken with actual unbinding events.

Polymerization problem can be alleviated by depositing silanes in a highly controlled three-step process (Krasnoslobodtsev & Smirnov, 2002) or by functionalization of surfaces with compounds that do not polymerize in solution. Possible alternative to APTES include silanes with only one group that reacts with silanols (e.g., 3-aminopropyl dimethylethoxysilane is commercially available) (Kuhner et al. 2004; Jiang et al. 2010). In addition, silatrane compounds (Table 4.2) do not polymerize in solution and has been used to functionalize surfaces (Shlyakhtenko et al. 2003). However, amino-functional silatranes are not commercially available and this limits their application in DFS experiments.

Amination of surfaces with ethanolamine shown in Figures 4.4b is an alternative to silanization (Willemsen et al. 1998; Baumgartner et al. 2000; Hinterdorfer et al. 2002). This reaction is based on esterification of silanols with hydroxyl group of organic compounds according to mechanism $\equiv Si-OH + R-OH \leftrightarrow \equiv Si-OR + H_2O$ (Ossenkamp et al. 2002). Advantages of this approach are that ethanolamine does not polymerize, it is commercially available and produces surfaces with relatively low coverage of attached molecules (Hinterdorfer et al. 2002). This is an important advantage because it helps to decrease probability of multiple interactions during the probe-surface contact. Moreover, surfaces prepared by silanization often show large adhesion forces with the AFM probe. Consequently, the off-contact jump of the AFM probe might mask recognition events. Force plots collected using bare surfaces aminated using ethanolamine method show no adhesion. Therefore, this amination approach is very attractive in DFS experiments (Baumgartner et al. 2000; Hinterdorfer et al. 2002; Bonanni et al. 2005; Kienberger et al. 2005; Ray & Akhremitchev, 2005; Ray et al. 2006; Guo et al. 2008; Bizzarri & Cannistraro, 2009). It should be noted that this approach requires anhydrous conditions because water is a product of reaction; moreover, formed covalent bonds can hydrolyze in less than an hour in boiling water (Ossenkamp et al. 2002; Ballard et al. 1961). Consequently, reaction chamber with moisture control (dry box) is the most suitable environment for this reaction (Ray & Akhremitchev, 2005).

Converting silanols to chemical groups other than amine can be applied in DFS experiments. Surfaces exhibiting thiols or epoxy functional groups can be prepared using 3-mercaptomethyl diethylethoxysilane or 3-glycidoxypropyl methyldiethoxysilane (Fritz et al. 1998; Neuert et al. 2006; Kim et al. 2009). However, these surface modification reactions are used relatively rarely in DFS experiments, probably because of chemical convenience of the amino group. More elaborate immobilization approach involves functionalization of surfaces with custom synthesized dendron molecules (Jung et al. 2007; Kim et al. 2009). This procedure allows for better control of spacing between immobilized molecules, thus, providing method for avoiding multiple bond ruptures.

4.4.2.3 Immobilization of Molecules to Functionalized Surfaces

Functionalization of AFM probe and substrate surfaces is followed by immobilization reactions described in Section 4.2.1. Surfaces that display amine or thiol functional groups are often functionalized using bifunctional cross linkers using chemical reactions shown in Figure 4.2. For example, aminated surface can be made reactive with thiols by activation surface with cross linker that has NHS ester and maleimide groups. Additional reactions that can be used to connect epoxy and aldehyde functional group to amines are shown in Figure 4.5. Here epoxy group is first converted to diol group with 0.5 M sulfuric acid, the diol is subsequently oxidized to form aldehyde using periodic acid. Aldehydes react with amines and form Schiff base.

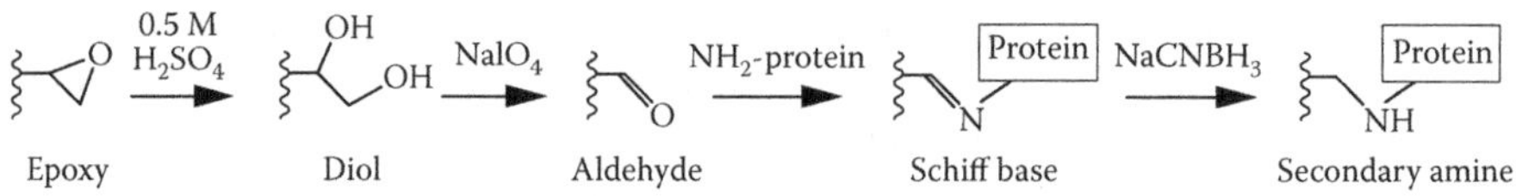

FIGURE 4.5 Immobilization of molecules with amine functional group onto surface with epoxy or aldehyde functional group. Reaction proceeds through formation of Schiff base that is subsequently reduced into a stable secondary amine bond.

Schiff base is susceptible to hydrolysis and thus it is reduced into a stable bond using sodium cyanoborohydride (Larsson, 1984; Hermanson, 1996; Gong et al. 2000).

Many different cross linkers with different functional groups and with different spacer lengths (including polymeric cross linkers) are commercially available. Table 4.3 lists several typical reactive cross-linkers and reactive polymeric linkers that can be used to couple to amines and thiols. Coupling reactions are shown in Figures 4.2 and 4.5. In addition to bifunctional tethers listed in Table 4.3, monofunctional tethers can be mixed with bifunctional tethers to block reactive groups on the surface. Therefore, surface density of attached molecules decreases facilitating single-molecule detection. Some molecules listed in Table 4.3 are homo-bifunctional. Therefore, bridging between functional groups on surfaces is possible. However, this is not detrimental to use of these tethers because this effect helps decreasing nonspecific adhesion to surfaces (Gong et al. 2000).

Functionalized polymeric tethers that can be used with activated substrates and biomolecules are listed in Table 4.4. These tethers are also commercially available in a wide range of molecular weights. In this table, possible uses of such tethers are listed instead of providing chemical formula that is evident from the name. Aforementioned comments about mono-functional tethers and about bridging of homo-bi-functional tethers between surface groups also apply here.

Although there are many more commercially available chemical coupling reagents and cross-linking tethers than are mentioned in this section, those indicated here are likely to be sufficient to immobilize biorecognition partners in DFS experiments. In the following sections, two sample preparations for DFS measurements are described, one method uses gold-thiol chemistry and another method uses silanol chemistry.

4.4.2.4   Example of Chemical Immobilization on Gold Surfaces

Here we describe sample preparation for investigation of "A-a" knob-hole interaction of fibrinogen. This interaction is involved in the early stages of blood clotting and provides mechanical stability of fibrinogen network before chemical cross linking of fibrin occurs (Averett et al. 2008; Averett et al. 2009). Fibrinogen consists of linear array of three globular-like units held together by thin connecting threads (Hall & Slayter, 1959). The "a" holes are located at both terminal units that are spaced ~ 50 nm apart. The "A" knobs are located in the central unit of fibrinogen

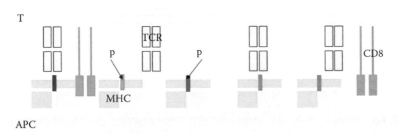

FIGURE 1.3 T cell activation. A CD4 or CD8 (shown here) T lymphocyte (T) interacting with an antigen-presenting cell (APC) is endowed with several tens of thousands of *identical* T cell receptors (TCR) specific for a unique combination of a MHC molecule and an oligopeptide resulting from the degradation of a particular protein. There may be only a few tens, or less, of specific ligands of a TCR on an APC.

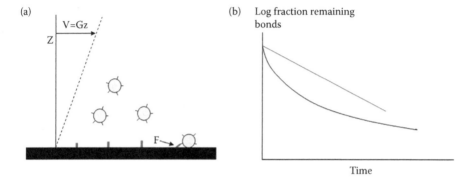

FIGURE 1.5 Studying molecular interactions with a flow chamber. (a) Optimal information can be obtained by studying the motion of receptor-coated microspheres near ligand-coated surfaces in presence of a wall shear rate of a few s^{-1}. Using microspheres of a few μm diameter, trajectories can be monitored with an accuracy of several tens of nm and time resolution of 20 ms with standard video equipment. The force exerted on a particle of 1.4 μm radius may be a fraction of a piconewton, and the force on the bond may be estimated at a few pN when the wall shear rate is on the order of several s^{-1}, which provides high sensitivity. The possibility to scan extensive contact areas is well suited to the use of low surface density coatings and determination of association rates. (b) Bond rupture may be studied by plotting the logarithm of the fraction of surviving bond versus time after initial arrest. In some cases, curves are straight lines and the slope represents the off rate, which may depend on the wall shear rate (red line). In many cases, the curve is more complex (black line) and may be analyzed to estimate some quantitative properties of energy landscapes. The capacity of the flow chamber to measure the kinetic and mechanical properties of weak bonds is described in a recent review [156].

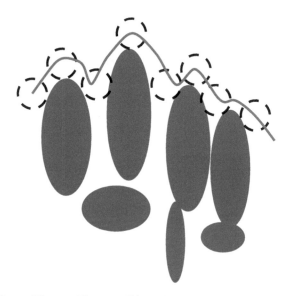

FIGURE 1.7 Accessible area. The accessible surface (red line) may be defined as the surface spanned by the center of a sphere representing a water molecule (broken contour) moving in contact with atoms constituting the protein and modeled as hard spheres with a known van der Walls radius (dark blue areas).

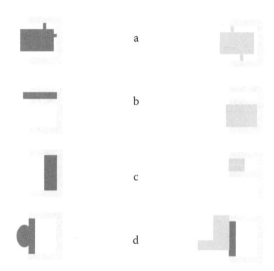

FIGURE 1.8 Possible mechanisms of cross reactivity. Several different mechanisms were shown to result in receptor promiscuity. (a) Different ligands may form different elementary bonds in a same binding site [110]. (b) A molecule may display several unrelated binding sites on its surface. (c) A binding site may be flexible and may accommodate different ligands [95]. (d) Two unrelated molecules may display some local similarity [103].

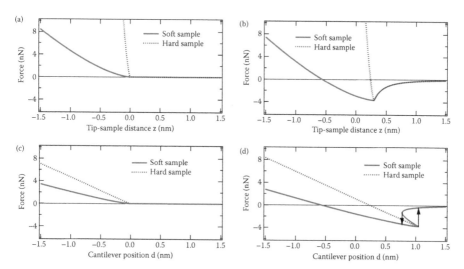

FIGURE 2.3 Tip-sample forces using the (a) Hertz and (b) DMT-M model for a hard ($E_{hard} = 100\,\text{GPa}$) and soft ($E_{soft} = 1\,\text{GPa}$) sample assuming a tip radius of 10 nm. The other parameters are $z_0 = 0.3\,\text{nm}$, $\mu_t = \mu_s = 0.3$, $E_t = 130\,\text{GPa}$, $A_H = 0.2\,\text{aJ}$. (c) If these forces are measured with an atomic force microscope using a cantilever with a spring constant of 5 N/m, the resulting force versus curves show significantly reduced slope due to the elasticity of the cantilever. Without the presence of adhesion forces (Hertz model), the curves are continuous. (d) Adhesion results in a hysteresis between forward and backward movement of the cantilever as marked by the arrows. (See Section. 2.2 for more details on this effect.)

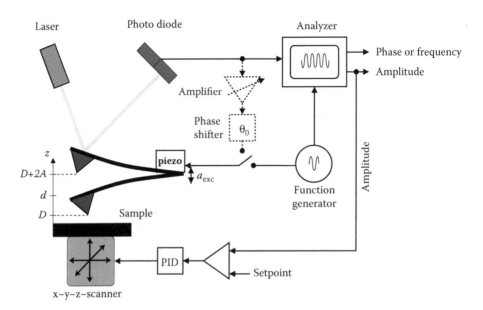

FIGURE 2.8 Schematic drawing of the experimental setup of a dynamic force microscope where the driving of the cantilever can be switched between amplitude-modulation (AM) mode (solid lines) or frequency-modulation (FM) mode (dashed lines). While the cantilever in the AM-mode is externally driven with a frequency generator, the FM-mode exhibits a feedback loop consisting of a time ("phase") shifter and an amplifier. In both cases, we assume that the laser beam deflection method is used to measure the oscillation of the tip, which oscillates between the nearest tip-sample position D and $D+2A$. The equilibrium position of the tip is denoted as d.

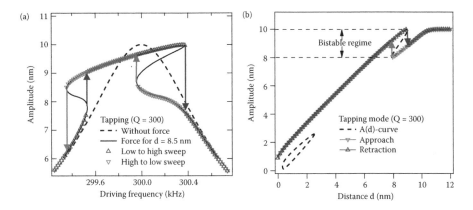

FIGURE 2.11 (a) Resonance curve for AM-mode operation if the cantilever oscillates near the sample surface with $d = 8.5$ nm and $A_0 = 10$ nm, thereby experiencing the model force field given by Equation 2.9. The solid lines represent the analytical result of Equation 2.24a, while the symbols are obtained from the numerical solution of the equation of motion Equation 2.12. The dashed lines reflect the resonance curves without tip-sample force and are shown purely for comparison. The resonance curve exhibits instabilities ("jumps") during a frequency sweep. These jumps take place at different positions (marked by arrows) depending on whether the driving frequency is increased or decreased. (b) A hysteresis is also observed for amplitude versus distance curves. The dashed line shows the analytical result, and the symbols show the numerical solutions for approach and retraction using a driving frequency of 300 kHz and the same parameters as in (a).

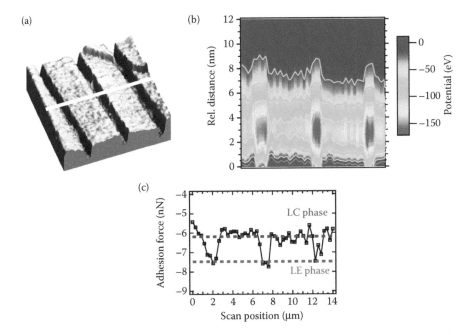

FIGURE 2.14 (a) Surface plot (scan size: $14 \times 14\,\mu\text{m}^2$) of the topography of the L-α-dipalmitoyl-phosphatidycholine (DPPC) film prepared by the Langmuir–Blogdett technique. The monolayer shows alternating stripes and channels that consists of DPPC adsorbed in the liquid condensed and liquid expanded phase, respectively. The white line marks the position where we recorded the frequency shift and amplitude versus distance curves for the construction of the contour map of the tip-sample interaction potential shown in (b). The graph in (c) displays the corresponding adhesion force obtained from the data shown in (b). The parameters of the cantilever were $f_0 = 170,460\,\text{Hz}$, $k_{\text{cant}} = 39.6\,\text{N/m}$, and $Q = 492$.

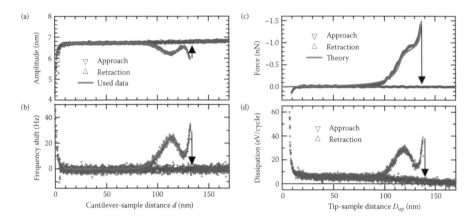

FIGURE 2.16 An application of the introduced dynamic force spectroscopy technique. (a) Amplitude (A) and (b) frequency shift (Δf) curves measured during approach to and retraction from the surface covered with dextran molecules. During the retraction, one dextran molecule bound to the tip as revealed by the change of the frequency and amplitude signal. At a position of about 135 nm (see arrows), the maximum binding force was exceeded and the cantilever oscillated freely again. Only the data before this jump is used for the subsequent analysis. (c) Using Equation 2.40, the force acting on the dextran molecule (symbols) is reconstructed as a function of the actual tip position D_{up}. The experimental result is well described by a "single-click" model using only the number of molecules ($N = 266$) as fitting parameter (solid line). (d) The energy dissipated per oscillation cycle can be calculated from Equation 2.39 for approach and retraction. The zero of the x-axes is arbitrarily set to the left side of the graphs.

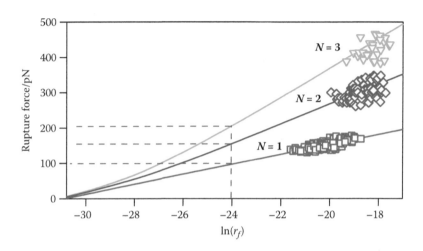

FIGURE 3.8 Rupture of multiple parallel bonds for 1, 2, and 3 multivalent clusters bewteen the cancer marker Mucin-1 and its antigen fragment. Data points reproduced with permission from Sulchek, T. et al. 2005. *Proc. Nat. Acad. Sci. USA*, 102 (46), 16638–16643. Solid lines are numerical solutions of Equation 3.86. Only the single-bond, $N = 1$, curve is fit to the data to determine the kinetic unbinding rate and transition state, while the $N = 2$ and $N = 3$ curves are predictions for bivalent and trivalent bonding. The dashed lines illustrate that, for a given loading rate, including two or three bonds to the cluster does not add a significant increase in total rupture force. In fact, at the chosen loading rate, the rupture force for $N = 3$ bonds is just twice that of the single $N = 1$ case. This is due to the increased probability of rupture when multiple bonds are present. The enhancement over a single bond improves for increased loading rate, but even at extremely fast loading, the rupture force for N bonds is limited to less than N times the single-bond rupture force. As discussed in the text, binding enhancement due to multivalency is most prominent in the equilibrium regime where rebinding and entropic effects stabilize the bound state.

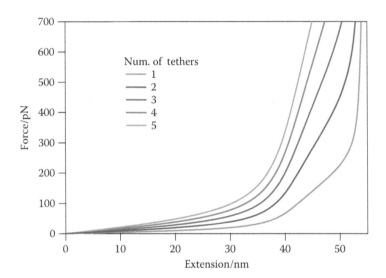

FIGURE 3.10 Theoretical force-extension curves of multiple, parallel PEG polymers from numerical solution of Equation 3.102. (Adapted from Sulchek, T. et al. 2006. *Biophys. J.*, 90 (12), 4686–4691.)

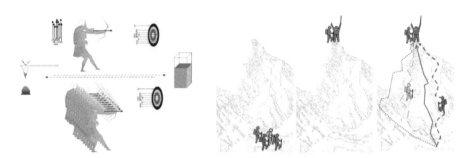

FIGURE 5.1 Energy landscape and single molecules. (a) To obtain results comparable to those provided using batch measurements (symbolized by the spectroscopy cuvette on the right), experiments on single molecules request a very large sampling represented by the bunch of arrows (shot by a single archer), which stochastically hit the target (top). Ergodic hypothesis is illustrated showing that the score obtained by the single archer shooting many arrows is exactly the same as the one obtained by a troop of archers shooting each only one arrow. (b) In biophysical or biochemical experiments, most of the results are often given taking only account of half time measurements. Starting (on the far left panel) from a population of climbers at the bottom of the mountain (the background represents K2, which was attempted by George Bell in 1953 with a well-knit team; the fascinating story can be found in *K2, The Savage Mountain*) the experiment will be considered as ending at time $t_{1/2}$ when half of the population have changed their states (50% of the climbers reached the top of the mountain, middle panel). This approach obviously masks the states of all the other climbers (green and red) still in progress and definitely miss all the different ways individually used.

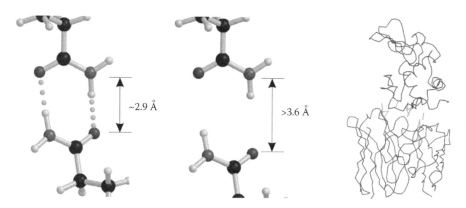

FIGURE 5.2 Representation of hydrogen bonds in proteins. On the left panel, a glutamine residue is making a double hydrogen bond with another glutamine residue. Ideal bond length for a hydrogen bond is about 2.9 Å. Covalent bonds are represented by yellow tubes, whereas noncovalent bonds are represented by green dots. Carbon atoms are in black, oxygen atoms in red, nitrogen atoms in blue, and hydrogen atoms in white. The middle panel shows the same two residues as in the left panel except that their distance has increased by about 0.7 Å. In such a conformation, hydrogen bonds are broken and the two residues are no more in contact. All the potential 14 hydrogen bonds (green dots) found in the complex between an antigen (magenta tubes) and an antibody (orange tubes) are represented in the right panel. Many other interactions occur in protein–protein complexes, such as salt bridges, van der Waals contact, π-cation interactions (Chen et al. 2009) for a recent description of interactions in antibody-antigen complexes and (Lin et al. 2011) for π-cation interactions). These interactions are not represented here for clarity. The figure was drawn with Molscript (Kraulis, 1991) and rendered using Raster3D (Merritt & Bacon, 1997).

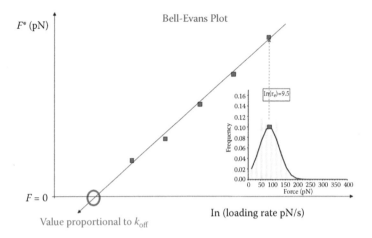

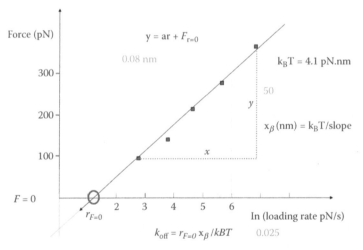

FIGURE 5.8 Theoretical Bell-Evans plot. The upper graph indicated how the Bell-Evans plot is built from the most probable rupture forces (F^*). These forces are obtained from a Gaussian fit of the force distribution obtained at a given effective loading rate. The lower graph indicates how to extract the energy barrier properties: the width (x_β in nm) and the kinetic dissociation rate (k_{off} in s^{-1}). The width x_β is obtained from the slope: $x_\beta = k_B T/\text{slope}$. The k_{off} value is obtained by extrapolating the loading rate value at $F^* = 0$: $k_{off} = r_{F=0}\, x_\beta\, /k_B T$.

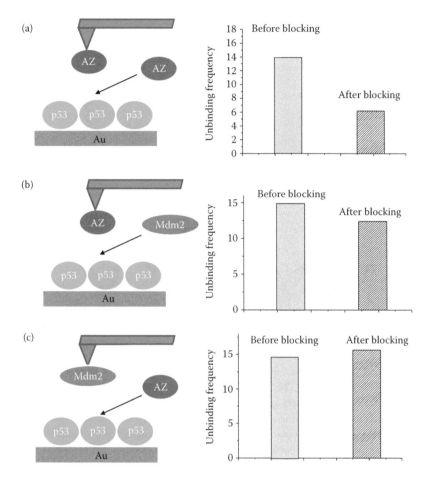

FIGURE 6.8 Competitive blocking experiments on the p53–mdm2–azurin ternary complex. (a) Azurin is used as a competitor for the p53-azurin complex (left); unbinding frequencies before and after blocking the substrate with a solution of free azurin (right). (b) mdm2 is used as a competitor for the p53-azurin complex (left); unbinding frequencies before and after blocking the substrate with a solution of free mdm2 (right). (c) Azurin is used as a competitor for the p53-mdm2 complex (left); unbinding frequencies before and after blocking the substrate with a solution of free azurin (right). (Adapted from Bizzarri, A. R. and Cannistraro, S. 2009. *J. Phys. Chem. B*, 113, 16449–16464. Adapted from Funari, G. et al. 2010. *J. Mol. Recognit.*, 23, 343–351.)

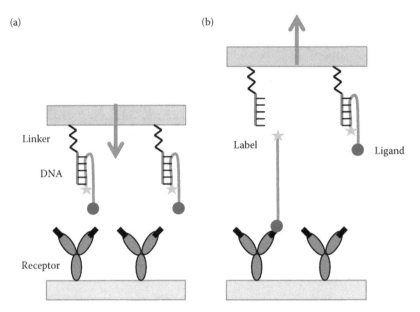

FIGURE 6.12 Sketch of the experimental setup used for a force-based biosensor. A DNA strand is bound to the top plate through a linker. Both a fluorescent label and the ligand molecule are bound to DNA. The receptor molecules are immobilized on the bottom plate. (a) The top and the bottom plates approached each other to promote a biorecognition process between the partners. (b) During the retraction, one of the two DNA strands is expected to remain attached to the bottom plate, with a concomitant deposition of the fluorophore label upon the interaction between the ligand and the receptor overcomes that between the DNA strands. (Adapted from Blank, K. et al. 2003. *Proc. Nat. Acad. Sci. USA*, 100:11356–11360.)

TABLE 4.3
Examples of Compounds for Functionalization of Silanols

Coupling Groups	Name of Compound	Chemical Formula
Amine to amine	Disuccinimidyl glutarate (DSG)	
	Glutaraldehyde[a]	
	NHS-PEG-NHS[b]	
Amine and create protected amine	NHS-PEG-NH-FMOC[c]	
Amine to thiol	N-[γ-maleimidobutyryloxy] succinimide ester	
	NHS-PEG-Mal[b]	
Amine to avidin (or strepta-vidin)	NHS-PEG-Biotin	

[a] Coupling is followed by the Schiff base reduction with $NaCNBH_3$, see Figure 4.5.

[b] Various tether lengths are commercially available, from several ethoxy monomer groups in the linker chain to long polymers with number of monomers ranging from ~ 25 to ~ 200.

[c] This compound can be used to move amine functional group away from the surface. After reacting with the surface amines, FMOC protection can be removed (20%–50% piperidine in DMF) exposing PEG-tethered amines.

TABLE 4.4

Functionalized Polymeric Tethers

Terminal Groups	Name of Compound	Typical Use
Bis-amine	Amine-PEG-Amine	To attach amine-reactive molecules to amine-reactive surfaces (e.g., prepared by activation of carboxyls with EDCI)
Bis-thiol	Thiol-PEG-Thiol	To attach thiol-reactive molecules to thiol-reactive surfaces (including gold)
Amine and thiol	Amine-PEG-Thiol	To attach amine-reactive molecules to thiol-reactive surfaces (including gold) and vice versa
Thiol and carboxyl	Thiol-PEG-Carboxymethyl	Attaches to gold surfaces and allows to couple molecules with amines using EDCI coupling

(NDSK fragment) (Averett et al. 2008). In this work, the entire fibrinogen molecule with inactive knobs is immobilized on the gold-coated substrate, and the central unit with active "A" knobs is attached to the gold-coated AFM probe. Sample preparation consists of the following steps:

1. **Cleaning of glass microscope slides and silicon nitride probes:** Glass slides were cleaned by soaking in Piranha solution (3:1 H_2SO_4/H_2O_2 mixture) for at least 20 min, were sonicated in deionized water, then in ethanol. After this, they are rinsed with deionized water and dried under a stream of nitrogen. AFM probes were rinsed with chloroform, dried with a gentle stream of nitrogen, and cleaned with ozone for 30 min using a BioForce TipCleaner (Ames, IA).

2. **Coating of probes and substrates with gold:** Both probes and slides were coated with 3-nm adhesion layer of chromium and then with 45-nm layer of gold using a magnetron sputtering system (Kurt Lesker Inc.). (If commercial gold-coated probes and substrates are used, then the first two steps can be replaced by cleaning surfaces, for example, with UV ozone, rinsing with organic solvent, and drying either under nitrogen or in vacuum).

3. **Functionalization with carboxylic acid terminated SAM:** The gold-coated probes and slides were immersed in 11-mercaptoundecanoic acid solution (2 mM in absolute ethanol) overnight immediately before use. After coating probes and substrates, they were rinsed with 100% ethanol and water and dried with nitrogen before use.

4. **Attachment of proteins to carboxylic acid terminated surfaces:** EDCI (see Figure 4.2c) coupling procedure was employed. The surfaces were immersed in solution composed of equal volumes of 0.1 M NHS and 0.4 M EDCI in water for 30 min and then rinsed with water. After activation, protein was coupled to the surfaces by incubating the surface with protein solutions in HEPES buffer for 10 min. Concentration of proteins under study was 75 μg/ml. Fibrinogen incubated on the substrate was mixed with

BSA in 1:1 mass ratio. Remaining active sites on surfaces were blocked by rinsing and 10-min incubation in AFM buffer (20 mM HEPES, 150 mM NaCl, 3 mM CaCl$_2$, 2 mg/ml BSA, 0.1% Triton X-100, pH 7.4).

5. **Removing noncovalently attached proteins:** The surfaces were rinsed alternately and incubated for 1 min each with a high salt (50 mM HEPES, 1 M NaCl, pH 7.4) and low pH buffer (50 mM NaOAc, 300 mM NaCl, pH 4.0). This procedure was repeated five times. Before loading into the AFM instrument, the surfaces were rinsed copiously with AFM buffer.

Significant fraction of DFS data collected using this sample preparation protocol consisted of single-molecule rupture events that exhibit a characteristic pattern (Averett et al. 2008). Typical force curve exhibiting this pattern is shown in Figure 4.6. Number of test experiments have been performed in this and in the following study (Averett et al. 2009) to confirm that measured interactions are indeed single-molecule events that come from breaking the bond under study. This might be somewhat surprising because multiple interactions are usually detected with molecules immobilized directly to surfaces. It is likely that two factors are of importance here. Distance between holes in fibrinogen molecule is rather large (~ 50 nm). Therefore, it seems unlikely that two holes will be in proximity of "A" knobs on the probe. This promotes single-molecule detection. The second factor is that surfaces were blocked with BSA and AFM buffer contained a surfactant. This reduced nonspecific adhesion that is essentially absent in force curve is shown in Figure 4.6.

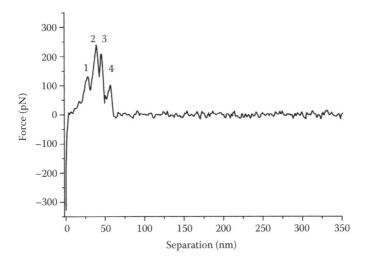

FIGURE 4.6 Typical pattern of bond breaking in "A-a" knob-hole interaction. It can be noted that the final bond breaking occurs at the tip-sample separation that is close in dimension to the size of interacting molecules (although somewhat larger probably due to unfolding of the D domain). (Reproduced from Averett, L. E. et al. 2008. *Langmuir* 24 (9):4979–4988.)

4.4.2.5   Example of Chemical Immobilization Using Silanol Chemistry

Example of sample preparation using silanol chemistry is taken from a study that aims to explain why specific blocking of active sites in DFS experiments is by far less efficient than binding in solution (Guo et al. 2010b). This is an important question pertaining to DFS methodology because such measurements are often used to test specificity of recognition in DFS. This study has employed typical biorecognition partners: biotin and streptavidin. Large widths of the distributions of rupture forces indicate that biotin–streptavidin bond breaking by DFS has large contribution of multiple ruptures (Guo et al. 2008). Therefore, it is desirable to decrease surface density of molecules by using polymeric linkers. Also, rupture of a molecular bond that is tethered to the surface occurs after characteristic stretching of an entropic spring and happens at the tip-sample separation that is close to the length of the linker employed. These considerations add certainty that measured rupture events come from recognition events under study and not from various artifacts.

Immobilization of molecules was performed on silicon nitride probes and glass cover slips. For the majority of experiments, streptavidin was covalently attached to the surface directly, however, test experiments involved attaching streptavidin to the substrate by PEG linker. This sample preparation for double-tether experiments is described below. Steps of the probe and substrate modification were as follows:

Cleaning of substrates and probes: Substrates and probes are cleaned by soaking in 2% Hellmanex II (Hellma, Müllheim, Germany) aqueous solution for 3 h, then rinsed by deionized water (18 MΩ·cm), anhydrous ethanol, and dried under vacuum.

Amination of surfaces: The cleaned and dried probes and glass substrates were aminated in a saturated ethanolamine hydrochloride solution of anhydrous dimethylsulfoxide (DMSO) for 72 h at room temperature in a nitrogen-filled glove box. After amination, surfaces were rinsed with DMSO.

Tethering biotin to silicon nitride probes: The aminated probes were covalently attached by biotin-poly (ethylene glycol)-succinimidyl carboxymethyl (Biotin-PEG-SCM) linkers with a mass-average molecular mass of 3400 Da (Laysan Bio Inc., Arab, AL). This reaction was carried out for 24 h in anhydrous dimethylformamide (DMF) with 5% pyridine (v/v). Next, the probes were treated by acetic anhydride, pyridine, and DMF mixture solution (v/v, 5:4:3) to block the remaining amines. Chemically modified probes were cleaned by immersing in preheated (50°C–60°C) hexanes/i-propanol (3/2) for 1 h and then in warm deionized water (50°C–60°C) for 10 min. And then, the probes were washed successively in toluene, DMF, and ethanol for 15 min each with a platform shaker. Finally, the probes were dried under vacuum and used immediately for data collection.

Tethering streptavidin to the glass substrates via PEG linkers: The aminated substrates were first incubated in DMF solution containing fluorenylmethoxycarbonyl-poly(ethylene glycol)-succinimidyl carboxymethyl

(FMOC-PEG-SCM) with a mass-average molecular mass of 3400 Da (Laysan Bio Inc.) for 24 h (10 mg of FMOC-PEG-SCM were dissolved in 300 µl of DMF with 10% pyridine [v/v]). Next, the substrates were treated with poly(ethylene glycol)mono-methyl ether mono(succinimidyl succinate) ester (PEG-SS) (Polysciences Inc., Warrington, PA) with a mass-average molecular mass of 1900 Da for 24 h (10 mg of PEG-SS were dissolved in 300 µl of DMF with 10% pyridine [v/v]). This step was aimed to block unreacted amines and hinder adsorption of streptavidin on surface of glass substrate. After that, the substrates were thoroughly rinsed with DMF and activated in DMF and piperidine mixture solution (v/v, 4:1) for 30 min to remove FMOC group. After removing FMOC group, the substrates were further reacted with solution of 10 mg of homo-bi-functional amine-reactive cross-linker 1,4-phenylenediisothiocyanate (PDITC) in 300 µl of DMF with 5% pyridine (v/v) for 2 h. Activated slides were cleaned by ultrasonication in DMF and ethanol twice for 10 min each. Then, 200 µl of 100 µg/ml solution of streptavidin in phosphate buffer (PBS, VWR international, 0.05 M pH 7 and pH 10 PBS were mixed together to reach pH 8) was deposited on the PDITC-activated substrates. The covalent attachment of streptavidin was performed in a dark environment for 2 h. Here, pH 8 PBS solution was chosen to optimize the PDITC-streptavidin reaction but keep streptavidin stable (Weber et al. 1989; Hermanson, 1996). Substrates with attached streptavidin were thoroughly rinsed with pH 7 PBS buffer solution and used immediately for data collection.

Figure 4.7 compares force curves with rupture events measured when streptavidin is attached directly to the surface and when it is tethered using PEG linker. In the double-tether experiment, ruptures occur consistently at larger tip-sample separation when in the single-tether experiment. This is consistent with streptavidin attachment to the end of polymeric tether.

4.5 MULTIPLE INTERACTIONS BETWEEN IMMOBILIZED MOLECULES

Covalent immobilization of molecules with polymeric tethers removes problems associated with potentially weak physical attachment, avoids effects of nonspecific probe adhesion to the substrate, and decreases steric hindrance. However, using tethers does not eliminate possibility of multiple interactions as illustrated in Figure 4.1. Feasibility of multiple bonds forming during the probe-sample contact is illustrated in Figure 4.8. Simple estimate of possible number of bonds between the probe and the substrate can be obtained assuming high density of attachments of interacting molecules to surfaces. If we further assume that biomolecules tethered to the probe can bind to the partners immobilized on the substrate at separation that is similar in magnitude to the end-to-end distance l_{rms} of a free tether in solution, then the largest number of bonds during one tip-sample contact would be approximately equal to the ratio of the surface area of spherical cap shown in Figure 4.8 to the area occupied by

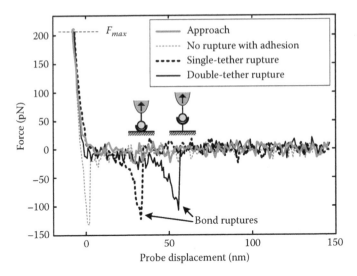

FIGURE 4.7 Three withdraw parts force curves are shown and a typical approach part in measurements of interactions between biotin and streptavidin is indicated by the legend in the figure. One force curve has adsorption at the surface. Two other curves show rupture events that are removed from the surface of substrate as illustrated by a cartoon in the figure. (Reproduced from Guo, S. L. et al. 2010b. *Journal of the American Chemical Society.* 132 (28):9681–9687.)

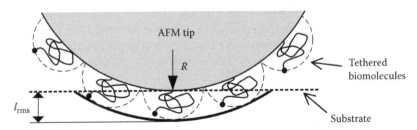

FIGURE 4.8 Cartoon illustrating formation of multiple bonds between molecules tethered by polymeric linkers to the atomic force spectroscopy tip and recognition partners that are closely packed on the substrate.

one molecule:

$$N_{\max} \approx \frac{2R}{l_{\mathrm{rms}}} \tag{4.1}$$

where R is the radius of the tip curvature. Thus, if molecules are tethered to the probe with radius of curvature of 25 nm with polymeric linkers with the end-to-end distance of 5 nm then upto 10 bonds might be formed during one probe-sample contact (assuming that recognition partners on the substrate are relatively small and densely packed). Such large number of bonds impedes accurate data analysis. In this section, problem of multiple interactions is addressed by considering the following questions: (1) What is the expected probability of the single-molecule interactions in comparison to the multiple interactions and (2) How to distinguish effects of multiple interactions in the measured data.

4.5.1 PROBABILITY OF SINGLE-MOLECULE EVENTS

It might be expected that if the radius of the probe curvature were so low so that only one molecule can be attached at the apex, then there will be no problem of multiple bond ruptures. It has been proposed that AFM probe equipped with carbon nanotube at the apex is suitable for single-molecule measurements by covalent functionalization with biological molecule of interest (Wong et al. 1998). However, preparation of such probes is difficult and did not find wide spread use in DFS. Moreover, there is still no guarantee that only one molecule will reside at the end of the nanotube (Wong et al. 1998).

Probes with gold-coated tips usually have higher radius of curvature than the bare probes (according to manufacturers' specification by 10–30 nm). Therefore, according to Equation 4.1, it might be expected that such probes are more prone to the multiple bond artifacts than the uncoated probes. Thus, dilution of functionalized linkers with chemically inert linkers is necessary to reduce number of multiple interactions and such dilution should be more substantial for gold-coated probes. If probability of forming bonds between multiple recognition partners on the tip and on the substrate follows Poisson distribution, then fraction of events that involve rupture of N bonds is (Tees et al. 2001)

$$P(N) = \frac{\left(\frac{1}{P_{tot}} - 1\right)\left[\ln\left(\frac{1}{1-P_{tot}}\right)\right]^N}{N!} \qquad (4.2)$$

where P_{tot} is the overall probability of detecting rupture events. Consequently, fractional probability of rupturing more than one bond P_{mult} is

$$P_{mult} = 1 - \left(\frac{1}{P_{tot}} - 1\right)\ln\left(\frac{1}{1-P_{tot}}\right) \qquad (4.3)$$

This equation indicates that multiple bond ruptures are expected to occur rather often in DFS measurements if P_{tot} is high. For example, if $P_{tot} = 0.5$, then the multiple bond ruptures occur only approximately twice less as often as the single bond ruptures. According to Equation 4.3, for multiple bond ruptures to constitute less than 5% fraction of all bond ruptures the overall probability P_{tot} should be less than 10%.

If formation of bonds between recognition partners is not independent of each other, then Poisson statistics does not apply and Equation 4.3 breaks down. This might occur when recognition partners are multivalent or cluster during sample preparation (Guo et al. 2010a). In this case, multiple interactions will occur more often than predicted by Equation 4.3. Another assumption in estimation of fraction of multiple bonds is that all bond ruptures are detected. However, if detection is limited by noise, then significant fraction of single bond ruptures might miss the detection, particularly at low and high probe velocities. This might result in considerable contribution of multiple bond ruptures that is not described by Equation 4.3 (Guo et al. 2008). Therefore, additional experiments are usually necessary for verification of the single bond nature of measured interactions. Several recommendations are given in the section below.

4.5.2 EFFECTS OF MULTIPLE INTERACTIONS ON MEASURED DATA

In the data analysis, only the last rupture event is usually included in statistics of bond ruptures, thus, facilitating removal of multiple bond ruptures from analysis. However, it is possible that the apparently single rupture transition (like shown in Figure 4.7) corresponds to simultaneous rupture of more than one bond (Guo et al. 2008; Guo et al. 2010a; Mayyas et al. 2010). Such events might occur if significant fraction of force that was holding two bonds gets transferred onto the remaining bond that will in turn rupture quickly. If rupture of the last bond occurs on the time scale shorter than the rupture detection time, then the last bond rupture remains undetected.

According to Equation 4.3, if $P_{tot} < 0.25$, then among multiple bond ruptures, the most prevalent will be ruptures, of two bonds (more than 90% of all multiple bond ruptures). Therefore, in this subsection, the focus will be on two-bond ruptures. It might be expected that if two bonds are loaded simultaneously by external force, then measured force of bond rupture will be higher than the force of a single bond rupture. It has been shown that if a large number of bonds are loaded simultaneously and the force is distributed evenly between these bonds, then total rupture force is proportional to the number of bonds (Seifert, 2000). However, if only two bonds are loaded simultaneously, then the most probable rupture force will be less than two times the rupture force of a single bond (Tees et al. 2001; Williams, 2003). Nonetheless, if forces are distributed evenly between bonds, then in the histogram of rupture forces, well-separated peaks are expected. However, if the load is not distributed evenly between the tethers the total rupture force decreases further and peaks in the histogram of rupture forces might significantly overlap (Gu et al. 2008; Guo et al. 2008). The unequal distribution of forces is expected in experiments that utilize polymeric linkers to immobilize molecules because tethers are often polydisperse, also tethers might attach at different heights along the tip and at different positions on the sample surface. Even small difference in contour lengths of linkers might result in substantial shift in the distribution of rupture forces toward lower values as illustrated in Figure 4.9 (Gu et al. 2008; Guo et al. 2010a).

If distributions of rupture forces show a shoulder at the high force side of the distribution, then multiple bond ruptures might be suspected. Therefore, to avoid ambiguity in the data analysis, measurements should be repeated with lower grafting density of molecules on the substrate. However, this approach might not significantly change distribution of rupture forces if interactions are multivalent or if molecules cluster on the surface during immobilization. In this case, probability to form multiple bonds might be affected by changing grafting density of biomolecules on the

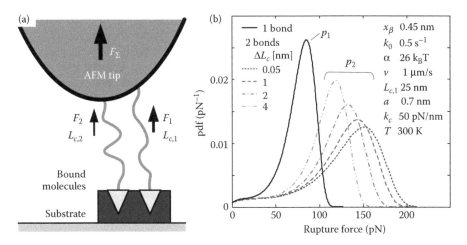

FIGURE 4.9 Panel (a) shows two-bond rupture geometry. Panel (b) shows corresponding probability of rupture forces for one- and two-bond ruptures for tethers of different lengths. Significant shift in position of peak for two-bond ruptures can be noticed even for less than 10% difference in contour lengths of two linkers. In calculations, parabolic potential with the cusp barrier was used in the kinetic model and the freely-jointed chain model extended to fit stretching of PEG linkers in water was used for calculating dynamics of loading. The legend and calculation parameters are shown in the graph. Here, x_β is the distance from the potential minimum to the transition state along the pulling direction, k_0 is the zero-force rate of transition from the bound state, α is the depth of potential from the minimum to the transition state in units of thermal energy k_B T, v is the probe velocity, $L_{c,1}$ is the contour length of a shorter linker, a is the Kuhn length of polymeric linker, k_c is the spring constant of the AFM cantilever, T is the absolute temperature, and ΔL_c is the difference between length of longer and shorter linkers. (Reproduced from Guo, S. L. et al. 2010a. *Journal of Physical Chemistry C* 114 (19):8755–8765.)

probe. Therefore, performing experiments by varying attachment density on the probe might reveal contribution of multiple bonds to the data. Maximum number of possible bonds obtained from Equation 4.1 can be used as a molar dilution factor of biomolecules on the probe. Using this factor implies that dilution is performed with polymers of approximately the same length. If dilution is performed with shorter polymers, then the molar dilution factor should be higher. Assuming that polymer attached to the surface occupies area that is approximately equal to πl_{rms}^2, then by using Equation 4.1, the dilution factor of the back-filling polymer to the polymeric linkers that hold recognition partner can be estimated by

$$\frac{C_b}{C_1} = \frac{2l_{rms,1}R}{l_{rms,b}^2} \tag{4.4}$$

where C_b and $l_{rms,b}$ are the concentration and the rms of the end-to-end distance of the back-filling molecules, respectively; C_l and $l_{rms,l}$ are the concentration and the rms of the end-to-end distance of polymeric linker that holds recognition partner molecule, respectively; and R is the radius of the tip curvature. For example, if polymer linker and the back-filling polymer have 40 and 15 statistical units, respectively, with the statistical (Kuhn) length of 0.7 nm, and the probe with 40 nm radius of curvature the dilution ratio of the linker to the back-filling polymer factor is $\sim$1:25. For back-filling molecules significantly shorter than the linkers Equation 4.4 might underestimate the dilution factor because of the entropy of mixing effects: Fraction of longer polymeric molecules that chemically attach to surfaces is higher than the fraction of longer polymers in solution (Al-Maawali et al. 2001). Therefore, Equation 4.4 provides only the starting point for experiments that aim to significantly reduce multiple bond artifacts. It should be noted that the aspect of avoiding multiple interactions remains poorly researched in DFS literature.

4.6 CONCLUSION

In this chapter, we have considered methods of attaching molecules to surfaces in DFS experiments primarily focusing on the most widely used approaches. The main conclusion of comparing physical and chemical methods of attaching molecules is that in spite of the simplicity of sample preparation using physical attachment method, this method requires substantial verification for various artifacts. Chemical methods of attaching molecules, particularly approaches that utilize polymeric tethers, remove majority of limitations pertinent to the physical methods of attachment. Chemical methods discussed in this chapter focus on approaches that have been employed in DFS literature. Also, we have indicated that specific requirements of DFS technique are considerably different from other techniques that employ immobilized biomolecules. Commercially available cross-linking compounds and functionalized tethers enable wide spread use of chemical attachment in DFS experiments and facilitate application of this methodology by biophysics scientists who do not have extensive chemistry training. Arsenal of available reactions to attach molecules to interfaces is considerably more extensive than reactions discussed here. It might be expected that newly developed reactions and experimental methodologies will help solve the remaining problem of multiple interactions in DFS and facilitate widespread use of DFS methodology as an analytical tool in biophysical research.

REFERENCES

Ahmad, S. F., L. A. Chtcheglova, B. Mayer, S. A. Kuznetsov, and P. Hinterdorfer. 2011. Nanosensing of Fc gamma receptors on macrophages. *Analytical and Bioanalytical Chemistry* 399 (7):2359–2367.

Al-Maawali, S., J. E. Bemis, B. B. Akhremitchev, R. Leecharoen, B. G. Janesko, and G. C. Walker. 2001. Study of the polydispersity of grafted poly(dimethylsiloxane) surfaces using single-molecule atomic force microscopy. *Journal of Physical Chemistry B* 105 (18):3965–3971.

Aswal, D. K., S. Lenfant, D. Guerin, J. V. Yakhmi, and D. Vuillaume. 2006. Self-assembled monolayers on silicon for molecular electronics. *Analytica Chimica Acta* 568 (1–2):84–108.

Averett, L. E., C. B. Geer, R. R. Fuierer, B. B. Akhremitchev, O. V. Gorkun, and M. H. Schoenfisch. 2008. Complexity of 'A-a' Knob-Hole fibrin interaction revealed by atomic force spectroscopy. *Langmuir* 24 (9):4979–4988.

Averett, L. E., M. H. Schoenfisch, B. B. Akhremitchev, and O. V. Gorkun. 2009. Kinetics of the multistep rupture of fibrin 'A-a' polymerization interactions measured using atomic force microscopy. *Biophysical Journal* 97 (10):2820–2828.

Ballard, C. C., E. C. Broge, R. K. Iler, John D. S. St, and J. R. McWhorter. 1961. Esterification of the surface of amorphous silica. *Journal of Physical Chemistry* 65 (Copyright (C) 2011 American Chemical Society (ACS). All Rights Reserved.):20–5.

Bartels, F. W., B. Baumgarth, D. Anselmetti, R. Ros, and A. Becker. 2003. Specific binding of the regulatory protein ExpG to promoter regions of the galactoglucan biosynthesis gene cluster of Sinorhizobium meliloti - a combined molecular biology and force spectroscopy investigation. *Journal of Structural Biology* 143 (2):145–152.

Baumgartner, W., P. Hinterdorfer, W. Ness, A. Raab, D. Vestweber, H. Schindler, and D. Drenckhahn. 2000. Cadherin interaction probed by atomic force microscopy. *Proceedings of the National Academy of Sciences of the United States of America* 97 (8):4005–4010.

Bizzarri, A. R., and S. Cannistraro. 2009. Atomic force spectroscopy in biological complex formation: Strategies and perspectives. *Journal of Physical Chemistry B* 113 (52):16449–16464.

Bizzarri, A. R., and S. Cannistraro. 2010. The application of atomic force spectroscopy to the study of biological complexes undergoing a biorecognition process. *Chemical Society Reviews* 39 (2):734–749.

Bonanni, B., A. S. M. Kamruzzahan, A. R. Bizzarri, C. Rankl, H. J. Gruber, P. Hinterdorfer, and S. Cannistraro. 2005. Single molecule recognition between Cytochrome C 551 and gold-immobilized Azurin by force spectroscopy. *Biophysical Journal* 89 (4):2783–2791.

Chilkoti, A., T. Boland, B. D. Ratner, and P. S. Stayton. 1995. The relationship between ligand-binding thermodynamics and protein-ligand interaction forces measured by atomic force microscopy. *Biophysical Journal* 69 (5):2125–2130.

Chtcheglova, L. A., A. Haeberli, and G. Dietler. 2008. Force spectroscopy of the fibrin(ogen) - Fibrinogen interaction. *Biopolymers* 89 (4):292–301.

Dammer, U., M. Hegner, D. Anselmetti, P. Wagner, M. Dreier, W. Huber, and H. J. Guntherodt. 1996. Specific antigen/antibody interactions measured by force microscopy. *Biophysical Journal* 70 (5):2437–2441.

Dammer, U., O. Popescu, P. Wagner, D. Anselmetti, H. J. Guntherodt, and G. N. Misevic. 1995. Binding strength between cell-adhesion proteoglycans measured by atomic-force microscopy. *Science* 267 (5201):1173–1175.

Dettmann, W., M. Grandbois, S. Andre, M. Benoit, A. K. Wehle, H. Kaltner, H. J. Gabius, and H. E. Gaub. 2000. Differences in zero-force and force-driven kinetics of ligand dissociation from beta-galactoside-specific proteins (plant and animal lectins, immunoglobulin G) monitored by plasmon resonance and dynamic single molecule force microscopy. *Archives of Biochemistry and Biophysics* 383 (2):157–170.

Evans, E., A. Leung, D. Hammer, and S. Simon. 2001. Chemically distinct transition states govern rapid dissociation of single L-selectin bonds under force. *Proceedings of the National Academy of Sciences of the United States of America* 98 (7):3784–3789.

Evans, E., A. Leung, V. Heinrich, and C. Zhu. 2004. Mechanical switching and coupling between two dissociation pathways in a P-selectin adhesion bond. *Proceedings of the National Academy of Sciences of the United States of America* 101 (31):11281–11286.

Fadeev, A. Y., and T. J. McCarthy. 2000. Self-assembly is not the only reaction possible between alkyltrichlorosilanes and surfaces: Monomolecular and oligomeric covalently attached layers of dichloro- and trichloroalkylsilanes on silicon. *Langmuir* 16 (18):7268–7274.

Ferrer, J. M., H. Lee, J. Chen, B. Pelz, F. Nakamura, R. D. Kamm, and M. J. Lang. 2008. Measuring molecular rupture forces between single actin filaments and actin-binding proteins. *Proceedings of the National Academy of Sciences of the United States of America* 105 (27):9221–9226.

Florin, E. L., V. T. Moy, and H. E. Gaub. 1994. Adhesion forces between individual ligand-receptor pairs. *Science* 264 (5157):415–417.

Friddle, R. W., P. Podsiadlo, A. B. Artyukhin, and A. Noy. 2008. Near-equilibrium chemical force microscopy. *Journal of Physical Chemistry C* 112 (13):4986–4990.

Friedsam, C., A. D. Becares, U. Jonas, H. F. Gaub, and M. Seitz. 2004. Polymer functionalized AFM tips for long-term measurements in single-molecule force spectroscopy. *Chemphyschem* 5 (3):388–393.

Fritz, M. C., C. Carraro, and R. Maboudian. 2001. Functionalization of scanning force microscopy cantilevers via galvanic displacement technique. *Tribology Letters* 11 (3-4):171–175.

Fritz, J., A. G. Katopodis, F. Kolbinger, and D. Anselmetti. 1998. Force-mediated kinetics of single P-selectin ligand complexes observed by atomic force microscopy. *Proceedings of the National Academy of Sciences of the United States of America* 95 (21):12283–12288.

Gad, M., A. Itoh, and A. Ikai. 1997. Mapping cell wall polysaccharides of living microbial cells using atomic force microscopy. *Cell Biology International* 21 (11):697–706.

Gilbert, Y., M. Deghorain, L. Wang, B. Xu, P. D. Pollheimer, H. J. Gruber, J. Errington, B. Hallet, X. Haulot, C. Verbelen, P. Hols, and Y. F. Dufrene. 2007. Single-molecule force spectroscopy and imaging of the vancomycin/D-Ala-D-Ala interaction. *Nano Letters* 7 (3):796–801.

Gong, X., L. Dai, H. J. Griesser, and A. W. H. Mau. 2000. Surface immobilization of poly(ethylene oxide): Structure and properties. *Journal of Polymer Science Part B-Polymer Physics* 38 (17):2323–2332.

Grandbois, M., M. Beyer, M. Rief, H. Clausen-Schaumann, and H. E. Gaub. 1999. How strong is a covalent bond? *Science* 283 (5408):1727–1730.

Gu, C., A. Kirkpatrick, C. Ray, S. Guo, and B. B. Akhremitchev. 2008. Effects of multiple-bond ruptures in force spectroscopy measurements of interactions between Fullerene C60 molecules in water. *Journal of Physical Chemistry C* 112 (13):5085–5092.

Guo, S. L., N. Li, N. Lad, S. Desai, and B. B. Akhremitchev. 2010a. Distributions of parameters and features of multiple bond ruptures in force spectroscopy by atomic force microscopy. *Journal of Physical Chemistry C* 114 (19):8755–8765.

Guo, S. L., N. Li, N. Lad, C. Ray, and B. B. Akhremitchev. 2010b. Mechanical distortion of protein receptor decreases the lifetime of a receptor-ligand bond. *Journal of the American Chemical Society* 132 (28):9681–9687.

Guo, S., C. Ray, A. Kirkpatrick, N. Lad, and B. B. Akhremitchev. 2008. Effects of multiple-bond ruptures on kinetic parameters extracted from force spectroscopy measurements: Revisiting biotin-streptavidin interactions. *Biophysical Journal* 95 (8):3964–3976.

Hall, Cecil E., and Henry S. Slayter. 1959. The fibrinogen molecule: Its size, shape, and mode of polymerization. *The Journal of Biophysical and Biochemical Cytology* 5 (1): 11–27.

Hermanson, G.T. 1996. *Bioconjugate Techniques*. San Diego: Academic Press.

Hinterdorfer, P., and Y. F. Dufrene. 2006. Detection and localization of single molecular recognition events using atomic force microscopy. *Nature Methods* 3 (5):347–355.

Hinterdorfer, P., W. Baumgartner, H. J. Gruber, K. Schilcher, and H. Schindler. 1996. Detection and localization of individual antibody-antigen recognition events by atomic force microscopy. *Proceedings of the National Academy of Sciences of the United States of America* 93 (8):3477–3481.

Hinterdorfer, P., H. J. Gruber, F. Kienberger, G. Kada, C. Riener, C. Borken, and H. Schindler. 2002. Surface attachment of ligands and receptors for molecular recognition force microscopy. *Colloids and Surfaces B-Biointerfaces* 23 (2–3):115–123.

Howarter, J. A., and J. P. Youngblood. 2006. Optimization of silica silanization by 3-aminopropyltriethoxysilane. *Langmuir* 22 (26):11142–11147.

Hukkanen, E. J., J. A. Wieland, A. Gewirth, D. E. Leckband, and R. D. Braatz. 2005. Multiple-bond kinetics from single-molecule pulling experiments: Evidence for multiple NCAM bonds. *Biophysical Journal* 89 (5):3434–3445.

Jiang, Z. H., Y. H. Zhang, Y. Yu, Z. Q. Wang, X. Zhang, X. R. Duan, and S. Wang. 2010. Study on intercalations between double-stranded DNA and pyrene by single-molecule force spectroscopy: Toward the detection of mismatch in DNA. *Langmuir* 26 (17):13773–13777.

Jung, Y. J., B. J. Hong, W. Zhang, S. J. B. Tendler, P. M. Williams, S. Allen, and J. W. Park. 2007. Dendron arrays for the force-based detection of DNA hybridization events. *Journal of the American Chemical Society* 129 (30):9349–9355.

Kaas, R. L. , and J. L. Kardos. 1971. The interaction of alkoxy silane coupling agents with silica surfaces. *Polymer Engineering and Science* 11 (1):11–18.

Kienberger, F., G. Kada, H. Mueller, and P. Hinterdorfer. 2005. Single molecule studies of antibody-antigen interaction strength versus intra-molecular antigen stability. *Journal of Molecular Biology* 347 (3):597–606.

Kim, I. H., H. Y. Lee, H. D. Lee, Y. J. Jung, S. J. B. Tendler, P. M. Williams, S. Allen, S. H. Ryu, and J. W. Park. 2009. Interactions between signal-transducing proteins measured by atomic force microscopy. *Analytical Chemistry* 81 (9):3276–3284.

Kim, J., A. Opdahl, K. C. Chou, and G. A. Somorjai. 2003. Hydrophobic-interaction-induced alignment of polymers at the solid/liquid interface studied by infrared-visible sum frequency generation. *Langmuir* 19 (23):9551–9553.

Krasnoslobodtsev, A. V., and S. N. Smirnov. 2002. Effect of water on silanization of silica by trimethoxysilanes. *Langmuir* 18 (8):3181–3184.

Kuhner, F., L. T. Costa, P. M. Bisch, S. Thalhammer, W. M. Heckl, and H. E. Gaub. 2004. LexA-DNA bond strength by single molecule force spectroscopy. *Biophysical Journal* 87 (4):2683–2690.

Larsson, P. O. 1984. High-performance liquid affinity-chromatography. *Methods in Enzymology* 104:212–223.

Lee, G. U., D. A. Kidwell, and R. J. Colton. 1994. Sensing discrete streptavidin biotin interactions with atomic-force microscopy. *Langmuir* 10 (2):354–357.

Lomant, A. J., and G. Fairbanks. 1976. Chemical probes of extended biological structures - synthesis and properties of cleavable protein cross-linking reagent dithiobis(succinimidyl-s-35 propionate). *Journal of Molecular Biology* 104 (1):243–261.

Love, J. C., L. A. Estroff, J. K. Kriebel, R. G. Nuzzo, and G. M. Whitesides. 2005. Self-assembled monolayers of thiolates on metals as a form of nanotechnology. *Chemical Reviews* 105 (4):1103–1169.

Ludwig, M., M. Rief, L. Schmidt, H. Li, F. Oesterhelt, M. Gautel, and H. E. Gaub. 1999. AFM, a tool for single-molecule experiments. *Applied Physics a-Materials Science & Processing* 68 (2):173–176.

Ma, H. W., D. J. Li, X. Sheng, B. Zhao, and A. Chilkoti. 2006. Protein-resistant polymer coatings on silicon oxide by surface-initiated atom transfer radical polymerization. *Langmuir* 22 (8):3751–3756.

Mallik, R., J. Tao, and D. S. Hage. 2004. High-performance affinity monolith chromatography: Development and evaluation of human serum albumin columns. *Analytical Chemistry* 76 (23):7013–7022.

Mallik, R., C. L. Wa, and D. S. Hage. 2007. Development of sulfhydryl-reactive silica for protein immobilization in high-performance affinity chromatography. *Analytical Chemistry* 79 (4):1411–1424.

Mayyas, E., M. Bernardo, L. Runyan, A. Sohail, V. Subba-Rao, M. Pantea, R. Fridman, and P. M. Hoffmann. 2010. Dissociation kinetics of an enzyme-inhibitor system using single-molecule force measurements. *Biomacromolecules* 11 (12):3352–3358.

Moy, V. T., E. L. Florin, and H. E. Gaub. 1994. Intermolecular forces and energies between ligands and receptors. *Science* 266 (5183):257–259.

Neuert, G., C. Albrecht, E. Pamir, and H. E. Gaub. 2006. Dynamic force spectroscopy of the digoxigenin-antibody complex. *Febs Letters* 580 (2):505–509.

Neuman, K. C., and A. Nagy. 2008. Single-molecule force spectroscopy: Optical tweezers, magnetic tweezers and atomic force microscopy. *Nature Methods* 5 (6):491–505.

Nguyen, T. H., L. J. Steinbock, H. J. Butt, M. Helm, and R. Berger. 2011. Measuring single small molecule binding via rupture forces of a split aptamer. *Journal of the American Chemical Society* 133 (7):2025–2027.

Odorico, M., J. M. Teulon, T. Bessou, C. Vidaud, L. Bellanger, S. W. W. Chen, E. Quemeneur, P. Parot, and J. L. Pellequer. 2007. Energy landscape of chelated uranyl: Antibody interactions by dynamic force spectroscopy. *Biophysical Journal* 93 (2):645–654.

Oesterhelt, F., M. Rief, and H. E. Gaub. 1999. Single molecule force spectroscopy by AFM indicates helical structure of poly(ethylene-glycol) in water. *New Journal of Physics* 1:6.1.

Ossenkamp, G. C., T. Kemmitt, and J. H. Johnston. 2002. Toward functionalized surfaces through surface esterification of silica. *Langmuir* 18 (Copyright (C) 2011 American Chemical Society (ACS). All Rights Reserved.):5749–5754.

Pagac, E. S., D. C. Prieve, Y. Solomentsev, and R. D. Tilton. 1997. A comparison of polystyrene-poly(ethylene oxide) diblock copolymer and poly(ethylene oxide) homopolymer adsorption from aqueous solutions. *Langmuir* 13 (11):2993–3001.

Patel, A. B., S. Allen, M. C. Davies, C. J. Roberts, S. J. B. Tendler, and P. M. Williams. 2004. Influence of architecture on the kinetic stability of molecular assemblies. *Journal of the American Chemical Society* 126 (5):1318–1319.

Pincet, F., and J. Husson. 2005. The solution to the streptavidin-biotin paradox: The influence of history on the strength of single molecular bonds. *Biophysical Journal* 89 (6):4374–4381.

Radmacher, M., M. Fritz, H. G. Hansma, and P. K. Hansma. 1994. Direct observation of enzyme-activity with the atomic-force microscope. *Science* 265 (5178):1577–1579.

Ratto, T. V., K. C. Langry, R. E. Rudd, R. L. Balhorn, M. J. Allen, and M. W. McElfresh. 2004. Force spectroscopy of the double-tethered concanavalin-A mannose bond. *Biophysical Journal* 86 (4):2430–2437.

Ray, C., and B. B. Akhremitchev. 2005. Conformational heterogeneity of surface-grafted amyloidogenic fragments of alpha-synuclein dimers detected by atomic force microscopy. *Journal of the American Chemical Society* 127 (42):14739–14744.

Ray, C., J. R. Brown, and B. B. Akhremitchev. 2006. Single-molecule force spectroscopy measurements of "hydrophobic bond" between tethered hexadecane molecules. *Journal of Physical Chemistry B* 110 (35):17578–17583.

Ros, R., F. Schwesinger, D. Anselmetti, M. Kubon, R. Schafer, A. Pluckthun, and L. Tiefenauer. 1998. Antigen binding forces of individually addressed single-chain Fv antibody molecules. *Proceedings of the National Academy of Sciences of the United States of America* 95 (13):7402–7405.

Schmitt, L., M. Ludwig, H. E. Gaub, and R. Tampe. 2000. A metal-chelating microscopy tip as a new toolbox for single-molecule experiments by atomic force microscopy. *Biophysical Journal* 78 (6):3275–3285.

Schwesinger, F., R. Ros, T. Strunz, D. Anselmetti, H. J. Guntherodt, A. Honegger, L. Jermutus, L. Tiefenauer, and A. Pluckthun. 2000. Unbinding forces of single antibody-antigen complexes correlate with their thermal dissociation rates. *Proceedings of the National Academy of Sciences of the United States of America* 97 (18):9972–9977.

Seifert, U. 2000. Rupture of multiple parallel molecular bonds under dynamic loading. *Physical Review Letters* 84 (12):2750–2753.

Sekiguchi, H., H. Arakawa, H. Taguchi, T. Ito, R. Kokawa, and A. Ikai. 2003. Specific interaction between GroEL and denatured protein measured by compression-free force spectroscopy. *Biophysical Journal* 85 (1):484–490.

Shar, J. A., T. M. Obey, and T. Cosgrove. 1999. Adsorption studies of polyethers - Part II: Adsorption onto hydrophilic surfaces. *Colloids and Surfaces a-Physicochemical and Engineering Aspects* 150 (1–3):15–23.

Shlyakhtenko, L. S., A. A. Gall, A. Filonov, Z. Cerovac, A. Lushnikov, and Y. L. Lyubchenko. 2003. Silatrane-based surface chemistry for immobilization of DNA, protein-DNA complexes and other biological materials. *Ultramicroscopy* 97 (1–4):279–287.

Staros, J. V., R. W. Wright, and D. M. Swingle. 1986. Enhancement by N-hydroxysulfosuccinimide of water-soluble carbodiimide-mediated coupling reactions. *Analytical Biochemistry* 156 (1):220–222.

Sulchek, T. A., R. W. Friddle, K. Langry, E. Y. Lau, H. Albrecht, T. V. Ratto, S. J. DeNardo, M. E. Colvin, and A. Noy. 2005. Dynamic force spectroscopy of parallel individual Mucin1-antibody bonds. *Proceedings of the National Academy of Sciences of the United States of America* 102 (46):16638–16643.

Tees, D. F. J., J. T. Woodward, and D. A. Hammer. 2001. Reliability theory for receptor-ligand bond dissociation. *Journal of Chemical Physics* 114 (17):7483–7496.

Tees, D. F. J., R. E. Waugh, and D. A. Hammer. 2001. A microcantilever device to assess the effect of force on the lifetime of selectin-carbohydrate bonds. *Biophysical Journal* 80 (2):668–682.

Todd, B. A., V. A. Parsegian, A. Shirahata, T. J. Thomas, and D. C. Rau. 2008. Attractive forces between cation condensed DNA double helices. *Biophysical Journal* 94 (12):4775–4782.

Ulman, A. 1991. *An Introduction to Ultrathin Organic Films*. New York: Academic Press.

Vandenberg, E. T., L. Bertilsson, B. Liedberg, K. Uvdal, R. Erlandsson, H. Elwing, and I. Lundstrom. 1991. Structure of 3-aminopropyl triethoxy silane on silicon-oxide. *Journal of Colloid and Interface Science* 147 (1):103–118.

Weber, P. C., D. H. Ohlendorf, J. J. Wendoloski, and F. R. Salemme. 1989. Structural origins of high-affinity biotin binding to streptavidin. *Science* 243 (4887):85–88.

Willemsen, O. H., M. M. E. Snel, K. O. van der Werf, B. G. de Grooth, J. Greve, P. Hinter-dorfer, H. J. Gruber, H. Schindler, Y. van Kooyk, and C. G. Figdor. 1998. Simultaneous height and adhesion imaging of antibody-antigen interactions by atomic force microscopy. *Biophysical Journal* 75 (5):2220–2228.

Williams, P. M. 2003. Analytical descriptions of dynamic force spectroscopy: Behaviour of multiple connections. *Analytica Chimica Acta* 479 (1):107–115.

Wojcikiewicz, E. P., M. H. Abdulreda, X. H. Zhang, and V. T. Moy. 2006. Force spectroscopy of LFA-1 and its ligands, ICAM-1 and ICAM-2. *Biomacromolecules* 7 (11):3188–3195.

Wong, L. S., F. Khan, and J. Micklefield. 2009. Selective covalent protein immobilization: Strategies and applications. *Chemical Reviews* 109 (9):4025–4053.

Wong, S. S., E. Joselevich, A. T. Woolley, C. L. Cheung, and C. M. Lieber. 1998. Covalently functionalized nanotubes as nanometre-sized probes in chemistry and biology. *Nature* 394 (6688):52–55.

Yip, C. M., C. C. Yip, and M. D. Ward. 1998. Direct force measurements of insulin monomer-monomer interactions. *Biochemistry* 37 (16):5439–5449.

Zhang, X. H., and V. T. Moy. 2003. Cooperative adhesion of ligand-receptor bonds. *Biophysical Chemistry* 104 (1):271–278.

Zhang, X. H., S. E. Craig, H. Kirby, M. J. Humphries, and V. T. Moy. 2004. Molecular basis for the dynamic strength of the integrin alpha(4)beta(1)/VCAM-1 interaction. *Biophysical Journal* 87 (5):3470–3478.

5 Biomolecular Recognition: Analysis of DFS Data

*Michael Odorico, Jean-Marie Teulon,
Yannick Delcuze, Shu-wen W. Chen,
Pierre Parot, and Jean-Luc Pellequer*

CONTENTS

5.1 INTRODUCTION TO DYNAMIC FORCE SPECTROSCOPY

One of the first meanings of *dynamic force spectroscopy* (DFS) refers to analyzing the dynamic interaction of the scanning force microscope tip with the sample surface. To our knowledge, this expression first appeared in the literature from the group of Harald Fuchs (Anczykowski et al. 1996) followed by a dissertation, *Kraftspectro-scopie an einzelnen Molekülen* of Matthias Rief, in Munich on 1997. Afterward, the term was reused by the group of Evan Evans (Evans & Ritchie, 1997; Merkel et al. 1999) *to probe the complex relation between force-lifetime and chemistry in single molecular bonds.* The first review on the topic was published in 2000 (Janshoff et al. 2000) and the acronym DFS appeared in 2001 (Evans, 2001). Now, more than 100 publications directly refer to DFS in their title.

In DFS, the word *spectroscopy*, despite its widely use, is a misleading term because, unlike other true spectroscopy, there is no matter-radiation interaction involved in DFS experiments. DFS mainly consists in measuring the behavior of single complexes under stretching mechanical force and more especially the rupture of bonds between two molecules in solution: a ligand and a receptor attached either to the substrate or to the tip of the cantilever. Individual bond rupture in a complex consists in a stochastic event and consequently a large collection of rupture events should be recorded. The measure of bond ruptures are performed using force-displacement (FD) experiments that consist in bringing the ligand-coated tip at the surface of the receptor-coated support, then retracting the tip and observing the deflection of the cantilever. Such deflections obtained at a given speed of retract is called the bond loading rate or simply *the loading rate*. Basically, rupture forces are obtained by calculating the product of cantilever deflection and cantilever spring constant. A *force spectrum* or *the force distribution* is obtained by plotting the frequency of rupture forces. In DFS, several force distributions are obtained depending on the applied loading rate which should be as large as that permitted by atomic force microscopy (AFM). These distributions are usually fitted by Gaussian or Poisson functions for determining the most probable rupture forces ($F*$).

The diagram of the most probable rupture forces ($F*$) versus the natural logarithm of the bond loading rate is usually called the *standard plot* or *Bell-Evans plot*. In the simplest conceivable case, where a single bond between ligand and receptor is measured, the Bell-Evans plot exhibits a single linear fit showing the increase of the most probable rupture forces as function of a logarithm of the loading rate. The slope of the fit is equal to $k_B T / x_\beta$, where x_β is the distance from the energy minimum to the transition state, k_B is the Boltzmann constant, and T is the temperature.[*]

When the measured interactions involve multiple bonds, the Bell-Evans plot could exhibit several linear fits corresponding to multiple parallel bonds. Multiple bonds may originate from multivalent systems (e.g., antibodies) or from multiple-ligand–receptor interactions depending on the density of ligands and receptors.

Which biological systems can be studied with DFS? Since 1997, DFS has been used for studying properties of many biological questions including: ligand–receptor interactions, DNA base pair interactions, polymer conformations, supramolecular interactions, and hydrophobic interactions, as well as protein folding pathways, bimolecular substitution reactions (Friddle et al. 2007; Neuman & Nagy, 2008; Bizzarri & Cannistraro, 2009; Puchner & Gaub, 2009).

In particular, DFS has been applied to study protein–protein interactions on proteins such as leukocyte rolling with P-selectin and P-selectin glycoprotein ligand-1 (Fritz et al. 1998); the leukocyte function-associated antigen-1 (LFA-1) with its cognate ligand the intercellular adhesion molecule-1 (ICAM-1) (Zhang et al. 2002); penta-repeats of the Mucin-1 20-residue epitope with an antibody single-chain variable fragment (scFv) (Sulchek et al. 2005); redox partners azurin and cytochrome c551 (Bonanni et al. 2005); platelet membrane receptor glycoprotein Ib-IX (GP Ib-IX) with the A1 domain of von Willebrand factor (Arya et al. 2005); antilysozyme Fv fragment and lysozyme (Berquand et al. 2005); ternary complex involved in neurotransmitters release by soluble N ethylmaleimide-sensitive fusion

---

[*]$k_B T \approx 4.1$ pN. nm at 300K.

protein attachment protein receptor (SNARE) complex, which is made of syntaxin, synaptosome-associated protein of 25 kDa (SNAP25), and synaptobrevin 2 (Sb2) (Liu et al. 2006) or with SNAP23 (Montana et al. 2009); iron transporter protein transferring (Tf) and its cell surface receptor (TfR) (Yersin et al. 2008); human tumor suppressor p53 with the bacterial redox protein azurin (Taranta et al. 2008); actin and heparin-binding hemagglutinin (HBHA) from *Mycobacterium tuberculosis* (Verbelen et al. 2008); *Staphylococcus aureus* surface protein IsdA with human proteins present on cornified envelope of desquamated epithelial cells: involucin, loricin, and cytokeratin K10 (Clarke et al. 2009); signal-transducing proteins with the Phox homology domain of mammalian phospholipase D1 (PLD1) and the Src homology domain (SH3) of phospholipase c-γ1 (PLC-γ1), and Munc-18-1 (Kim, 2009). DFS has also been used to study protein–protein homophilic interactions such as the vascular endothelial (VE)-cadherin (Baumgartner et al. 2000a); quadruple H-bonded ureido-4[1H]-pyrimidinone (UPy) dimmers (Zou et al. 2005); HBHA-HBHA (Verbelen et al. 2007); extracellular fragment of nectin-1 (nef-1) with the wild-type L-fibroblasts that express nectin-1 (Vedula et al. 2007); tight-junction proteins Claudin-1 (Lim et al. 2008b) and Claudin-2 (Lim et al. 2008a); negatively-charged aggregan molecules extracted from bovine cartilage extracellular matrix (Harder et al. 2010). DFS was used to study interaction of proteins with transmembrane (TM) proteins such as the multivalent *Psythyrella velutina* lectin (PVL) and the most abundant TM protein in human erythrocyte glycophorin A (Yan et al. 2009). DFS was applied on the study of protein–peptide interactions such as the GCN4(7P14P) peptide and a scFv at different maturation level (Morfill et al. 2007); PDZ protein Tax-interacting protein-1 (TIP-1) and its recognition peptide derived from β–catenin (Maki et al. 2007).

DFS has also been applied on various types of molecules other than peptides or proteins. In particular, DFS has been used to study interactions between DNA and proteins such as the regulatory DNA-binding protein ExpG from *Sinorhizobium meliloti 2011* and its target *exp* gene sequences (Bartels et al. 2003); BsoBI and XhoI restrictions enzymes with their specific DNA sites using a clever unzipping approach (Koch & Wang, 2003); various length of aptamers (22-nt, 29-nt, and 37-nt) with IgE antibodies (Yu et al. 2007); DNA-binding domain of the PhoB transcription factor from *Escherichia coli* and DNA-binding sequences (Wollschlager et al. 2009). Other types of ligand studied by DFS include small molecules such as fluorescein and anti-fluorescein scFv (Schwesinger. et al. 2000); digoxigenin and antidigoxigenin antibodies (Neuert et al. 2006); uranyl-dicarboxy-phenanthroline chelate with monoclonal antibodies (Odorico et al. 2007a; Teulon et al. 2007; Teulon et al. 2008). DFS was used to study protein-carbohydrate interactions such as AlgE4 epimerase and mannuronan (Sletmoen et al. 2004; Sletmoen et al. 2005); extracellular matrix network between fibronectin and heparin (Mitchell et al. 2007); *Helicobacter pylori* adhesion–receptor complex BabA and fucosylated ABO/Lewis b blood group antigen (Bjornham et al. 2009). DFS showed applications for studying the interactions between SNARE proteins and egg L-α-phosphatidylcholine bilayers (Abdulreda et al. 2008). Application of DFS also covers polymers with the study of *Ralstonia pickettii T1* poly[(R)-3-hydroxybutyrate] (P(3HB)) depolymerase with P(3HB) and poly(L-lactic acid) (PLLA) (Matsumoto et al. 2008).

Regarding entire cells or viruses, DFS obtained promising results such as the study of LFA-1 expressed on Jurka T cells and intercellular adhesion molecule-1 and -2 (ICAM-1 and ICAM-2) (Wojcikiewicz et al. 2006); human rhinovirus 2 (HRV2) with very low-density lipoprotein receptor (VLDLR)1-8 (Rankl et al. 2008); transferring with HeLa cells (Yersin et al. 2008); heregulin β1 (HRG) with cells expressing growth factor receptors (HER2 and HER3) (Shi et al. 2009); junctional adhesion molecules (JAM) with murin L929 cells expressing JAM-A (Vedula et al. 2008). DFS has also been attempted on cell–cell interactions such as the study of E-cadherin expressed on the surface of live human parental breast cancer cells and cells reexpresseing a-catenin (Bajpai et al. 2009).

Finally, DFS has been useful not only in the study of unfolding and extraction of consensus sequence for TM protein segments (WALP), with a N-terminal Cys residue, from gel-phase DPPC and DSPC bilayers (Contera et al. 2005); unfolding membrane proteins such as the Na+/H+ antiporter NhaA (Kedrov et al. 2008), bovine rhodopsin and bacteriorhodopsin (Sapra et al. 2008); myomesin the most prominent structural component of the sarcomeric M-band (Bertoncini et al. 2005); but also in the study of nonbonded rupture such as in the interaction between a silicon nitride tip and 1-nonanethiol self-assembled monolayer on gold (Ptak et al. 2009).

What is the strength of AFM over other techniques for measuring kinetics parameters? Undoubtedly, the advantage of AFM is the possibility to measure biological system with apparent ultra-high affinities in solution that would take weeks or years with classical techniques such as surface plasmon resonance. The avidin–biotin complex, which is widely used in biotechnology, epitomizes the need for alternative techniques due to its femtomolar affinity (Green, 1963). Such affinity is due to the tetrameric quaternary structure of avidin, that is, with the presence of four binding sites toward biotin. The importance of multiple bonds will be emphasized later in this chapter. The first study of avidin–biotin complex using AFM was made in 1994 (Florin et al. 1994) and about 20 reports followed on this system or its bacterial equivalent of streptavidin–biotin complex (Teulon et al. 2011). Avidin–biotin systems have been studied using at least 10 different experimental setups that explained the current lack of consensus in the results (Walton et al. 2008) although a reconciling hypothesis has been recently proposed (Teulon et al. 2011). Avidin–biotin system has also been chosen for performing molecular simulations (Zhou et al. 2006; Walton et al. 2008), for measuring the energy landscape roughness (Rico & Moy, 2007), or to test concepts related to multiple-bond systems (Guo et al. 2008). It should be emphasized that DFS can be performed with other techniques different from AFM such as the biomembrane force probe with micropipettes (Merkel et al. 1999), optical tweezers (Arya et al. 2005; Andersson et al. 2006; Bjornham et al. 2009), or the laminar flux chamber (Robert et al. 2007).

Why write a chapter on DFS data analysis? The first thing that comes to mind is probably the lack of details in publications. Due to the restriction in page length, it is not possible to display even a small fraction of experimental data. Beside, there is no repository system for DFS data, yet. Data analysis covers all the technical details involved between the physical measurement and the interpretation of the data. The major requirement in DFS is the availability of an automated FD curve analysis. This is of vital importance as the number of recording may reach 10,000 or more depending on the range of loading rates or on the ratio of specific and nonspecific events. Beside, an automatization will reduce the operator bias compared with the computational bias that will be homogeneously applied all over the analysis. The computational method should aim at identifying rupture events that are necessary for building the cumulative probability distribution (CPD) of rupture forces. As always in digital processing, while the human eye often perceives immediately the essential information, artifacts or erratic points, coding this perception as well as its selection process into a computer program is always a great challenge. Two strategies have been employed: One proposes a quick and easy identification of most obvious rupture events hoping that errors will be damped by the sheer number of data, and second aims to analyze in detail the geometry of rupture events. None of these strategies is superior to the other; but the experience shows that FD curves can be utterly complicated and simple identification procedures often fail. Finally, data analysis includes the fitting of the CPD using an analytical function (usually a Gaussian or a beta-law fit). The fits in turn will provide the most probable rupture events, the required values for making Bell-Evans plots.

As will be seen in Section 5.2, working at the single molecule level usually require a large collection of measurements. Measurements may characterize single or multiple bond ruptures (Section 5.3) as well as specific and nonspecific interactions (Section 5.4). A key parameter in DFS is the force-induced bond rupture between the two molecules characterized by the loading rate (Section 5.5). In Sections 5.6 through 5.7, we will illustrate in detail a methodology to characterize energy barriers along the rupture pathways also known as energy landscape. We will emphasize on the rupture event detection (Section 5.6) as well as the calculation of the CPD using histograms and Gaussian fits necessary to obtain the most probable rupture forces (Section 5.7). In Sections 5.8 and 5.9, we will provide classical interpretations of Bell-Evans plots. Finally, we will give a brief description of published methodology for DFS data analysis as well as available commercial products (Section 5.10).

5.2 ENERGY LANDSCAPE AND SINGLE MOLECULES

Paradoxically, treatment of DFS data must be taken into consideration before their acquisition for several reasons. First to meet certain criteria purely dictated by statistical thermodynamic considerations (Evans & Williams, 2002; Tinoco & Bustamante, 2002) and then to meet problems posed by the assumption of ergodicity. As previously developed in Chapters 1 (Biorecognition Processes) and 3 (Theoretical Models in Force Spectroscopy), the theory on molecular interaction derives mainly from work on cell adhesion starting with the seminal paper by George Bell (Bell, 1978). This approach has permitted the understanding of the importance of different experimental parameters associated with the statistical treatment of the data, such as the loading rate, the contact time, the contact energy between ligand and receptor as well as the energy transferred to the system (Evans, 2001). Recent results show that the application of the ergodic hypothesis is not enough to correlate results coming from traditional bulk experiments. This is especially true with those obtained on single molecules when the loading rate dependence of measured rupture forces was not taken into account. It is believed that the understanding of dynamic strength of molecular bonds starts with the control of the loading rate, and thus most historical papers published before 1999 did not adequately portray an accurate kinetic characterization of bond rupture.[*]

We are aware that despite significant progress in ergodic and chaos theories, using the ergodic hypothesis to justify the use of the microcanonical ensemble in statistical mechanics remains controversial till date. However, we continue using[†] the expression *ergodic hypothesis* that could be understood by saying that for microscopic quantities, average, and fluctuations over time are the same as average and fluctuations over space or in other words after a sufficiently long time a system explores all of its microscopic states. This very important condition explains the capacity to reveal energy landscapes for single molecules that are different (in their details) than those obtained from bulk experiments (Figure 5.1). The counterpart is that exploring single molecule properties requires recording a huge number of experimental rupture events.

The energy landscape of a bond rupture explored by DFS characterizes the force-driven pathway along the pulling direction until bond rupture. A classical representation of energy landscape is made in a one-dimension plot representing the energy of the system versus the reaction coordinates (Kramers, 1940). The shape of the energy landscape is thus constituted by the height of energy barriers (mountains in Figure 5.1) and the energy barrier width (flat distance) between the valley and the summit of the mountain. In this chapter, the height of the energy barrier is characterized by a k_{off} value, whereas the energy barrier width is called x_β.

5.3 SINGLE AND MULTIPLE BONDS

Most of the early works in force spectroscopy attempted to simplify their experimental models for considering interactions limited to single bonds (Florin et al. 1994; Moy et al. 1994) using linkers in their biochemical setup (Hinterdorfer et al. 1996) or by filtering and selecting experimental results (Dammer et al. 1996; Baumgartner et al. 2000a). Actually from a biological point of view, it is very difficult to design such simple setups in protein–protein, protein–membrane, protein–DNA, or other biological interactions (Mammen et al. 1998). Besides, several important biological reactions involve multiple interactions as evidenced by antibody molecules

---

[*]To our knowledge, the first mention that the rupture force vary with rupture speed was indicated in 1996 in a modeling paper by Grubmüller et al. (Grubmüller et al. 1996) citing a relevant experimental observation done by Gil Lee et al. in 1993 (Lee et al. 1994).

[†]As we do with physical principles, they cannot be demonstrated but experimentally verified anytime.

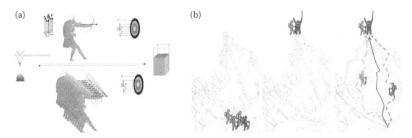

FIGURE 5.1 **(See color insert.)** Energy landscape and single molecules. (a) To obtain results comparable to those provided using batch measurements (symbolized by the spectroscopy cuvette on the right), experiments on single molecules request a very large sampling represented by the bunch of arrows (shot by a single archer), which stochastically hit the target (top). Ergodic hypothesis is illustrated showing that the score obtained by the single archer shooting many arrows is exactly the same as the one obtained by a troop of archers shooting each only one arrow. (b) In biophysical or biochemical experiments, most of the results are often given taking only account of half time measurements. Starting (on the far left panel) from a population of climbers at the bottom of the mountain (the background represents K2, which was attempted by George Bell in 1953 with a well-knit team; the fascinating story can be found in *K2, The Savage Mountain*) the experiment will be considered as ending at time $t_{1/2}$ when half of the population have changed their states (50% of the climbers reached the top of the mountain, middle panel). This approach obviously masks the states of all the other climbers (green and red) still in progress and definitely miss all the different ways individually used.

that are usually divalent, while others such as avidin are tetravalent. Multiple interactions tend to be stronger than monovalent ones, a critical biological parameter in regulation of physiological processes (Badjic et al. 2005).

Very few experiments have tackled the study of multiple-bond ruptures (Hukkanen et al. 2005; Levy & Maaloum, 2005; Sulchek et al. 2005; Odorico et al. 2007a; Teulon et al. 2007; Tsukasaki et al. 2007; Erdmann et al. 2008; Guo et al. 2008; Teulon et al. 2008). A likely reason is the difficulty to interpret the complex data distribution that was collected and analyzed from the FD curves obtained at different loading rates. In particular, the evaluation of the effective loading rate (r_e) remains a critical step (Section 5.5).

Till now, the term "bond" was used to depict some kind of specific link between two entities. These entities could be molecules or cells. It is of upmost importance to make it clear what this term "bond" means. In DFS, one "bond" indicates the sum of all noncovalent interactions between two entities; that is, when a "bond" breaks, all the noncovalent interactions between the two partners are broken. In chemistry, noncovalent interactions, such as salt bridges or hydrogen bonds, are specific chemical contacts between electron acceptor and electron donor atoms. The length of these interactions is usually short (<4 Å) compared with the size of proteins or cells (nanometer or micrometer). Most importantly, the distance required to break such specific noncovalent contacts is even smaller, usually below 1Å (Figure 5.2).

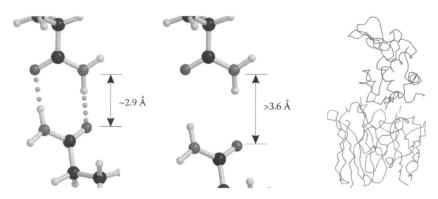

FIGURE 5.2   (See color insert.) Representation of hydrogen bonds in proteins. On the left panel, a glutamine residue is making a double hydrogen bond with another glutamine residue. Ideal bond length for a hydrogen bond is about 2.9 Å. Covalent bonds are represented by yellow tubes, whereas noncovalent bonds are represented by green dots. Carbon atoms are in black, oxygen atoms in red, nitrogen atoms in blue, and hydrogen atoms in white. The middle panel shows the same two residues as in the left panel except that their distance has increased by about 0.7 Å. In such a conformation, hydrogen bonds are broken and the two residues are no more in contact. All the potential 14 hydrogen bonds (green dots) found in the complex between an antigen (magenta tubes) and an antibody (orange tubes) are represented in the right panel. Many other interactions occur in protein–protein complexes, such as salt bridges, van der Waals contact, π-cation interactions (Chen et al. 2009) for a recent description of interactions in antibody-antigen complexes and (Lin et al. 2011) for π-cation interactions). These interactions are not represented here for clarity. The figure was drawn with Molscript (Kraulis, 1991) and rendered using Raster3D (Merritt & Bacon, 1997).

5.4   SPECIFIC AND NONSPECIFIC INTERACTIONS

As previously mentioned, this chapter only concerns DFS using AFM. With this technique, the chemical grafting of molecules may alter the property of the cantilever, thus the calibration of a newly mounted tip should be systematically performed before DFS experiments. It is also assumed that users are equipped with standard modern AFM instruments, allowing automatic spring constant measurement of the cantilever such as the thermal tune module of Nanoscope software or any equivalent software. Otherwise, reader should refer to information about thermal and other calibrating methods (Odorico et al. 2007a; te Riet et al. 2011).

FD curves are the primary output of DFS measurements, also known as force-distance or approach-retract curves. The term displacement describes the motion in Z direction of the piezoelectric tube on which the tip or the surface is attached to. These curves are obtained by bringing the ligand-coated tip in contact with the receptor-coated surface. Two possibilities follow: First, there is no interaction between molecules attached on the tip surface and molecules attached on the substrate, thus, upon retraction of the tip, no rupture events is observed (Figure 5.3a); Second, there are interactions, thus upon withdrawing of the tip, a rupture in the

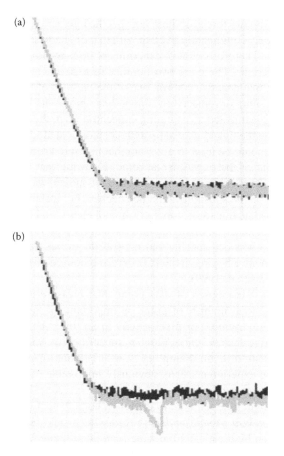

FIGURE 5.3 Force-displacement (FD) curve represented by the approaching step (black) and the retraction step (light gray). (a) FD curve without rupture event, (b) FD curve with a single interaction event.

experimental setup is observed (Figure 5.3b). Repeating FD curves and observing the same result may indicate that no modification appears on the receptor-coated surface or on the tip.

Recording FD curves and visually observing rupture events can be very operator dependent. As previously mentioned, it is easy to favor the presence of rupture events simply by increasing the contact duration between the tip and the substrate. Unfortunately, we know that it drastically increases the percentage of nonspecific events. It is important to define the term *nonspecific* as it is often a misleading expression. Nonspecific interactions in DFS concerns all kind of physical or chemical interactions that we do not want to measure. For instance, the gold-coated tip with the gold-coated glass, or the tip-coated ligand with molecules used to saturate the gold-coated substrate, or the polyethylene glycol linker atoms with atoms from the substrate-coated receptor, etc. The rupture forces of these nonspecific interactions are often weak, but their distribution may significantly overlap with that of single specific bond ruptures at low loading rate. Therefore, control experiments are mandatory to avoid recording nonspecific interactions. There are two families of experimental controls: negative and positive ones. In a negative control, one of the partners of the complex is either removed from the setup or replaced by a nonrelated molecule. In a positive control, one of the partners is injected at saturating concentration in the solution to block the interaction between the ligand-coated tip and receptor-coated substrate. Ideally, both types of controls should be used to ascertain that measured rupture events show the expected specificity of the experimental setup. Knowing that it is extremely diffi-cult to control tip functionalization in a simple or economic way, only the quality of the film deposited on the substrate could be checked. For instance, it is important to determine the substrate thickness, topology, homogeneity, and the apparent concen-tration of biological material. Changing the concentration of receptor could allow the control of the ratio between specific and nonspecific events. In special cases, ther-mal fluctuations create a hoping in the FD curve and nonspecific interactions may be detected (Willemsen et al. 1999).

Each FD curve is composed of data points that measure the cantilever deflec-tion at time t. The total number of points is important for the data processing of the FD curve. A low number of pixel results in an undersampling leading to the likely absence of significant events. A high number of points will make the measure-ment at high velocity difficult. Sampling is an important aspect of data collection in single molecule experiments. It is obviously impossible to experimentally sur-vey the entire population of single molecules for two reasons: First, it will take for-ever (for instance, the time required to explore a homogeneous surface of a sample [$\sim$25 mm^2] at *standard protein molecule resolution* would be about 10 millenar-ies!*); second, even by drastically reducing scanning duration, the sample stability in liquid environment will likely change over time. Intuitively, an optimal sampling is necessary to reach a requested experimental accuracy and should take into account two aspects: possibility (event happening) and specificity of the interaction (ensure that measured forces correspond to the complex we wish to target).

We have seen that two kinds of FD curves can be observed (Figure 5.3). Those with interactions (N_i) and those without any interactions (N_{ni}) for a total number of measures (N_t), where $N_t = N_i + N_{ni}$. This ratio N_i/N_{ni} is experimenter dependent and can be modulated (as mentioned above) by changing the concentrations of biological material or by adjusting experimental parameters such as the approach-retract veloc-ity (approach or retract speed of the cantilever) or the duration of contact between the tip and the substrate. One FD curve selection criterion is the experimentally accept-able ratio N_i/N_{ni}. Poisson statistics can provide an important estimate of the likeli-hood of multiple interactions amongst a series of tests of bond strength as explained by Evans and Williams (Evans & Williams, 2002). However, when multiple bonds are the main targets of the experimenter, it is only necessary to consider that each

---

*Even reduced at a field like the surface of cell (100 μm^2), it will take years, at the same resolution!

rupture event is a particular sample from the complex distributions of hundred forces needed to describe the process at each loading rate. Besides, if one calculates the effective bond loading rate for each rupture event, all the data collected at various retraction speeds can be combined into a single analysis.

5.5 LOADING RATE

DFS experiments on single molecules is meaningless when the loading rate dependence of measured forces is not taken into account (Evans & Ritchie, 1997; Merkel et al. 1999; Evans & Ludwig, 2000; Gergely et al. 2000; Fantner et al. 2006; Prakasam et al. 2006). The nominal loading rate is the product of the spring constant of the cantilever (N/m) and the retract velocity (m/s) and it is often assumed that it remains constant from one measure to the other. However, in the presence of multiple bonds, it is more appropriate to use the effective loading rate, which is the loading rate measured before bond rupture (Friedsam et al. 2003) using the following relationship:

$$r_e = \frac{k_{samp} \cdot v}{1 + \frac{k_{samp}}{k_{cant}}}, \quad \text{where} \quad k_{samp} = \frac{k_{eq} \cdot k_{cant}}{k_{cant} - k_{eq}}$$

and k_{eq} is the slope of the retract curve before the unbinding event. The identification of rupture events (Odorico et al. 2007a), as computed in YieldFinder (Odorico et al. 2007b), requires the estimation of the stiffness constant (k_{samp}) of the study system based on the series-parallel spring model and the effective loading rates (Erdmann, 2005), which is the loading rate really experienced by the molecular bonds (Figure 5.4).

In a multiparallel systems $k_{samp} = nk_{bond}$, where k_{bond} remains unknown. The number of parallel bonds can be extracted from Bell-Evans or Williams' plots of all combined experimental rupture force events. In case of multiple parallel bonds, a distribution of loading rates can be obtained. It is then possible to obtain the most frequent loading rate as peaks in that distribution. For each most frequent loading rate, it is then possible to determine the distribution of rupture forces necessary to build the Bell-Evans plot.

To study the ligand–receptor rupture force dependence with the loading rate, the probe approach and retract velocity can be varied typically within the range of 100 nm/s to 1 μm/s (adjusted by changing the scan rate, the retract velocity, or the ramp size). For biological samples, useful spring constants (k_{cant}) of commercial cantilever are in the range of 6–200 pN/nm, leading to nominal loading rates between 600 pN s^{-1} and 200,000 pN s^{-1} (6.4–12.2 in log scale). With advanced instruments, it is also possible to adjust the contact time when recording FD curves. The effect of the piezo velocity (i.e., loading rate) is easily seen on FD curves by observing a shift toward higher rupture forces at high velocity. Usually recording nondistorted FD curves require reducing piezo speed and in all cases not to exceed 1–2 μm/s. Otherwise, shorter or stiffer cantilever should be used (Kim et al. 2010). While the

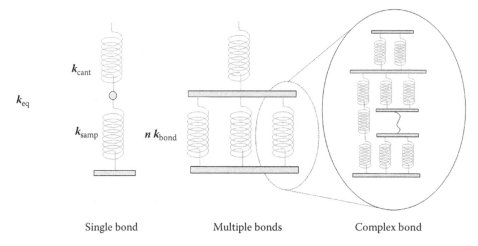

Single bond   Multiple bonds   Complex bond

FIGURE 5.4   Spring model for interactions between the functionalized AFM tip and the biological sample. In a series-parallel spring model, a complicate system of molecular interactions is considered as a sophisticated assembly of connected springs. Each "simple" single bond is modeled by a single spring. If an interaction involves multiple bonds, then this interaction can be represented by multiple parallel springs, and so on. Accordingly, the rupture force F_{rupt} is a function of the number of parallel bonds, n and $F_{rupt} = nF_1$ where F_1 is the rupture force of a single bond. Hypothetically, the modeled spring constant of the biological sample $k_{samp} = nk_{bond}$ where k_{bond} is the spring constant of a single bond corresponding to individual interaction.

effective loading rates of each event will be extracted from FD curves, it is not necessary in this specific approach to especially target nominal loading rates by combining well-defined k_{cant} with specific piezo speed values.

5.6   DETECTING RUPTURE EVENTS

FD curves contain a set of data representing distance on the x-axis and cantilever deflection on the y-axis. These discrete curves are too complicated to be straightforwardly analyzed without prior processing. Small cumulative errors in the analysis lead to erroneous quantification of the energy landscape parameters (Björnham & Schedin, 2009).

The first step in FD curve processing is to import an often proprietary-owned formatted file. Therefore, a conversion into a human-readable (ASCII) format is often required as file formats may change from one software version to another (as it is often seen). A standardization of such intermediate ASCII files is essential so that FD data sets could be exchanged more easily as suggested in the COST Action TD1002 (COST Action TD1002, 2011).

The second step deals with the calibration of FD curves. For accurate interpretation, it is necessary to recalibrate the Y-axis of FD curves such that the approach region (on the right of the curve, Figure 5.5) is set to 0. Furthermore, it is critical to recalibrate the entire retract curve so that the slope of the hard contact zone (the most left part of the curve, Figure 5.5) match the experimentally determined cantilever spring constant (k_{cant}). This step is important for the appropriate calculation of the effective loading rate (see Section 5.5).

The third step concerns the identification of rupture events. It varies from one method to another. However, we have found it very difficult to analyze rupture events directly from data points of the retract curve. The use of a derivative of the retract curve (Sugisaki and Nakagini, 1999) or a sliding window approach was not concluding in our hand to handle the large variety of observed retract curves at various loading rates. Instead, a modified version of YieldFinder (Odorico et al. 2007b) consists in modeling pulling events by a succession of small straight lines (Figure 5.6). There are two main advantages: first, rupture events are defined by a change in the sign of slope in two consecutive segments, and second, it makes it straightforward to measure the loading slope before each rupture event.

The last step in FD curve analysis is the selection (or validation) of rupture events and the building of an event database for further treatments (see below). To select putative rupture events, several authors used different criteria such as

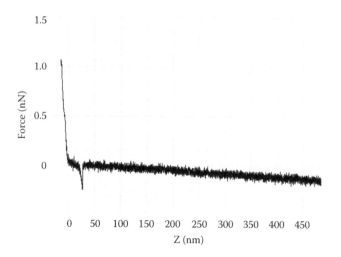

FIGURE 5.5 Raw experimental retract curve obtained from a force-displacement curve. The piezo displacement (Z) is shown in nanometer (nm). The Y-axis is a transformation of the cantilever deflection into forces in nN. The tip displacement (piezo) of the retract curve starts from the hard contact zone on the left, where the cantilever is bending due to the contact with the hard substrate. Then, when the tip retracts, the cantilever stops bending when it crosses the Y-axis at $F = 0$. Continuing the tip retraction shows a binding event represented by a negative peak followed by an abrupt rupture and the return to the baseline toward the right of the curve. To automatically detect rupture events, it is clear that a normalization of retract curves is necessary.

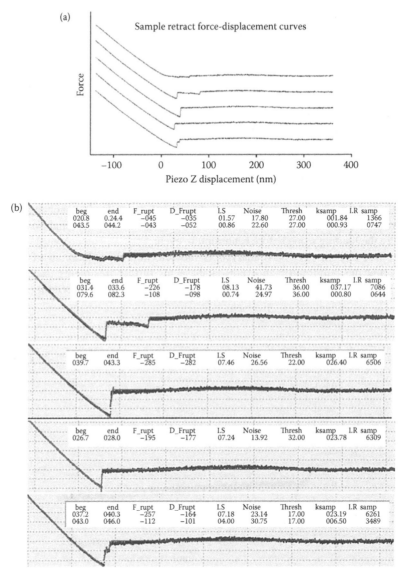

FIGURE 5.6 Identification of unbinding events. (a) Five experimental force-displacement (FD) curves showing only the retract section. Each of these retract curves has been automatically analyzed by YieldFinder to identify rupture events. (b) Modeling the five FD curves shown in (a) using YieldFinder. Raw data is shown in black, and fitted model is drawn in gray segments. An insert is shown on each curve and provides detected values by YieldFinder (variable names will be shown in italic). *Beg* and *end* values identify the beginning and the end of rupture events in nanometer. *F_rupt* and *D_Frupt* are the maximum and the event rupture forces, respectively. *LS* is the loading slope before the rupture (known as k_{eq} in Section 5.5). *Noise* is the observed noise on the approach curve. *Thresh* is the force threshold used to automatically select rupture events. k_{samp} and LR_{samp} are the k_{samp} value (see Section 5.5) and the effective loading rate value (r_e), respectively. In YieldFinder, the threshold value is controlled by the user, but it applied on every FD curves loaded.

the selection of the last event only (Yuan et al. 2000), some select events only if a significant stretching is observed before the rupture (Baumgartner et al. 2000b), and others select events when the rupture force is greater than a user-defined threshold (usually based on the noise observed in the approach curve) such as in YieldFinder.

5.7   CPD AND MOST PROBABLE RUPTURE FORCES

To make a Bell-Evans plot, we need to obtain the most probable rupture forces from the CPD of rupture forces. Usually, the CPD is constructed as the sum of N Gaussian distributions when multiple parallel-bond ruptures occur (Williams, 2008). However, the first obstacle is to build the probability distribution histogram. The shape of the histogram is sensitive to the number of selected bins. With wide bins, important information might be lost; for instance, a bimodal distribution could appear as a mono modal one. With narrow bins, apparent information may be caused by random fluctuations (noise). To determine whether the bin size is appropriately set, different bin size should be tested. Unfortunately, there are no simple rules for determining the optimal bin size of a histogram. Ideally, finer bins seem more appropriate than larger ones, but it follows that a large number of observations is required. This balance is a pure matter of experimenter judgment.

Some theoreticians have attempted to determine an optimal number of bins, but usually these methods make strong assumptions about the shape of the distribution. One should always experiment with various bin sizes before choosing one (or more) method that illustrate the salient features in the data. A distribution is defined by the number of events (n), the number of bins (k), and the size of each bin (h) according to the relation: k = (max x − min x)/h, where max x and min x are the largest and smallest values on the x-axis. One of the most encountered formula (in excel for instance) is the square root choice that defines the number of bins as the square root of the number of events n (Table 5.1). The Sturges formula (Table 5.1) implicitly defines the bin size on the basis of the range of the data but perform poorly when n < 30. The Scott's choice assumes that the bin size is related to σ, the sample standard deviation (Table 5.1). Finally, the Freedman–Diaconis choice defines the bin size in function of the interquartile range, which determines the length of the interval that divides data points such that 25% of the data is left on both sides (Table 5.1). Using a test case of 100 data points, we show results of each of these methods in Figure 5.7. The operator choice is also indicated in the table and shows that the closest method is that of Freedman–Diaconis.

The most probable rupture forces F^* correspond to the maximum of the rupture force distribution (indicated with * in Figure 5.7). The most common method to determine F^* is to fit the CPD using an analytical formula such as a Gaussian or a beta law curves (Odorico et al. 2007b). In most cases, these fits are performed in spreadsheet environments and are visually adjusted. In practice, the operator set the position of the maximum value of the fit and varies the standard deviation and the size of the population (number of events) until the fitting curve matches the distribution.

TABLE 5.1

Various Methods for Determining Bin Sizes of a Distribution and Their Effect on F^* Determination[a]

Choice	Sturges	Scott	Square Root	Freedman Diaconis	Operator
Reference	(Sturges, 1926)	(Scott, 1979)		(Freedman & Diaconis, 1981)	
Formula[b]	$k = [\log_2 n + 1]$	$h = \dfrac{3.5\sigma}{n^{1/3}}$	$k = \sqrt{n}$	$h = 2\dfrac{IQR(x)}{n^{1/3}}$	visual
h	16.9	15.1	12.9	9.6	5
k	7.6	8.6	10	13.4	16
F^{*1} (in pN)	21 ± 18	18 ± 16	17 ± 9	12 ± 9	16 ± 7
F^{*2} (in pN)	ND	ND	35 ± 8	33 ± 11	38 ± 9

[a] Data set: $n = 100$, ave $= 26$, stdev $= 20$, min $= 2$, max $= 131$, Q1 $= 12.75$, Q3 $= 35$

[b] k is the number of bins, h is the size of each bin, n is the total number of values, and $IQR(x)$ is the interquartile range defined as Q3-Q1.

The spreadsheet is used to calculate numerical points that can be used to draw the fit line over the distribution.

Since most relevant biological interactions involve multiple parallel bonds, we decide to combine all rupture events into a single analysis. Thus, all the rupture events collected at various retraction speeds were pooled together. However, it is thus necessary to build a distribution of effective loading rate (LR*). Determining bin sizes is again a key element. In the end, we obtained a complex 2D plot that shows the distribution of rupture forces in function of effective loading rates.

Finally, with a data set (F^*, ln(LR*)), it is possible to build the Bell-Evans plot (Fig. 5.8) to obtain the energy barrier width from the slope of the fitting line and the dissociation constant k_{off} by extrapolating the fitting line until $F^* = 0$ (Figure 5.8). It is noteworthy that other approaches do not rely on the identification of most probable rupture forces to compute the kinetic dissociation constant $k_{off}(F)$ (Serpe et al. 2008).

The interpretation of the energy landscape using energy barriers as described by Bell-Evans (Bell, 1978; Evans & Ritchie, 1997) provides useful information related to the chemistry of binding. For instance, the width of an energy barrier measured at 1 Å or less likely involves the rupture of hydrogen bonds or salt bridges from a rigid ligand (Teulon et al. 2008; Teulon et al. 2011). However, a putative energy barrier width of 10 Å likely implies a stretching or deformation of one or both partners before the rupture.

5.8   SINGLE AND MULTIPLE SLOPES IN BELL-EVANS PLOTS

Bell-Evans plots are obtained by measuring the dependence of rupture forces on the loading rate. The energy landscape, as defined in the framework of the one-

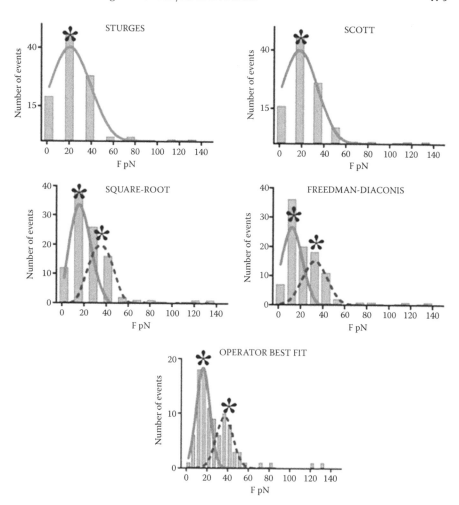

FIGURE 5.7 Effect of various methods used to assemble the cumulative probability distribution. The width of histogram bins has been obtained from Table 5.1. The total number of event is 100. Gaussian fits are overlaid on histograms; lines and dashed lines correspond to the fitting of the first and second population, respectively. The localization of most probable rupture forces is shown as * located on the maximum of the Gaussian fit.

dimensional kinetic reaction rate theory, characterizes the dissociation reaction (energy barriers) of a complex. From a Bell-Evans plot, the energy barrier width (x_β) and the energy barrier height (related to k_{off}) can be extrapolated (Figure 5.8). The theory predicts a linear correlation between the most probable rupture force (F^*) and the logarithm of the most probable loading rate (LR^*). However, the presence of two or three regimes in loading rates has been observed and was initially attributed to the presence of multiple energy barriers (Merkel et al. 1999). This observation

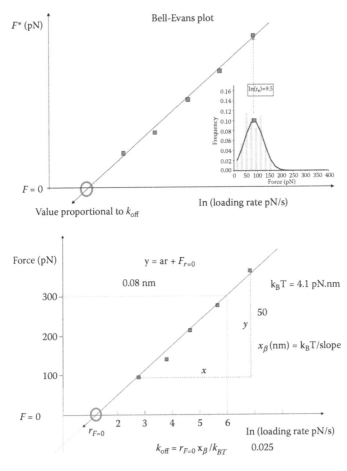

FIGURE 5.8 (See color insert.) Theoretical Bell-Evans plot. The upper graph indicated how the Bell-Evans plot is built from the most probable rupture forces ($F*$). These forces are obtained from a Gaussian fit of the force distribution obtained at a given effective loading rate. The lower graph indicates how to extract the energy barrier properties: the width (x_β in nm) and the kinetic dissociation rate (k_{off} in s^{-1}). The width x_β is obtained from the slope: $x_\beta = k_B T/slope$. The k_{off} value is obtained by extrapolating the loading rate value at $F* = 0$: $k_{off} = r_{F=0} x_\beta /k_B T$.

is still under intense debate and several explanations have been proposed to explain this apparent nonlinearity of data (Dudko et al. 2003; Williams, 2003; Derenyi et al. 2004; Neuert et al. 2006; Li et al. 2010). In several studies of our group, such as antibody–hapten (Odorico et al. 2007a; Teulon et al. 2007; Teulon et al. 2008) and avidin–biotin (Teulon et al. 2011), such nonlinearity has been observed (Figure 5.9). In these cases, two loading rate regimes were demonstrated to indicate two different structural events in the dissociation pathway (Teulon et al. 2008; Teulon et al.

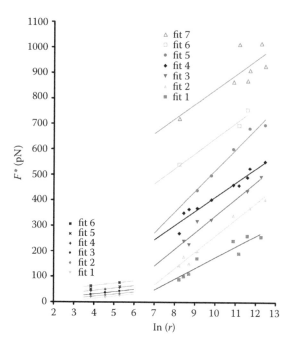

FIGURE 5.9 Bell-Evans plot. Experimental plot showing multiple fits obtained from antibody–uranyl chelator complexes (Odorico et al. 2007a). The two loading rate regimes are clearly visible: one at low loading rates from 44 to 200 pN/s and one at high loading rates from 3641 to 242,802 pN/s. Using various antibodies and different metals, it was shown that the low loading rate regime depicts the rupture between the antibody and the metal, whereas the high loading rate regime depicts the rupture between the antibody and the metal-chelator (Teulon et al. 2008).

2011). In the case of two loading rate regimes, an inner energy barrier is crossed out first followed by a second outer energy barrier (Yuan et al. 2000). The formalism of Evans (Evans & Ritchie, 1997) explains that by applying a linear force on a bond, the energy landscape is tilted such that the outer energy barrier is reduced. Consequently, when present the inner energy barrier can be observed by applying high loading rates.

5.9 MULTIPLE PARALLEL UNBINDING IN BELL-EVANS PLOTS

The presence of multiple parallel unbinding events is first observed in the distribution of rupture forces with the presence of multiple Gaussian fits (Figure 5.10a). Early single molecule force spectroscopy studies with AFM identified this distribution as quanta of forces (Florin et al. 1994). This has been widely verified since then. Now, this appears as a difficulty for DFS treatment since at a given loading rate we obtained not only one but two or more most probable rupture forces. Consequently, the Bell-

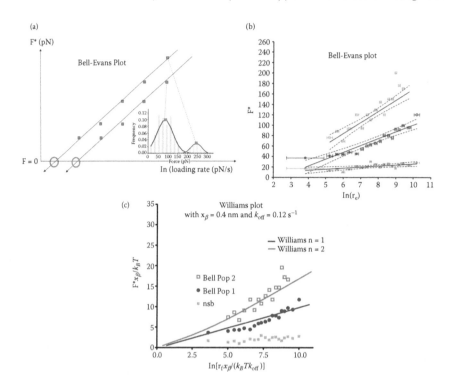

FIGURE 5.10 Bell-Evans and Williams plots for multiple parallel bonds. (a) Theoretical Bell-Evans plot of the most probable rupture forces (F^*) versus the logarithm of the loading rate. Two fitted line are obtained when multiple Gaussian distributions are observed (lower inset) at a given loading rate. (b) Experimental Bell-Evans plot showing multiple fits (lines) and their standard deviations (dotted lines); these fits are well separated from one fit to another. The standard deviation of most probable loading rates is shown by horizontal bars. (c) Williams plot representing the correlation between normalized most probable rupture forces versus normalized logarithm of the loading rate. Data points from (b) are normalized. To obtain the fit line, a simulation is performed by manually adjusting the energy barrier width (x_β) and the natural kinetic dissociation rate (k_{off}). From the fit, it can be seen that single-bond rupture corresponds to filled circle symbols whereas that of double bond ruptures corresponds to open symbols. Note that the light gray square symbols at the bottom of the plot could not be fitted and thus are attributed to nonspecific rupture events.

Evans plot becomes more complicated (Figure 5.10b) due to the presence of multiple fits. These fits should not be confused with the observed nonlinearity in Bell-Evans plots (see above) that is commonly attributed to the presence of multiple energy barriers in the dissociation pathway.

The main difficulty in the interpretation of multiple fitting lines in a Bell-Evans plot is the correct attribution of the elementary interaction (n = 1). The presence of rupture forces corresponding to nonspecific events often pollutes the Bell-Evans plot near the low rupture forces. To tackle the identification of elementary events, a mathematical treatment has been developed by Williams (Williams, 2003). It was shown that the kinetic equations for the uncorrelated parallel bond rupture cannot be solved analytically. However, a simplified numerical solution can be successfully used as the one developed by Sulchek et al. (Sulchek et al. 2005) that presents a direct experimental verification of the Markovian model described by Williams. This model uses normalized forces (F) and normalized loading rates (R) such as

$$F = \frac{f}{f_\beta} \quad \text{and} \quad R = \frac{r}{f_\beta k_{\text{off}}}$$

where f_β is the thermal force scale defined as $k_B T / x_\beta$ and the equivalent single-bond approximation is defined as

$$R = \left[\frac{1}{k_{\text{off}} f_\beta} \sum_{n=1}^{N} \frac{1}{n^2} \exp\left(-\frac{F*}{n} \right) \right]^{-1}$$

where N is the number of bonds and $F*$ is the normalized most probable rupture force. To operate a Williams plot, first it is necessary to normalize most probable rupture forces and loading rates and to plot these values according to $F*$ versus $\ln[R]$. Then, it is necessary to model the N-attachments by computing theoretical values of R in function of given $F*$ and n. Adjusting k_{off} and γ parameters allow the theoretical curve to match the normalized experimental data points (Figure 5.10c). In the end, one obtains a normalized plot with fitting lines that correspond to the description of N-attachment. At each curve, a set of k_{off} and x_β values are also obtained.

5.10 EXISTING SOFTWARE AND METHODS FOR DFS DATA ANALYSIS

In this last section, we would like to provide a nonexhaustive selection of existing software or website that performs DFS analysis. First, we should recommend a very didactical website on data processing and analysis in force spectroscopy (technical aspects of single molecule force spectroscopy) from Boris Akhremitchev lab: (http://www.chem.duke.edu/~boris/research/force_spectroscopy/force_spectroscopy .htm). Readers will find a very interesting and educative presentation, where most of the different aspects of DFS are pictured in detail with many classical and very useful references. Boris A is now here: http://www.chem.duke.edu/~boris

It is clear now that using a program for automatic detection of rupture events is a major improvement for two reasons. This detection is a time saver and automated analysis eliminates measurement bias related to subjective choices of the person performing the analysis. Even if, as previously explain, choosing the number of bin and their width remains operator's skill dependant!

Several home-made softwares have been developed in AFM groups during the past years and can be found in the following papers or websites.

In S. Kasas' data analysis, a program has been coded under MATLAB® 5.1. FD curves are analyzed using the fuzzy logic toolbox of MATLAB. Rupture events are detected by making a convolution between the FD curve and a geometrical shape of ideal event (a v-shape feature followed by a steep vertical segment and terminated by a right angle turn). Grades are attributed to rupture events for easy selection (Kasas et al. 2000).

In P. Hinterdorfer's data analysis, the experimental FD curve is least-squared fitted and unbinding events are characterized by a nonlinear delay preceding the jump. Nonspecific rupture events are eliminated after identifying the point of contact and the point of separation and the stretching of the spacer molecule (Baumgartner et al. 2000b).

Gergely et al. (Gergely et al. 2001) developed a semiautomatic method for detecting minima in the FD curves after a polynomial of degree 2 fitting procedure for p consecutive points.

In H. Gaub's data analysis, force-extension curves, the rupture force, the rupture length, and the corresponding loading rate were determined using the program Igor Pro 5.0 (Wavemetrics, Lake Oswego, OR) and a custom-written set of procedures (Morfill et al. 2007). Dissociation rate and the energy barrier width are obtained with a probability density function as a fit function of rupture forces and loading rates (Friedsam et al. 2003).

Levy and Maaloum (Levy & Maaloum, 2004; Levy & Maaloum, 2005) developed an algorithm and program written using Labview. The algorithm calculates the standard deviation in a test window (including a few experimental points) that moves along the curve. A rupture is a peak found outside the standard deviation. The algorithm detects the peaks whose height and width exceed the thresholds set by the user. Several criteria are defined to eliminate false positives.

In D. Müller's data analysis, they convert FD curves into force-extension curves (Kuhn et al. 2005). An automated alignment and pattern recognition of single molecule force spectroscopy data was developed in C++ using the GNU Scientific Library and the Wavemetrics Igor Pro. Force-extension curves are fitted using the worm-like chain model and rupture events are determined based on contour lengths. The binaries are available at http://fskit.blogspot.com/.

YieldFinder was developed as an integrated tool that reads, treats, and displays FD curves (Odorico et al. 2007b). In a recent development, FD curves have been fit by short successive linear segments that greatly improve the quality of analysis, especially the extraction of the effective loading rate values (Figure 5.11).

In R. Ros' data analysis, two steps are required: First, rupture events are modeled using a second-degree polynomial fit from the FD curve; then, the fit is used as a master curve to select appropriate experimental curve in the final analysis. The program is developed in MATLAB and C programming language (Fuhrmann et al. 2008).

HOOKE is an open software platform for force spectroscopy and was developed in Samori's lab (Sandal et al. 2009). Hooke is written in Python and runs on most operating systems. It can be extended by anyone by mean of simple plug ins.

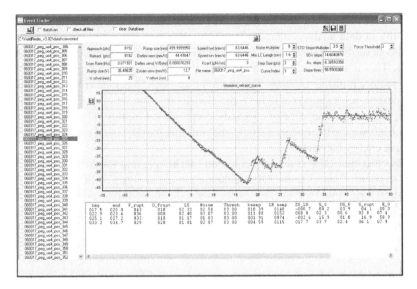

FIGURE 5.11 YieldFinder event-finder window. It displays a force-displacement (FD) curve from experimental records (left-side panel) and setup parameters (top panel). The plotting box (central panel) shows the fitted FD curve as lines, while the raw retract curve is displayed in points. Automatic identification of rupture events are shown in the bottom panel. Selected rupture event parameters will be saved in a local database for determining the cumulative probability distribution of forces before assembling the Bell-Evans plot.

Hooke contains utilities for automation of data selection and analysis and supports the most common commercial file formats. Hooke is under active development at http://code.google.com/p/hooke.

A few commercial products are available on the market. They almost all pretend offering unlimited access to automated, precise, and fast analysis of experimental data based on DFS. Most of the time, despite of their indisputable qualities, neither can solve the crucial step of rupture events distribution fitting. Among them, we can mention (probably nonexhaustive list):

PUNIAS (Protein Unfolding and Nano-Indentation Analysis Software) has been developed for single molecule force spectroscopy (Carl et al. 2001). It is capable of treating large set of data in batch mode (http://punias.voila.net/).

SPIP is a commercial image processing package for microscopy. The software includes several modules, including one in force curve analysis (http://www.imagemet.com/index.php?id=14&main=products&sub=modules).

5.11 CONCLUSIONS AND PERSPECTIVES

This chapter attempted to illustrate all the necessary steps in data analysis required in DFS measurements. Experimentally, we showed that one of the major challenges in DFS was to ascertain the level of bond attachment (single or multiple) since the difficulty in data analysis increased with the presence of multiple bonds. However, we also described methods and tools that allow the adequate treatment of multiple bonds. This is important since the most important recognition processes in tissues often require multiple binding. The second take home message from this chapter is that, despite all its approximations, the Bell-Evans' model is well adapted to analyzing the energy landscape of a ligand–receptor bond rupture. The Bell-Evans' formalism allows the treatment of single and multiple parallel bond ruptures and that of multiple energy barriers.

It is clear that DFS is a relatively young technique and improvements will continue to occur. Central to the DFS technique is the analysis of FD curves. Most of the laboratories involved in DFS experiments developed their own set of tools for interpreting the retract trace from the FD curve. Unfortunately, no software comparison has been performed to assess the performance of each method. Undoubtedly, this is a tedious work. Nevertheless, it is an important step for strengthening the pertinence of DFS in single molecule interactions. A central point to such assessment is the availability of well-characterized test sets of FD curves. Rules will have to be defined to characterize what is an adequate FD curve? For instance, a data file format should be defined; units will need to be homogenized; negative and positive controls will need to be present; nonspecific FD curves should also be included. The availability of DFS test sets will allow in-depth study of several tasks in data analysis. For instance, the shape of rupture events or the position of rupture events along the dissociation can be assessed and their effects on the energy landscape quantified. In addition, effects of experimental parameters such as the contact time, the trigger threshold, or the time delay between two consecutive measures could be evaluated. It is the purpose of the European COST Action TD1002 (AFM4NanoMed&Bio) to tackle some of these challenges so that the field of DFS will continue to grow and application of DFS to biologically relevant system will flourish.

ACKNOWLEDGMENTS

This work has been supported by the program for environmental nuclear toxicology of the Commissariat l'énergie atomique et aux énergies alternatives (CEA), France.

ABBREVIATIONS

AFM   Atomic force microscopy
AFS   Atomic force spectroscopy
CPD   Cumulative probability distribution
DFS   Dynamic force spectroscopy
F^* Most probable rupture force
FD   Force-displacement
F_{rupt} Rupture force
k_{cant} Spring constant of the cantilever
k_{eq} Slope of the retract curve before the unbinding event

k_{off} Kinetic dissociation rate
k_{samp} Stiffness constant of the studied sample
$LR/LR*$ Loading rate/Most probable loading rate
r_e Effective loading rate
SFM   Scanning force microscopy
x_β Energy barrier width

REFERENCES

Abdulreda, M.H., A. Bhalla, E.R. Chapman & V.T. Moy. 2008. Atomic force microscope spectroscopy reveals a hemifusion intermediate during soluble N-ethylmaleimide-sensitive factor-attachment protein receptors-mediated membrane fusion. *Biophys. J.* 94:648–55.

Anczykowski, B., D. Krüger, K.L. Babcock & H. Fuchs. 1996. Basic properties of dynamic force spectroscopy with the scanning force microscope in experiment and simulation. *Ultramicroscopy* 66:251–9.

Andersson, M., E. Fallman, B.E. Uhlin & O. Axner. 2006. Dynamic force spectroscopy of E. coli P pili. *Biophys. J.* 91:2717–25.

Arya, M., A.B. Kolomeisky, G.M. Romo, et al. 2005. Dynamic force spectroscopy of glycoprotein Ib-IX & von Willebrand factor. *Biophys. J.* 88:4391–401.

Badjic, J.D., A. Nelson, S.J. Cantrill, W.B. Turbull & J.F. Stoddart. 2005. Multivalency and cooperativity in supramolecular chemistry. *Acc. Chem. Res.* 38:723–32.

Bajpai, S., Y. Feng, R. Krishnamurthy, G.D. Longmore & D. Wirtz. 2009. Loss of alphacatenin decreases the strength of single E-cadherin bonds between human cancer cells. *J. Biol. Chem.* 284:18252–9.

Bartels, F.W., B. Baumgarth, D. Anselmetti, R. Ros & A. Becker. 2003. Specific binding of the regulatory protein ExpG to promoter regions of the galactoglucan biosynthesis gene cluster of Sinorhizobium meliloti - a combined molecular biology and force spectroscopy investigation. *J. Struct. Biol.* 143:145–52.

Baumgartner, W., P. Hinterdorfer, W. Ness, et al. 2000a. Cadherin interaction probed by atomic force microscopy. *Proc. Natl. Acad. Sci. USA* 97:4005–10.

Baumgartner, W., P. Hinterdorfer & H. Schindler. 2000b. Data analysis of interaction forces measured with the atomic force microscope. *Ultramicroscopy* 82:85–95.

Bell, G.I. 1978. Models for the specific adhesion of cells to cells. *Science* 200:618–27.

Berquand, A., N. Xia, D.G. Castner, et al. 2005. Antigen binding forces of single antilysozyme Fv fragments explored by atomic force microscopy. *Langmuir* 21:5517–23.

Bertoncini, P., R. Schoenauer, I. Agarkova, et al. 2005. Study of the mechanical properties of myomesin proteins using dynamic force spectroscopy. *J. Mol. Biol.* 348:1127–37.

Bizzarri, A. & S. Cannistraro. 2009. Atomic force spectroscopy in biological complex formation: Strategies & perspectives. *J. Phys. Chem. B* 113:16.

Björnham, O., J. Bugaytsova, T. Boren & S. Schedin. 2009. Dynamic force spectroscopy of the Helicobacter pylori BabA-Lewis b binding. *Biophys. Chem.* 143:102–5.

Björnham, O. & S. Schedin. 2009. Methods & estimations of uncertainties in single-molecule dynamic force spectroscopy. *Eur. Biophys. J.* 38:911–22.

Bonanni, B., A.S. Kamruzzahan, A.R. Bizzarri, et al. 2005. Single molecule recognition between cytochrome C 551 & gold-immobilized azurin by force spectroscopy. *Biophys. J.* 89:2783–91.

Carl, P., C.H. Kwok, G. Manderson, D.W. Speicher & D.E. Discher. 2001. Forced unfolding modulated by disulfide bonds in the Ig domains of a cell adhesion molecule. *Proc. Natl. Acad. Sci. USA* 98:1565–70.

Chen, S.-w.W., M.H.V. Van Regenmortel & J.-L. Pellequer. 2009. Structure-activity relationships in peptide-antibody complexes: Implications for epitope prediction and development of synthetic peptide vaccines. *Curr. Med. Chem.* 16:953–64.

Clarke, S.R., G. Andre, E.J. Walsh, et al. 2009. Iron-regulated surface determinant protein A mediates adhesion of Staphylococcus aureus to human corneocyte envelope proteins. *Infect. Immun.* 77:2408–16.

Contera, S.A., V. Lemaitre, M.R. de Planque, A. Watts & J.F. Ryan. 2005. Unfolding and extraction of a transmembrane alpha-helical peptide: Dynamic force spectroscopy and molecular dynamics simulations. *Biophys. J.* 89:3129–40.

COST Action TD1002. 2011. AFM4NanoMed&Bio, http://afm4nanomedbio.eu/

Dammer, U., M. Hegner, D. Anselmetti, et al. 1996. Specific antigen/antibody interactions measured by force microscopy. *Biophys. J.* 70:2437–41.

Derenyi, I., D. Bartolo & A. Ajdari. 2004. Effects of intermediate bound states in dynamic force spectroscopy. *Biophys. J.* 86:1263–9.

Dudko, O.K., A.E. Filippov, J. Klafter & M. Urbakh. 2003. Beyond the conventional description of dynamic force spectroscopy of adhesion bonds. *Proc. Natl. Acad. Sci. USA* 100:11378–81

Erdmann, T. 2005. Stochastic dynamics of adhesion clusters under force. *Theoretische Physik.* Potsdam, Universität Potsdam, Germany. Thesis:155.

Erdmann, T., S. Pierrat, P. Nassoy & U.S. Schwarz. 2008. Dynamic force spectroscopy on multiple bonds: experiments and model *EPL* 81:48001.

Evans, E. 2001. Probing the relation between force—lifetime—and chemistry in single molecular bonds. *Annu. Rev. Biophys. Biomol. Struct.* 30:105–28.

Evans, E. & L. Ludwig. 2000. Dynamic strengths of molecular anchoring and material cohesion in fluid biomembranes. *J. Phys.: Condens. Matter* 12:A315–A20.

Evans, E. & K. Ritchie. 1997. Dynamic strength of molecular adhesion bonds. *Biophys. J.* 72:1541–55.

Evans, E. & P. Williams. 2002. Dynamic force spectroscopy. In *Physics of Bio-Molecules and Cells*, ed. H. Flyvbjerg, F. Jülicher, P. Ormos and F. David, 145–204. Berlin: Springer-Verlag.

Fantner, G.E., E. Oroudjev & G. Schitter, et al. 2006. Sacrificial bonds and hidden length: unraveling molecular mesostructures in tough materials. *Biophys. J.* 90:1411–8.

Florin, E.L., V.T. Moy & H.E. Gaub. 1994. Adhesion forces between individual ligand-receptor pairs. *Science* 264:415–7.

Freedman, D. & P. Diaconis. 1981. On the histogram as a density estimator: L2 theory. *Zeit. Wahrscheinlichkeit.* 57:453–76.

Friddle, R.W., T.A. Sulchek, H. Albrecht, S.J. De Nardo & A. Noy. 2007. Counting and breaking individual biological bonds: force spectroscopy of tethered ligand-receptor pairs. *Curr Nanoscience* 3:41–8.

Friedsam, C., A.K. Wehle, F. Kühner & H. Gaub. 2003. Dynamic single-molecule force spectroscopy: bond rupture analysis with variable spacer length. *J. Phys. Cond. Mat.* 15:1709–23.

Fritz, J., A.G. Katopodis, F. Kolbinger & D. Anselmetti. 1998. Force-mediated kinetics of single P-selectin/ligand complexes observed by atomic force microscopy. *Proc. Natl. Acad. Sci. USA* 95:12283–8.

Fuhrmann, A., D. Anselmetti, R. Ros, S. Getfert & P. Reimann. 2008. Refined procedure of evaluating experimental single-molecule force spectroscopy data. *Phys. Rev. E Stat. Nonlin. Soft Matter Phys.* 77:031912.

Gergely, C., B. Senger, J.-C. Voegel, et al. 2001. Semi-automatized processing of AFM force-spectroscopy data. *Ultramicroscopy* 87:67–78.

Gergely, C., J. Voegel, P. Schaaf, et al. 2000. Unbinding process of adsorbed proteins under external stress studied by atomic force microscopy spectroscopy. *Proc. Natl. Acad. Sci. USA* 97:10802–7.

Green, N.M. 1963. Avidin. 1. The use of (14-C)biotin for kinetic studies and for assay. *Biochem. J.* 89:585–91.

Grubmüller, H., B. Heymann & P. Tavan. 1996. Ligand binding: Molecular mechanics calculation of the streptavidin-biotin rupture force. *Science* 271:997–9.

Guo, S., C. Ray, A. Kirkpatrick, N. Lad & B.B. Akhremitchev. 2008. Effects of multiple-bond ruptures on kinetic parameters extracted from force spectroscopy measurements: Revisiting biotin-streptavidin interactions. *Biophys. J.* 95:3964–76.

Harder, A., V. Walhorn, T. Dierks, X. Fernandez-Busquets & D. Anselmetti. 2010. Single-molecule force spectroscopy of cartilage aggrecan self-adhesion. *Biophys. J.* 99: 3498–504.

Hinterdorfer, P., W. Baumgartner, H.J. Gruber, K. Schilcher & H. Schindler. 1996. Detection and localization of individual antibody-antigen recognition events by atomic force microscopy. *Proc. Natl. Acad. Sci. USA* 93:3477–81.

Hukkanen, E.J., J.A. Wieland, A. Gewirth, D.E. Leckband & R.D. Braatz. 2005. Multiple-bond kinetics from single-molecule pulling experiments: Evidence for multiple NCAM bonds. *Biophys. J.* 89:3434–45.

Janshoff, A., M. Neitzert, Y. Oberdorfer & H. Fuchs. 2000. Force spectroscopy of molecular systems-single molecule spectroscopy of polymers and biomolecules. *Angew. Chem. Int. Ed. Engl.* 39:3212–37.

Kasas, S., B.M. Riederer, S. Catsicas, B. Cappella & G. Dietler. 2000. Fuzzy logic algorithm to extract specific Interaction forces from atomic force microscopy data. *Rev. Sci. Instrum.* 71:2082–6.

Kedrov, A., M. Appel, H. Baumann, C. Ziegler & D.J. Muller. 2008. Examining the dynamic energy landscape of an antiporter upon inhibitor binding. *J. Mol. Biol.* 375:1258–66.

Kim, I.H. 2009. Interactions between signal-transducing proteins measured by atomic force microscopy. *Anal. Chim. Acta* 81:3276–84.

Kim, Y., Y.I. Yang, I. Choi & J. Yi. 2010. Dependence of approaching velocity on the force-distance curve in AFM analysis. *Korean J. Chem. Eng.* 27:324–7.

Koch, S.J. & M.D. Wang. 2003. Dynamic force spectroscopy of protein-DNA interactions by unzipping DNA. *Phys. Rev. Lett.* 91:028103.

Kramers, H.K. 1940. Brownian motion in a field of force and the diffusion model of chemical reactions. *Physica* VII:284–304.

Kraulis, P.J. 1991. MOLSCRIPT: a program to produce both detailed and schematic plots of protein structures. *J. Appl. Cryst.* 24:946–50.

Kuhn, M., H. Janovjak, M. Hubain & D.J. Muller. 2005. Automated alignment and pattern recognition of single-molecule force spectroscopy data. *J. Microsc.* 218:125–32.

Lee, G.U., D.A. Kidwell & R.J. Colton. 1994. Sensing discrete streptavidin–biotin interactions with atomic force microscopy. *Langmuir* 10:354–357.

Levy, R. & M. Maaloum. 2004. Probing adsorbed polymer chains using atomic force microscopy: Interpretation of rupture distributions. *J.Phys.: Condens. Matter* 16: 7199–208.

Levy, R. & M. Maaloum. 2005. New tools for force spectroscopy. *Ultramicroscopy* 102: 311–5.

Li, N., S. Guo & B.B. Akhremitchev. 2010. Apparent dependence of rupture force on loading rate in single-molecule force spectroscopy. *Chemphyschem* 11:2096–8.

Lim, T.S., S.R. Vedula, W. Hunziker & C.T. Lim. 2008a. Kinetics of adhesion mediated by extracellular loops of claudin-2 as revealed by single-molecule force spectroscopy. *J. Mol. Biol.* 381:681–91.

Lim, T.S., S.R. Vedula, P.J. Kausalya, W. Hunziker & C.T. Lim. 2008b. Single-molecular-level study of claudin-1-mediated adhesion. *Langmuir* 24:490–5.

Lin, C., R. Chinnappan, K. Acharya, J.L. Pellequer & R. Jankowiak. 2011. On stabilization of a neutral aromatic ligand by p-cation interactions in monoclonal antibodies. *Biophys. Chem.* 154:35–40.

Liu, W., V. Montana, J. Bai, et al. 2006. Single molecule mechanical probing of the SNARE protein interactions. *Biophys. J.* 91:744–58.

Maki, T., S. Kidoaki, K. Usui, et al. 2007. Dynamic force spectroscopy of the specific interaction between the PDZ domain and its recognition peptides. *Langmuir* 23:2668–73.

Mammen, M., S.-K. Chio & G.M. Whitesides. 1998. Polyvalent interactions in biological systems: implications for design and use of multivalent ligands and inhibitors. *Angew. Chem. Int. Ed.* 37:2754–94.

Matsumoto, N., M. Fujita, T. Hiraishi, H. Abe & M. Maeda. 2008. Adsorption characteristics of P(3HB) depolymerase as evaluated by surface plasmon resonance and atomic force microscopy. *Biomacromolecules* 9:3201–7.

Merkel, R., P. Nassoy, A. Leung, K. Ritchie & E. Evans. 1999. Energy landscapes of receptor-ligand bonds explored with dynamic force spectroscopy. *Nature* 397:50–3.

Merritt, E.A. & D.J. Bacon. 1997. Raster3D: Photorealistic molecular graphics. *Meth. Enzymol.* 277:505–24.

Mitchell, G., C.A. Lamontagne, R. Lebel, M. Grandbois & F. Malouin. 2007. Single-molecule dynamic force spectroscopy of the fibronectin-heparin interaction. *Biochem. Biophys. Res. Comm.* 364:595–600.

Montana, V., W. Liu, U. Mohideen & V. Parpura. 2009. Single molecule measurements of mechanical interactions within ternary SNARE complexes and dynamics of their disassembly: SNAP25 vs. SNAP23. *J. Physiol.* 587:1943–60.

Morfill, J., K. Blank, C. Zahnd, et al. 2007. Affinity-matured recombinant antibody fragments analyzed by single molecule force spectroscopy. *Biophys. J.* 93:3583–90.

Moy, V.T., E.L. Florin & H.E. Gaub. 1994. Intermolecular forces and energies between ligands and receptors. *Science* 266:257–9.

Neuert, G., C. Albrecht, E. Pamir & H.E. Gaub. 2006. Dynamic force spectroscopy of the digoxigenin-antibody complex. *FEBS Lett.* 580:505–9.

Neuman, K.C. & A. Nagy. 2008. Single-molecule force spectroscopy: Optical tweezers, magnetic tweezers and atomic force microscopy. *Nat. Methods* 5:491–505.

Odorico, M., J.-M. Teulon, L. Bellanger, et al. 2007a. Energy landscape of chelated uranyl–antibody interactions by dynamic force spectroscopy. *Biophys. J.* 93:645–54.

Odorico, M., J.M. Teulon, O. Berthoumieu, et al. 2007b. An integrated methodology for data processing in dynamic force spectroscopy of ligand-receptor binding. *Ultramicroscopy* 107:887–94.

Prakasam, A.K., V. Maruthamuthu & D.E. Leckband. 2006. Similarities between heterophilic and homophilic cadherin adhesion. *Proc. Natl. Acad. Sci. USA* 103:15434–9.

Ptak, A., M. Kappl, S. Moreno-Flores, H. Gojzewski & H.J. Butt. 2009. Quantitative characterization of nanoadhesion by dynamic force spectroscopy. *Langmuir* 25:256–61.

Puchner, E.M. & H.E. Gaub. 2009. Force and function: Probing proteins with AFM-based force spectroscopy. *Curr. Opin. Struct. Biol.* 19:605–14.

Rankl, C., F. Kienberger, L. Wildling, et al. 2008. Multiple receptors involved in human rhinovirus attachment to live cells. *Proc. Natl. Acad. Sci. USA* 105:17778–83.

Rico, F. & V.T. Moy. 2007. Energy landscape roughness of the streptavidin-biotin interaction. *J. Mol. Recognit.* 20:495–501.

Robert, P., A.M. Benoliel, A. Pierres & P. Bongrand. 2007. What is the biological relevance of the specific bond properties revealed by single-molecule studies? *J. Mol. Recognit.* 20: 432–47.

Sandal, M., F. Benedetti, M. Brucale, A. Gomez-Casado & B. Samor. 2009. Hooke: An open software platform for force spectroscopy. *Bioinformatics* 25:1428–30.

Sapra, K.T., P.S. Park, K. Palczewski & D.J. Muller. 2008. Mechanical properties of bovine rhodopsin and bacteriorhodopsin: Possible roles in folding and function. *Langmuir* 24:1330–7.

Schwesinger, F., R. Ros, T. Strunz, et al. 2000. Unbinding forces of single antibody-antigen complexes correlate with their thermal dissociation rates. *Proc. Natl. Acad. Sci. USA* 97:9972–7.

Scott, D.W. 1979. On optimal & data-based histograms. *Biometrika* 66:605–10.

Serpe, M.J., F.R. Kersey, J.R. Whitehead, et al. 2008. A simple & practical spreadsheet-based method to extract single-molecule dissociation kinetics from variable loading-rate force spectroscopy data. *J. Phys. Chem. C* 112:19163–7.

Shi, X., L. Xu, J. Yu & X. Fang. 2009. Study of inhibition effect of herceptin on interaction between heregulin and erbB receptors HER3/HER2 by single-molecule force spectroscopy. *Exp. Cell Res.* 315:2847–55.

Sletmoen, M., G. Skjak-Braek & B.T. Stokke. 2004. Single-molecular pair unbinding studies of mannuronan C-5 epimerase AlgE4 and its polymer substrate. *Biomacromolecules* 5:1288–95.

Sletmoen, M., G. Skjak-Braek & B.T. Stokke. 2005. Mapping enzymatic functionalities of mannuronan C-5 epimerases and their modular units by dynamic force spectroscopy. *Carbohydr. Res.* 340:2782–95.

Sturges, H.A. 1926. The choice of a class interval. *J. Am. Stat. Assoc.* 21:65–6.

Sugisaki, K. 1999. Data processing in force curve mapping. *App. Surf. Sci.* 144–145:613–617.

Sulchek, T.A., R.W. Friddle, K. Langry, et al. 2005. Dynamic force spectroscopy of parallel individual mucin1-antibody bonds. *Proc. Natl. Acad. Sci. USA* 102:16638–43.

Taranta, M., A.R. Bizzarri & S. Cannistraro. 2008. Probing the interaction between p53 and the bacterial protein azurin by single molecule force spectroscopy. *J. Mol. Recognit.* 21: 63–70.

te Riet, J., A. Katan, C. Rankl, et al. 2011. Interlaboratory round robin on cantilever calibration for AFM force spectroscopy. *Ultramicroscopy* 111: doi:10.1016/j.ultramic.2011.09.012.

Teulon, J.M., M. Odorico, S.W. Chen, P. Parot & J.L. Pellequer. 2007. On molecular recognition of an uranyl chelate by monoclonal antibodies. *J. Mol. Recognit.* 20:508–15.

Teulon, J.-M., P. Parot, M. Odorico & J.-L. Pellequer. 2008. Deciphering the energy landscape of the interaction uranyl-DCP with antibodies using dynamic force spectroscopy. *Biophys. J.* 95:L63–5.

Teulon, J.-M., Y. Delcuze, & M. Odorico, et al. 2011. Single and multiple bonds in (strept)avidin–biotin interactions. *J. Mol. Recognit.* 24:472–83.

Tinoco, I., Jr. & C. Bustamante. 2002. The effect of force on thermodynamics and kinetics of single molecule reactions. *Biophys. Chem.* 101–102:513–33.

Tsukasaki, Y., K. Kitamura, K. Shimizu, et al. 2007. Role of multiple bonds between the single cell adhesion molecules, nectin and cadherin, revealed by high sensitive force measurements. *J. Mol. Biol.* 367:996–1006.

Vedula, S.R., T.S. Lim, S. Hui, et al. 2007. Molecular force spectroscopy of homophilic nectin-1 interactions. *Biochem. Biophys. Res. Commun.* 362:886–92.

Vedula, S.R., T.S. Lim, E. Kirchner, et al. 2008. A comparative molecular force spectroscopy study of homophilic JAM-A interactions and JAM-A interactions with reovirus attachment protein sigma1. *J. Mol. Recogn.* 21:210–6.

Verbelen, C., V. Dupres, D. Raze, et al. 2008. Interaction of the mycobacterial heparin-binding hemagglutinin with actin, as evidenced by single-molecule force spectroscopy. *J. Bacteriol.* 190:7614–20.

Verbelen, C., D. Raze, F. Dewitte, C. Locht & Y.F. Dufrene. 2007. Single-molecule force spectroscopy of mycobacterial adhesin-adhesin interactions. *J. Bacteriol.* 189: 8801–6.

Verbelen, C., V. Dupres, D. Raze, et al. 2008. Interaction of the mycobacterial heparin-binding hemagglutinin with actin, as evidenced by single-molecule force spectroscopy. *J. Bacteriol.* 190:7614–20.

Walton, E.B., S. Lee & K.J. Van Vliet. 2008. Extending Bell's model: How force transducer stiffness alters measured unbinding forces and kinetics of molecular complexes. *Biophys. J.* 94:2621–30.

Willemsen, O.H., M.M. Snel, L. Kuipers, et al. 1999. A physical approach to reduce non-specific adhesion in molecular recognition atomic force microscopy. *Biophys. J.* 76: 716–24.

Williams, P. 2008. Dynamic force spectroscopy with atomic force microscope. In *Handbook of Molecular Force Spectroscopy*, ed. A. Noy, 143–61. New York: Springer.

Williams, P.M. 2003. Analytical descriptions of dynamic force spectroscopy: Behaviour of multiple connections. *Anal. Chim. Acta* 479:107–15.

Wojcikiewicz, E.P., M.H. Abdulreda, X. Zhang & V.T. Moy. 2006. Force spectroscopy of LFA-1 and its ligands, ICAM-1 and ICAM-2. *Biomacromolecules* 7:3188–95.

Wollschlager, K., K. Gaus, A. Kornig, et al. 2009. Single-molecule experiments to elucidate the minimal requirement for DNA recognition by transcription factor epitopes. *Small* 5:484–95.

Yan, C., A. Yersin, R. Afrin, H. Sekiguchi & A. Ikai. 2009. Single molecular dynamic interactions between glycophorin A and lectin as probed by atomic force microscopy. *Biophys. Chem.* 144:72–7.

Yersin, A., T. Osada & A. Ikai. 2008. Exploring transferrin-receptor interactions at the single-molecule level. *Biophys. J.* 94:230–40.

Yu, J., Y. Jiang, X. Ma, Y. Lin & X. Fang. 2007. Energy landscape of aptamer/protein complexes studied by single-molecule force spectroscopy. *Chem. Asian J.* 2:284–9.

Yuan, C., A. Chen, P. Kolb & V.T. Moy. 2000. Energy landscape of streptavidin-biotin complexes measured by atomic force microscopy. *Biochemistry* 39:10219–23.

Zhang, X., E. Wojcikiewicz & V.T. Moy. 2002. Force spectroscopy of the leukocyte function-associated antigen-1/intercellular adhesion molecule-1 interaction. *Biophys. J.* 83:2270–9.

Zhou, J., L. Zhang, Y. Leng, et al. 2006. Unbinding of the streptavidin-biotin complex by atomic force microscopy: A hybrid simulation study. *J. Chem. Phys.* 125:104905.

Zou, S., H. Schonherr & G.J. Vancso. 2005. Force spectroscopy of quadruple H-bonded dimers by AFM: Dynamic bond rupture and molecular time-temperature superposition. *J. Am. Chem. Soc.* 127:11230–1.

6 Biological Applications of Dynamic Force Spectroscopy

Anna Rita Bizzarri and Salvatore Cannistraro

CONTENTS

6.1 INTRODUCTION

Interactions between biological molecules drive a large variety of cellular processes and span a wide range of strengths and complexity. Upon specific recognition mechanisms, biomolecules give rise to associations with different properties: from antigen–antibody complexes characterized by tight binding, long lifetime, and high specificity, to short-lived transient complexes involving molecules that recognize multiple partners, sometimes with a charge transfer capability (Janin, 1997; Crowley and Ubbink, 2003). The ability of biological molecules to undergo such highly controlled and hierarchical processes is regulated by forces at molecular scale based on a combination of noncovalent interactions (i.e., van der Waals, electrostatic, hydrophobic, hydrogen (H), and ionic bonds), which determine the strength and the characteristic time of the complexes. More generally, the instructions driving molecules to self-assemble into multicomponent structures are contained into their shape, chemical surface, and in their interaction with the environment in which the assembly takes place. Although many aspects of biorecognition have been elucidated, a full comprehension of the underlying mechanisms is far from being reached and many crucial questions are still debated; the biomolecular recognition processes having been described by progressively more refined theoretical frameworks (see Chapters 1 and 3).

Basically, a kinetic description of a biorecognition process between two partners is provided by the association rate k_{on}, which is essentially limited by the ligand diffusion and the geometric constraints of the binding site (k_{on} values range from 10^3 to 10^9 M s^{-1}), and by the dissociation rate k_{off}, which strongly depends on the interaction properties of the partners and determines the characteristics lifetime, τ, of the complex ($\tau = 1/k_{off}$) (Schreiber et al., 2009). Figure 6.1 shows the values of the dissociation rate k_{off} for some representative biological complexes. It comes out that k_{off}, and then the corresponding lifetime, spans an extremely wide range, consistent with the variety of functions played by the complexes. The experimental determination of k_{on} and k_{off} for free biomolecules in bulk solution, or when one of the partners is immobilized onto a surface, can be performed by several methods, with or without use of labels, such as optical spectroscopies, stop flow fluorimetry, surface plasmon resonance (SPR), nuclear magnetic resonance, isothermal titration calorimetry, and so on (Lauffenburger and Linferman, 1993; Morikis and Lambris, 2004; Lindorff-Larsen et al., 2005). However, since these techniques operate averagely on the molecule ensemble, some peculiar properties inherent to individual molecules, for example, rare events, transient phenomena, crowding effects, population heterogeneity, and so on, cannot be fully elucidated. With the advent of single molecule techniques, the study of these aspects has become accessible, offering novel and powerful tools for a more insightful investigation of biological processes (Neuman and Nagy, 2008). Among the rapidly expanding repertoire of single molecule techniques, including optical and magnetic tweezers, biomembrane force probe, and laminar flow chambers, dynamic force spectroscopy (DFS) represents a particularly valuable methodology to investigate interactions in biological systems, allowing to probe intra- and intermolecular forces with high sensitivity, without labeling and even in physiological conditions.

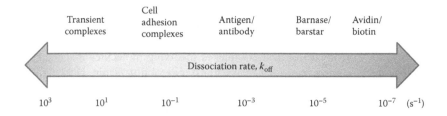

FIGURE 6.1 Typical k_{off} values for some representative biological complexes.

DFS can be performed by an atomic force microscopy (AFM) equipment, which is a high-resolution imaging tool based on force measurements, (for a complete description of AFM, see Chapter 2). Briefly, AFM imaging is obtained by scanning a very sharp tip, located at the end of a cantilever spring, over the sample that is placed onto a surface of a piezoelectric scanner, which is able to ensure a three dimensional positioning with subnanometer resolution (Jena and Hörber, 2002). The interaction force between tip and sample, optically measured from the cantilever deflection, is used to create a topographical image of the sample when the tip is raster-scanned in the horizontal x–y plane. In the DFS modality, the interaction forces between two biomolecules, one anchored to the tip and the other anchored to the substrate, can be probed during approaching and retraction cycles. From the analysis of the unbinding force data of biomolecular pairs, the kinetics and thermodynamics properties can be obtained in the framework of suitable theoretical models. The remarkable force sensitivity (down to pN), coupled with the small probe-sample contact area (as small as 10 nm^2), involving very few molecules, even down to only one, allows to investigate the subtle molecular features of biological systems (Janshoff et al., 2000; Zlatanova et al., 2000; Lee et al., 2007; Bizzarri and Cannistraro, 2009). In this respect, we remark that DFS requires only a very little amount of interacting species to carry out reliable experiments. Another interesting advantage of DFS measurements is that the application of an external force, yielding a reduction of the lifetime of the system, makes accessible the study of systems with long lifetimes. Since 1994, DFS has been applied to investigate a large variety of biomolecular complexes covering a wide range of different functions and biological processes, such as ligand–receptor or antibody–antigen pairs, protein unfolding, molecular stretching, conformational changes, cell deformation, membrane elasticity, cell adhesion, and so on. In this chapter, the results of some biological complexes investigated by DFS are reviewed and discussed in connection with the related data in bulk, when available. Selection of the cited articles has been done with the aim of providing an overview of the actual capabilities and potentialities of DFS to elucidate the molecular processes underlying some representative interactions of biological and medical interests. Due to the limited space, only some topics have been selected; therefore, our presentation is far from being exhaustive. We apologize for the omission of many key references, worth to be mentioned, that would have certainly enriched the present review.

This chapter is organized as follows: The main aspects involved in a DFS study of a biomolecular complex are overviewed in Section 6.2. In particular, the fundamental steps of a DFS experiment (immobilization strategies and data acquisition) are illustrated in Subsection 6.2.1, whereas the data analysis and the theoretical background are summarized in Subsection 6.2.2. A brief description of the computational methods useful when combined with DFS measurements is given in Subsection 6.2.3. Section 6.3 is devoted to describe and discuss the results of DFS applied to some representative biomolecular complexes. This section has been organized in subsections, each one of them is focused on a class of complexes playing a specific biological function; the experimental used setup, the applied data analysis, and the most relevant obtained information being outlined. Applications of DFS to the development of biosensors, together with some innovative approaches susceptible to emerge in the study of biological complexes, are presented in Section 6.4. Finally, Section 6.5 summarizes the results and outlines possible future developments of DFS applied to biorecognition processes.

6.2 SETUP AND ANALYSIS METHODS OF DFS EXPERIMENTS

6.2.1 DESCRIPTION OF A DFS EXPERIMENT

The setup of a DFS experiment commonly used to study the interaction of two biological molecular partners is sketched in Figure 6.2a. One partner is bound to the apex of an AFM tip (usually made of silicon, silicon nitride, sometimes with a gold coating), whereas the other one is anchored to the substrate, which can be made of silicon, glass, mica, gold, agarose bead, and so on. The biomolecular immobilization procedure often involves heterobifunctional linker molecules to covalently bind to one end the biomolecules and on the other the inorganic surfaces (Hinterdorfer and Dufrêne, 2006; for details see Chapter 4); in many cases, polyethylene glycol (PEG) polymer being used as linkers.

The introduction of these linkers can both prevent conformational distortions and denaturation due to a direct biomolecule-surface interaction and favor an optimized recognition between the partners. Figure 6.2b shows a typical trend (force curve) for the cantilever deflection Δz as a function of the piezo displacement, recorded in a DFS experiment by approaching and retracting the ligand-functionalized tip to and from the substrate covered with the receptors. Approaching and retraction are carried out with a constant loading rate given by $r_F = dF/dt$, where F is the applied force, which can be derived through the Hooke's formula by $F = -k_{cant}\Delta z$, where k_{cant} is the cantilever spring constant. At the beginning, the functionalized tip is far away from the substrate (Point A). As the tip is approached toward the substrate, the partners become closer and closer and they may undergo a biorecognition process, leading to a complex formation, provided that they are endowed with enough flexibility and reorientational freedom. Beyond the contact point (Point B), the cantilever begins to deflect upward due to the intermolecular repulsive forces. The approaching phase is stopped when the cantilever applies to the substrate a preset maximum contact force, usually kept below 1 nN to avoid damage to the sample (Point C). Next, the direction of motion is reversed and the cantilever retracts from the surface. During this retraction cycle, the cantilever reaches the baseline (Point D), after which it begins to bend downward, due to the adhesion forces, and/or bonds formed during the contact phase. When the applied force overcomes the interaction forces between the biomolecules, the cantilever pulls off sharply going to a noncontact position (jump-off-contact, Path FG), with a concomitant dissociation of the two partners. From the cantilever deflection d in correspondence of the jump (Figure 6.2), the unbinding force F_{unb} between the biomolecular partners is given by $F_{unb} = -k_{cant}d$. In addition, the piezo displacement corresponding to the nonlinear portion of the retraction curve before the unbinding event (Path EF) corresponds to the so-called unbinding length, which is somewhat correlated to the stretching of the linker and, sometimes,

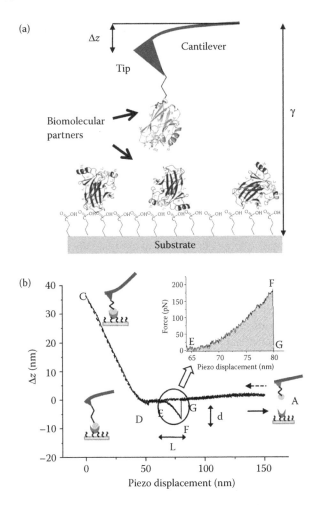

FIGURE 6.2 (a) Sketch of the experimental setup commonly used to investigate by DFS the unbinding process of two interacting molecules; Δz is the cantilever deflection, and γ is the piezo displacement. Both the partners are immobilized to the surfaces (tip or substrate) by linkers. (b) Representative experimental force curve (approach and retraction) for a specific unbinding event of a biomolecular complex. The unbinding event occurs at the jump-off (FG path); the corresponding unbinding force can be derived from $F = -k_{cant}\,d$, where d is the cantilever deflection and k_{cant} is the effective cantilever spring constant. L is the unbinding length given by the piezo displacement encompassing the nonlinear portion of the retraction curve before the unbinding event. For other details, see the text. *Inset:* A zoom of the region of the retraction curve corresponding to the unbinding process where the force F is given by $F = -k_{cant}\Delta z$. The area underlying the force versus piezo displacement (dashed region) provides the work done by the applied force (see Subsection 6.3.3).

of the biomolecules. Accordingly, a suitable modeling of the linker can be useful to discriminate between specific and nonspecific unbinding events (see below and Chapter 4). Successively, the tip is further retracted from the substrate to reach the initial position (Point A). In a DFS biorecognition experiment, a large number of force curves (hundreds or thousands) are usually acquired in a cyclic way at the same or at a different substrate location for a statistical sampling.

6.2.2 Analysis of the Force Curves

The approach of a biomolecule-functionalized tip toward a substrate, coated with the corresponding partner, does not necessarily result into the formation of a specific complex. An improper spatial contact between the biomolecules or an interaction between the tip and the substrate without the involvement of one, or both, of the biological partners, may give rise to nonspecific interactions between the tip and the substrate. Representative force curves, which are typically obtained in DFS experiments on biomolecular complexes are shown by Bizzarri and Cannistraro (2010). To reliably single out the force curves corresponding to specific unbinding events, different criteria have been followed. In general, curves exhibiting a linear trend after the contact point in the retraction curve are attributed to a nonspecific adhesion between the tip and the substrate, and then are discarded. Instead, curves showing a nonlinear trend after the contact point in the retraction curve are accepted and subjected to further analysis. When flexible linkers are used to immobilize the biomolecules, the unbinding length L is expected to match that one corresponding to the total stretching of the linkers (Hinterdorfer and Dufrêne, 2006). To further assess the specificity of the unbinding events, *ad hoc* experiments can be performed. Commonly, blocking experiments are carried out by a saturation of the functionalized tip, or substrate, with the corresponding free partner (Florin et al., 1994; Allen et al., 1997; Ros et al., 1998). In this way, the biorecognition process would be inhibited and the unbinding frequency (given by the ratio between the number of specific events and that of total recorded events) would be drastically reduced if the interaction is specific. While unbinding frequencies ranging from 15% to even 85% have been usually registered in standard experiments, a significant decrease (more than 50%) is expected after blocking (Bizzarri and Cannistraro, 2010). Other control experiments to verify the specificity of the detected unbinding events can be performed by replacing one of the two partners with a noninteracting molecule or by changing the biomolecule concentration. Once the specificity of the unbinding events has been assessed, the force curves are analyzed to obtain reliable quantitative information on the biorecognition process. For example, the histograms of the unbinding forces, which have been registered at a given value of the loading rate, are plotted to determine the most probable force (see below) and also the possible occurrence of multiple unbinding events. Indeed, these histograms have been generally found to be quite spread, even with the presence of multiple peaks. Such a variability could be generally ascribed to several factors such as the heterogeneity in the formation of the complex, slight differences in the relative arrangements of the partners, the existence of different binding sites, the occurrence of multiple unbinding processes, and so on. When the histogram displays a single-mode distribution, the unbinding force has been evaluated from the maximum of the distribution or through a fit with a Gaussian function. When multiple peaks are observed, the first peak has been commonly assumed as due to a single unbinding event. More specifically, if the peaks in the distribution are equally spaced, the distance between two subsequent peaks has been assumed as the quantum for the unbinding force. Alternatively, a Poisson statistical method has been developed to determine the unbinding force when a finite number of interacting biomolecular pairs are expected to be found within the tip–substrate contact area (see Chapters 3 and 4 and Lo et al., 2001). However, more accurate and refined procedures are generally required to analyze the force distributions in the presence of multiple unbinding events (see Chapters 3 through 5).

Notably, the first DFS studies on biomolecular systems have been focused on the unbinding force intensity at a fixed value of the loading rate. Successive investigations have put into evidence that the unbinding force alone does not necessarily reflect the effective strength of the interaction since this force strongly depends on the loading rate and then the equilibrium energy profile is altered (see Figure 6.3). As widely treated in Chapter 3, the determination of equilibrium properties of a biomolecular complex from DFS data obtained in nonequilibrium conditions can be achieved by applying suitable models. A phenomenological description of the effects of an applied force on the chemical reaction energy profile has been first provided by Bell and Evans-Ritchie (see Figure 6.3; Bell, 1978; Evans and Ritchie, 1997). According to the proposed model, the application of a weak external force F yields a lowering of the activation energy barrier ΔG^* given by $\Delta G^*(F) = \Delta G^* - F x_\beta$, where x_β is the width of the energy barrier (see dashed line in Figure 6.3). It should be noted that the possibility of an increase of this energy barrier upon the application of a force, with a concomitant prolonged lifetime of the complex, has also been predicted (see dotted line in Figure 6.3; Prezhdo and Pereverzev, 2009). Such an increase which gives rise to the so-called "catch bonds," has been later on confirmed experimentally (see Subsection 6.3.5).

Starting from the Bell model, and in the framework of the reaction rate theory for thermally activated processes, Evans and Ritchie described the unbinding process of a biomolecular complex in terms of crossing over a single energy barrier under the application of the force with a constant loading rate (Evans and Ritchie, 1997). They cast the probability distribution $P(F)$ of the unbinding force F into the equation:

$$P(F) = \frac{k_{off}}{r_F} e^{\left[\frac{F x_\beta}{k_B T} + \frac{k_{off} k_B T}{x_\beta r_F} \left(1 - e^{\frac{F x_\beta}{k_B T}} \right) \right]} \tag{6.1}$$

where k_{off} is the dissociation rate at equilibrium, k_B is the Boltzmann constant, and T is the absolute temperature (see Chapter 3). This distribution was found to be asymmetric and skewed toward low force values. Under the assumption of single unbinding events, the most probable unbinding force F^* can be generally derived

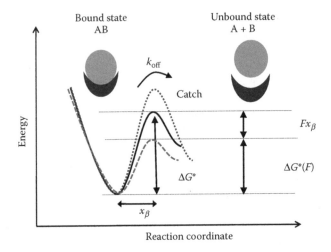

FIGURE 6.3 Diagram of the energy profile for an equilibrium biomolecular complex disso-ciation (continuous line) and under the application of an external force F (dashed line); ΔG^* and $\Delta G^*(F)$ being the corresponding activation free energies. The possibility of an increase of the energy barrier upon applying the external force is also represented (dotted line) (see the text).

from the maximum of the unbinding force distribution:

$$F^* = \frac{k_B T}{x_\beta} \ln\left(\frac{r_F x_\beta}{k_{off} k_B T} \right) \tag{6.2}$$

Equation 6.2 predicts a linear relationship between the most probable force F^* with the natural logarithm of the loading rate. Then, by plotting F^* as a function of $\ln(r_F)$, the equilibrium parameters k_{off} and x_β can be extracted from the slope and the intercept of a linear fit.

The Bell–Evans model has provided a successful description of the most proba-ble unbinding force trend with the loading rate for several biomolecular complexes. Indeed, it represents a landmark for describing unbinding processes as studied by DFS or by other single-molecule techniques (Lee et al., 2007; Bizzarri and Cannis-traro, 2009). When two or even three distinct linear regimes for the most probable unbinding force as a function of the natural logarithm of the loading rate have been observed, the existence of two or three energy barriers, respectively, between the bound and the unbound states have been hypothesized. On the other hand, models predicting a dependence for the unbinding force with the loading rate as $F^* \sim \ln(r)^\nu$, where ν can assume the value 1, 1/2 or 2/3 depending on the shape of the energy bar-rier, have been developed (Dudko et al., 2003; Hummer and Szabo, 2003; Friddle, 2008).

6.2.3 DFS AND COMPUTATIONAL METHODS

Both the design of DFS experiments and the corresponding data analysis can take large advantage of computational methods and in particular of docking and steered molecular dynamics (see also Chapter 1). To achieve effective and functional immobilization of the biomolecules to the substrates, the knowledge of the interaction sites between the two partners would be crucial to maximize the interaction probability by avoiding the involvement of the complex binding sites in the anchoring procedures (Bonanni et al., 2005; Bizzarri et al., 2007). If the molecular structure of the complex is not known, computational docking can be applied to predict the most probable arrangement of the complex starting from the 3D structures of the individual partners (Jones and Thornton, 1996). Briefly, the surfaces of each partner are probed by looking for all the possible binding modes that are then ranked according to a score function taking into account geometric, electrostatic, energy criteria, and so on (Camacho and Vajda, 2002). The predicted structure of the complex can also be used to evaluate the probability that the two partners may form a complex upon their immobilization on the surfaces, resulting in some help to estimate the unbinding frequency (Bizzarri et al., 2009). Notably, the predicted structure of the complex could be also extremely useful to analyze the unbinding process in connection with the structural properties at the complex interface (i.e., the H-bond network, the salt bridges, or the charge distribution).

DFS experiments can also be combined with steered molecular dynamics, a computational tool suitably developed from molecular dynamics simulation, to predict the unbinding dynamics of biomolecules under the application of an external force (Rief and Grubmüller, 2002). More specifically, the dynamics of the system can be followed at atomic resolution while one or both the biomolecular partners, bound to a spring, are pulled at a constant velocity to induce their unbinding. Although the pulling speed of the simulations (m/s range) is generally orders of magnitude higher than that used in DFS (μ/s range), steered molecular dynamics has been demonstrated to provide information on the sequence of the unbinding steps, thus helping to elucidate the molecular mechanisms regulating the unbinding processes (Grubmüller et al., 1996; Izrailev et al., 1997; Bayas et al., 2003).

6.3 DFS STUDIES OF BIOMOLECULAR COMPLEXES

The selected DFS studies of biomolecular complexes have been grouped into five subsections, each one of them having been focused on a specific class of biological systems. In each section, the most important aspects of the experimental setup (immobilization procedure, cantilever spring constant, and loading rate range) and the main results (unbinding force, dissociation rate, and energy barrier width) for a collection of DFS studies have been summarized in five tables. Some of the investigations reported in these tables are discussed into details with some emphasis on the relevance of the DFS approach for the biological functions in which the systems are involved. When available, the DFS results have been compared with those obtained by other single-molecule techniques, such as biomembrane force probe or optical tweezers, and with those coming from bulk experiments. Throughly, we make a warning to the readers about the fact that the interactions of ligands with receptors could display different properties when they are free in solution (3D) or covalently anchored to solid surfaces (2D; see also Chapter 1). On the other hand, it should be kept in mind that the zero-force extrapolation of k_{off} requires the existence of a single barrier between the bound and the unbound states (Janshoff and Steinem, 2001).

6.3.1   A Paradigmatic Complex with Long Lifetime: Avidin–Biotin

The interaction between the tetrameric protein avidin and the smaller ligand biotin is one of the strongest noncovalent interactions in nature, involving a highly stabilized network of polar and hydrophobic bonds. The study of the complexes that are characterized by an extremely long lifetime could present some difficulty when approached in a standard way since the dissociation process does not take place spontaneously within a reasonable observation time. However, an estimation of $10^{-8}-10^{-7}s^{-1}$ for the dissociation rate and of $7 \cdot 10^7 M^{-1}s^{-1}$ for the association rate has been obtained; the corresponding affinity constant ($K_a = k_{on}/k_{off}$) of about 10^{15} M^{-1} being one of the highest value detected for biomolecular systems. For this reason, the avidin–biotin complex has been object of extensive studies by a variety of experimental techniques and computer simulations (Izrailev et al., 1997; Merkel et al., 1999; Wilchek et al., 2006; Teulon et al., 2011). It represents a sort of benchmark in the investigation of the kinetic and thermodynamic properties of biomolecular complexes at single-molecule level. The avidin–biotin complex deserves a huge interest even in applicative fields, thanks to its strong affinity that allows it to be used as a molecular glue capable of strong, specific, and long-lasting interactions of biotechnological interest (Grunwald, 2008). Besides avidin, either the tetrameric protein streptavidin from the bacterium *Streptomyces avidini* with a structure very similar to that of avidin or the deglycosylated form of avidin (neutravidin) have an extraordinarily strong affinity for biotin (K_a of about $10^{14} M^{-1}$).

Also the avidin–neutravidin and streptavidin biotin complexes have been the object of extensive DFS investigations (Teulon et al., 2011). In the studies, the avidin–biotin and the neutravidin–biotin and, sometimes, the avidin–streptavidin pairs have been used interchangeably. In many cases, biotin covalently bound to bovine serum albumin (biotin-BSA), instead of bare biotin, has been used to overcome some experimental problems arising from the small dimension of this molecule. In DFS experiments (or in other related techniques, such as biomembrane force probe), the application of an external force yields a decrease of the energy barrier and consequently a reduction of the complex lifetime, making these systems accessible for investigation within reasonable experimental timescales. The first studies by DFS on biomolecular complex were carried out on avidin–biotin in 1994 (Florin et al., 1994; Lee et al., 1994b; Moy et al., 1994b). A sketch of the experimental setup that follows a rather simple architecture is shown in Figure 6.4a. Biotin was immobilized on an agarose bead that constitutes the substrate and directly adsorbed on the AFM tip. Avidin was, in turn, bound to the tip by taking advantage of its strong affinity for biotin and the presence of four equivalent binding sites. These pioneristic studies were devoted to measure the unbinding force between the partners at a fixed loading rate, whose value was not specified (Florin et al., 1994). At that time, indeed, it was believed that the unbinding force could be representative of the effective interaction strength between the biomolecular partners.

The histogram of the unbinding forces exhibits a large variability (Figure 6.4b); this being a common feature of single-molecule measurements and it was believed to reflect the heterogeneity of the analyzed system. Also, the presence of well-distinguishable peaks, almost equi-spaced, appears evident. These peaks have been correlated to the occurrence of multiple unbinding events, that is, a synchronous unbinding of one, two, three, and more pairs of biomolecules connected in parallel. The distance between two adjacent peaks in the histogram has been assumed to be the force quantum, that is, the force required for the unbinding of a single pair of interacting biomolecules (it was estimated to be about 160 pN).

Further, DFS data on avidin–biotin and steptavidin–biotin pairs are listed in Table 6.1. The values of unbinding force measured in these experiments for streptavidin–biotin fall in a wide range (from about 50 to 120 pN) at the loading rate of 1 nN/s for streptavidin–biotin, a similar range being observed for avidin–biotin. Such a variability has been confirmed by many independent studies. The significant variability in the unbinding force values as extracted by DFS from the avidin–biotin system has stimulated an extensive debate leading to several different claims and statements. Successive investigations on avidin–biotin and related pairs have been focused on the analysis of the unbinding forces measured at different loading rates

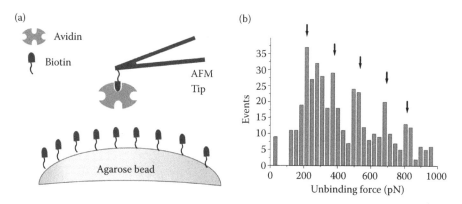

FIGURE 6.4 (a) Sketch of the experimental setup used for the first DFS experiments on the avidin–biotin complex. (b) Histogram of the unbinding force; the arrows indicate the multiple of the quantum force. (Adapted from Florin, E. L. et al. 1994. *Science*, 264:415–417.)

TABLE 6.1

DFS Results for a Collection of Avidin–Biotin and Streptavidin–Biotin Pairs: Unbinding Force at a Given Loading Rate (in parentheses); Dissociation Rate k_{off} and Energy Barrier Width x_β (in parentheses) as Extracted by Applying the Bell–Evans Model for Each Linear Trend Observed in the F^* versus $\ln(r_F)$ Plot (see Equation 6.2)

System: Substrate/ Tip	Immobilization: Substrate/ Tip	Cantilever Spring Constant k_{cant} (N/m)	Unbinding Force (pN) (Loading Rate (nN/s))	Loading Rate Range (nN/s)	k_{off} (s^{-1}) (x_β (nm))	Reference
Biotin-BSA/ Avidin	Agarose bead / Si$_3$N$_4$	0.06	160 (NA)	NA	NA	Florin et al. 1994
Avidin/ Biotin	Gold (Biotin) / Si$_3$N$_4$ (PEG)	0.034	38 (1)	0.01–10	10^{-3} (2) 100 (0.3)	De Paris et al., 2000
Biotin-BSA/ Streptavidin	Agarose bead / Si$_3$N$_4$	0.01–0.05	120 (1)	0.1–5	$1.71 \cdot 10^{-5}$ (0.5)	Yuan et al., 2000
Streptavidin/ Biotin-BSA	Silicon / Si$_3$N$_4$	0.12	83 (1)	0.1–1000	2.1 (0.05) 5.6 (0.12)	William et al., 2000
Biotin-BSA/ Streptavidin	Glass / Si$_3$N$_4$	0.039	167 (30)	30–6000	NA	Lo et al., 2001
Streptavidin/ Biotin-BSA	Silicon (Silane) / Si$_3$N$_4$	0.01–0.09	50 (1)	1–100	0.2 (0.6) 69 (0.14)	Patel et al., 2004
Biotin-BSA/ Streptavidin	Glass (Silane) / Si$_3$N$_4$ (Silane)	0.01–0.4	50 (1)	0.3–9.6	0.56 (0.081) 2.98 (0.024)	de Odrowaz et al., 2006
Biotin-BSA/ Streptavidin	Agarose bead / Si$_3$N$_4$ (Biotin)	0.01	50 (1)	0.03–200	0.1 (0.38) 23 (0.09)	Rico and Moy, 2007
Streptavidin/ Biotin	Glass (PEG) / Si$_3$N$_4$ (PEG)	0.05–0.16	50(1)	0.1–100	2.86 (0.28) 78 (0.073)	Guo et al., 2008
Streptavidin/ Biotin	Mica / Si$_3$N$_4$ (Silane)	0.01	15–25 (0.2)	0.1–2	$2.1 \cdot 10^{-7}$ (0.1) 2.1 (0.05)	Walton et al., 2008
Streptavidin/ Biotin	Gold / Gold (PEG)	0.03–0.32	68 (1.2)	1.8–960	0.07 (0.40) 4.38 (0.08)	Teulon et al., 2011

Note: Some details of the experimental setup: the immobilization strategies of the biomolecules to both the substrate and the AFM tip; the cantilever spring constant; the loading rate range. (NA = not available).

to determine the kinetic properties of the system in the framework of the Bell–Evans model.

A collection of unbinding force as a function of the logarithm of the loading rate from different works on the strepta(avidin)–biotin systems is shown in Figure 6.5. This figure shows that at the same loading rate, different values for the unbinding force have been found; this being in agreement with the variability observed in the previously mentioned experiments. For some experiments (e.g., Rico and Moy, 2007; Guo et al., 2008), the unbinding force data cluster on two distinct linear trends pointing out the existence of two distinct barriers in the energy landscape. In general, the values for k_{off} and x_β extracted by a fit through Equation 6.2, and reported in Table 6.1, exhibit a wide range of values. The dissociation rate values, ranging from about 10^{-7} to $100\ s^{-1}$, suggest the existence of both a slow and a fast processes. It is interesting to note that the lifetime of the slower process is substantially lower than the value in bulk ($10^7\ s$). Some variability, even if less marked, has been found for the energy barrier width x_β. This also has stimulated wide discussions and deeper investigations as witnessed by the abundant literature (see, e.g., Pincet and Husson, 2005; Rico and Moy, 2007; Teulon et al., 2011). A variety of possible explanations have been suggested, without, however, reaching a general consensus. For example, it has been hypothesized that the observed discrepancies could arise from differences in the used setup, for example, the kind of substrates (silicon, glass, agarose, gold, mica), the cantilever stiffness, the immobilization strategies, and so on (Patel et al., 2004; Walton et al., 2008). In other cases, more than one biomolecular pair has been postulated to be involved in the unbinding process, leading to significant errors in the

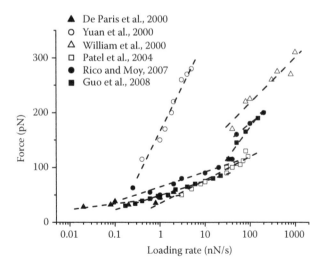

FIGURE 6.5 An overlay of the unbinding force as a function of the logarithm of the loading rate from a collection of DFS experimental data for the avidin–biotin complex. Dashed lines are obtained by a fit through Equation 6.2.

evaluation of the unbinding force. Accordingly, when multiple unbinding processes occur (in series or in parallel) suitable analysis procedures have been suggested to be used to avoid the introduction of systematic errors (Guo et al., 2008). Recently, a revisitation of the unbinding process of avidin–biotin pairs has taken into account the complexity of biomolecular systems, whose energy landscape may involve many nearly isoenergetic local minima, called *conformational substates*, according to the Frauenfelder glass-like picture of proteins (Frauenfelder et al., 1991; Rico and Moy, 2007). It has been postulated that during the binding and unbinding processes, the partners can explore different hierarchical regions of the energy landscape overcoming different barriers. Therefore, the biological system complexity gives rise to a huge variety of possible unbinding paths that would give rise to a significant spread in the unbinding force values. In this framework, the history of the experiments could play some role in guiding the system to preferentially enter a given path of the sample and/or of the measurements, and so on. Along this direction, it has been suggested that measurements performed at different temperatures make possible a direct estimation of the energy landscape roughness. Under the Hyeon and Thirumalai approximations, the roughness ϵ of the energy landscape can be calculated by (Hyeon and Thirumalai, 2003; Nevo et al., 2005):

$$\epsilon^2 = \frac{x_\beta(T_1)k_BT_1x_\beta(T_2)k_BT_2}{x_\beta(T_2)k_BT_2 - x_\beta(T_1)k_BT_1}$$

$$\left[\Delta G^* \left(\frac{1}{x_\beta(T_1)} - \frac{1}{x_\beta(T_2)} \right) + \frac{k_BT_1}{x_\beta(T_1)} \ln \frac{r_F(T_1)x_\beta(T_1))}{k_{\text{off}}(T_1)k_BT_1} \right.$$
$$\left. - \frac{k_BT_2}{x_\beta(T_2)} \ln \frac{r_F(T_2)x_\beta(T_2)}{k_{\text{off}}(T_2)k_BT_2} \right] \tag{6.3}$$

where ΔG^* is the energy barrier and $r_F(T_1)$ and $r_F(T_2)$ are the loading rates at two different temperatures T_1 and T_2, which give rise to the same unbinding force (Nevo et al., 2005). For streptavidin–biotin, a roughness of 5.6 and 7.5 k_BT has been evaluated for the inner and outer barriers, respectively (Rico and Moy, 2007); these values correspond to about 13%–18% of the energy barrier ΔG^*. Such a surface roughness would be connected with the unbinding force data scattering, resulting from the exploration of different conformational substates during the unbinding process. It has been moreover suggested that measurements performed for longer times would permit to reach a complete relaxation of the system, making possible to reconstruct the full energy landscape, reconciling, thus, the results among all the experiments. In this respect, the recent development of extremely stable, very low drift AFM apparata could allow to investigate the biological system for longer time, with the capability to reach a higher level of details for the unbinding processes (Junker et al., 2009).

To elucidate the molecular mechanisms regulating the escape of biotin from the binding sites of avidin, steered molecular dynamics has been applied (Grubmüller et al., 1996). The setup of the simulations has been designed to closely match that of used in the first DFS studies (Florin et al., 1994; Lee et al., 1994b; Moy et al., 1994b).

These simulations suggested the occurrence of a multiple-pathway rupture mechanism characterized by five unbinding steps. An overall unbinding force of about 250 pN has been estimated; such a value being slightly higher than the experimental one (see Table 6.1). In addition, the study has allowed the authors to elucidate the role of both the water bridges and the H-bond network within the binding pocket on the stability of the complex. Both binding force values and specificity are therefore attributed to a H-bond network between the biotin and the binding pocket residues of streptavidin, whereas the presence of additional water bridges has been suggested to be responsible for further enhancement of the stability of the complex toward the rupture.

A complex that can compete with avidin–biotin in terms of affinity and lifetime is formed between the ribonuclease Barnase and its inhibitor Barstar from the bacterium *Bacillus amyloliquefaciens* whose interaction needs to be very fast and strong to prevent cell death when Barnase is excreted by the bacterium. Indeed, the Barnase–Barstar complex formation, mainly driven by electrostatic interactions, is extremely fast, with an association rate of about 10^8 M s^{-1} and a dissociation rate of about 10^{-6} s^{-1}, as estimated in bulk (Schreiber and Fersht, 1993). The resulting affinity constant of about 10^{14}M is, therefore, close to that of avidin–biotin, although both k_{on} and k_{off} values are significantly different. Recent DFS measurements on the interaction between Barnase and Barstar have shown that the unbinding force as a function of the logarithm of the loading rate is characterized by a single, linear regime from which a dissociation rate k_{off} of 10^{-14} s^{-1} and an energy barrier width x_β of 0.12 nm have been extracted (Sekatskii et al., 2010). The much lower k_{off} value than that measured by standard biochemical methods in bulk has been attributed to the use of an oversimplified single energy barrier for describing the Barnase–Barstar interaction. Moreover, from the analysis of structural data, both a short and long electrostatic interaction is expected to regulate the Barnase–Barstar binding properties. Since DFS results were found not to be affected by the salt concentration, it has been hypothesized that the unbinding properties disclosed in DFS experiments could be mainly determined by the short-range interactions. Accordingly, the authors suggested the possibility to disentangle by DFS the short-range "contact" and long-range electrostatic interprotein interactions in a systematic way.

6.3.2 DFS STUDIES OF ANTIGEN–ANTIBODY COMPLEXES

Antigen–antibody systems are generally characterized by rather low dissociation rates, just below than those measured for the avidin–biotin and Barnase–Barstar complexes (see Figure 6.1). Antibodies (also known as immunoglobulins) are proteins used by the immune system of vertebrates to identify and neutralize foreign objects, called *antigens*, which can be molecules or molecular fragments, even belonging to bacteria, viruses, or cell surfaces. Recognition of an antigen by an antibody tags it for attack by other parts of the immune system. Antibodies are made of basic structural units, each one having two large heavy chains and two small light chains, giving rise to a Y-shaped structure, bearing a binding site on each arm of the Y. Although the general structure of all antibodies is very similar, the small binding region is extremely variable, allowing millions of different antigen binding sites; the variable domain being usually called *Fv* region. Antibodies can be derived from a single immune cell (monoclonal antibodies) or they can be made by the combination of genetic material from two or more sources (recombinant antibody). Both monoclonal and recombinant antibodies have become ubiquitous tools in biomedical science, since they are used to research, diagnose, and even as effective therapeutic treatments for cancer, various autoimmune disorders, and other diseases. Furthermore, they are widely used for self-assembling of nanostructures and for diagnostics assays to detect a wide range of analytes (Maynard and Georgiou, 2000). The large interest from both fundamental and applicative points of view has stimulated the study of the antigen–antibody interaction at single-molecule level, since from the beginning of DFS application to biological complexes (Moy et al., 1994a). The results obtained by DFS on some representative antigen–antibody pairs are summarized in Table 6.2. In general, the measured unbinding force values span a wide range from 30 to 250 pN, in agreement with those obtained for other biomolecular complexes. An analogous variability has been also observed for both the k_{off} and x_β parameters, consistently with the heterogeneous biological functions played by these systems.

Schwesinger et al. have investigated, by both DFS and stop-flow fluorescence spectroscopy, the interaction between fluorescein, a synthetic organic marker commonly used in microscopy, and different fragments derived from an antifluorescein antibody with punctual mutations (Schwesinger et al., 2000). In all these complexes, the unbinding force plotted as a function of the logarithm of the loading rate has shown a single linear trend with various intercepts and slopes; the dissociation rate values having been found to be different for each complex. In general, this indicates that DFS is extremely sensitive to even subtle details in the intermolecular interaction between very similar biomolecules. In addition, they found that the k_{off} values determined by DFS were almost the same as those obtained by fluorescence spectroscopy in bulk. Such an agreement, which is not very often observed, has been attributed to the low loading rates applied in the DFS measurements, a condition that better approximates the zero-force extrapolation of bulk data. Interestingly, an analysis of the plot of x_β versus logarithm of k_{off} has shown a high correlation between these two quantities, which could arise from an interplay between the energy landscape features (height and width) and the elasticity of the protein.

Berquand et al. have used an approach that combines DFS and SPR to study the interaction between the lysozyme protein and specific antilysozyme *Fv* fragments (Berquand et al., 2005). The two techniques have been applied under almost the same experimental conditions with the aim to investigate whether, and under which conditions, DFS results could be consistent with those obtained in bulk at zero force. An analysis of the unbinding force as a function of the logarithm of the loading rate has evidenced two linear trends with markedly different slopes, related to the existence of two distinct barriers in the energy landscape. In general, the involvement of two, or even more, barriers in the antigen–antibody unbinding process is believed to be a general feature of these systems, likely arising from the capability of achieving a fine tuning of their biological function. More specifically, the extracted dissociation rate

TABLE 6.2

DFS Results for a Collection of Antigen–Antibody Complexes: Unbinding Force at a Given Loading Rate (in parentheses); Dissociation Rate k_{off} and Energy Barrier Width x_β (in parentheses) as Extracted by Applying the Bell–Evans Model for Each Linear Trend Observed in the F^* versus $\ln(r_F)$ plot (see Equation 6.2)

System: Substrate/Tip	Immobilization: Substrate/Tip	Cantilever Spring Constant k_{cant} (N/m)	Unbinding Force (pN) (Loading Rate (nN/s))	Loading Rate Range (nN/s)	k_{off} (s^{-1}) (x_β (nm))	Reference
HSA/ Anti-HSA	Mica (PEG) Si_3N_4 (PEG)	0.27	244 (1)	0.1–5	$1.7 \cdot 10^{-5}$ (0.5)	Hinterdorfer et al., 1996
Antiferritin/ Ferritin	Silicon (Silane) Si_3N_4	0.043–0.057	50 (NA)	NA	NA	Allen et al., 1997
Antifluorescein/ Fluorescein	Gold Si_3N_4	0.008–0.014	140(1)	0.1–10	0.003 (0.4) 2.1 (0.05)	Schwesinger et al., 2000
Antilysozime/ Lysozime	Gold Gold–coated	0.01	50 (1)	1–100	0.001 (1) 150 (0.06)	Berquand et al., 2005
Mucin 1/ Anti Mucin 1	Gold Gold–coated (PEG)	0.04–0.18	120 (1)	0.4–10	$2.6\ 10^{-3}$ (2.8)	Sulchek et al., 2005
ICAM-1/ LF Antigen 1	Culture disk Si_3N_4 (Cells)	0.004–0.013	100 (1)	0.05–100	0.55 (2.6) 19 (0.49)	Wojcikiewicz et al., 2006
ICAM-2/ LF Antigen 1			60 (1)	0.05–100	0.31 (4.5) 0.015 (1.15)	
Anti-digoxigenin/ Digoxigenin	Glass (PEG) Si_3N_4 (PEG)	0.006–0.05	40 (2)	5–100	0.015 (1.15) 4.56 (0.35)	Neuert et al., 2006
Anti-GCN4/ GCN4	Glass (PEG) Si_3N_4 (PEG)	0.06–0.08	50(1.5)	0.081–40	$3.9\ 10^{-3}$ (0.88) $4.9\ 10^{-4}$ (0.90) $8.2\ 10^{-4}$ (0.92)	Morfill et al., 2007

Note: Some details of the experimental setup: the immobilization strategies of the biomolecules to both the substrate and the AFM tip; the cantilever spring constant; the loading rate range. (NA = not available.)

values of (10^{-3} s^{-1} and $1.5 \cdot 10^2$ s^{-1}) put into evidence the presence of a slow and a fast processes; with the switching between the two processes being evident from the abrupt change in the unbinding force plot when higher loading rates were applied. The k_{off} value corresponding to the slow process agrees with that measured by SPR ($3 \cdot 10^{-3}$ s^{-1}). Accordingly, they deduced that the DFS data could provide a good description of the zero-force SPR measurements when they are taken at very low loading rate and this should be kept in mind when comparing DFS and SPR data.

Sulchek et al. have used DFS to study the mucin 1 peptide in interaction with a recombinant antibody selected through a library screening (Sulchek et al., 2005). Mucin 1 is a transmembrane glycoprotein expressed in a variety of epithelial tissues, and it is involved in the adhesion process to neighboring cells. Since mucin 1 has been found to be overexpressed in some cancers, it may represent a suitable target for a family of immunotherapeutics for cancer treatment. Searching for information at single molecule level has been therefore conceived in the perspective of designing more effective drugs. A particular attention has been devoted to reliably discriminate between single and multiple unbinding events of mucin 1–involving complexes, in both monovalent and multivalent configurations. With such an aim, the authors have developed an immobilization strategy based on the mixing of two types of linkers, one possessing the ability to covalently bind the biomolecules and the other being unable to do it. Such an experimental setup, combined with an *ad hoc* analysis of the unbinding properties, even by varying the relative concentrations of these two linkers, has provided some general criteria to clearly distinguish between single and multiple events. In this respect, the authors demonstrated that when multiple events occur in parallel, the effective load applied to the system is reduced, and this should be taken into account in the analysis of DFS data. Such an approach led them to find out that the dissociation rate of the mucin 1–antibody complex drastically decreases when multivalent unbinding processes occur, thus supporting the importance for the biomolecules to preserve their multivalent capability for a more efficient binding. We would like to remark that the developed immobilization strategy and the related data analysis could be of rewarding help in the study of any ligand–receptor pairs.

Wojcikiewicz et al. have investigated the interaction of the leucocyte function-associated antigen-1 with two different intercellular adhesion proteins ICAM-1 and ICAM-2 involved in cell transmigration across endothelium (Wojcikiewicz et al., 2006). Antigen-1 is a transmembrane glycoprotein, while ICAM-2 and ICAM-2 are composed by immunoglobulin domains, a transmembrane domain and short cyto-plasmic domain. Under normal conditions, ICAM-2 is the dominant receptor for mediating leukocyte trafficking, whereas ICAM-1 becomes largely responsible for mediating the adhesion of leukocytes to the inflamed endothelium. In DFS experiments, antigen-1 molecules were directly embedded on the surface of Jurkat cells, which were, in turn, bound to the AFM tip, whereas the antibodies were immobi-lized on the substrate by a standard protocol (see Figure 6.6). Such an immobiliza-tion strategy has the advantage of approaching the single-molecule regime while the real physiological conditions are maintained. Both the complexes involving ICAM-1 or ICAM-2 show two distinct linear trends in the plot of the unbinding force as a

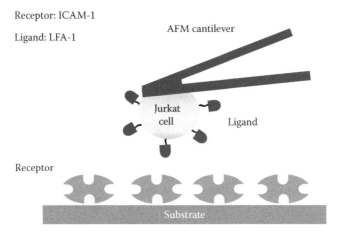

Receptor: ICAM-1

Ligand: LFA-1

AFM cantilever

Jurkat cell

Ligand

Receptor

Substrate

FIGURE 6.6 Sketch of the experimental setup used to investigate by DFS the interaction between the receptor, ICAM-1, and the ligand, LFA-1, located on the surface of Jurkat cells. (Adapted from Wojcikiewicz, E. P. et al. 2006. *Biomacromolecules*, 7:3188–3195.).

function of the logarithm loading rate, indicating that the unbinding process involves a double energy barrier. In particular, they have hypothesized the presence of an inner barrier associated with a fast process and an outer barrier responsible for the slower process (see Table 6.2). These DFS data, analyzed also in connection with the SPR data available in literature on the same systems, pointed out that the two complexes are characterized by similar energy heights but different energy barrier widths. Since ICAM-1 and ICAM-2 have a very similar structure, these differences have been traced back to small changes in the structural organization of the binding sites. The authors have also shown that the addition of Mg^{2+} to the solution can induce a strengthening of the interaction between the biomolecular partners, likely due to some changes in the binding site environment, arising from a modulation of the energy landscape. This work constitutes a remarkable example of how DFS experiments, carried out in single-molecule regime and in near-native environment, could be able to put into evidence even small differences in the binding properties of biological complexes; these differences being usually hidden in bulk measurements.

Morfill et al. have explored the possibility to use DFS to discriminate among variants of recombinant antibodies through the analysis of their binding properties with suitable ligands (Morfill et al., 2007). Indeed, selection of antibodies with high affinity for specific ligands is required in many applications. The interaction of the GCN4 peptide with three fragments, obtained from recombinant antibodies at different steps of the affinity maturation process, has been studied to determine their affinity. For all the three complexes, the unbinding force plotted as a function of the logarithm of the loading rate showed a linear trend with the same slope and slightly different intercept (see Table 6.2). Accordingly, they have hypothesized the existence of a single energy barrier whose height varies with the maturation degree of the antibody; such an effect being related to a modification in the antibody binding pocket. Interestingly, the standard Bell–Evans analysis has been combined with an alternative procedure to extract information on the energy barrier from DFS data. In particular, they have fitted the histograms of the unbinding force through a function, derived from Equation 6.1, to take into account for a distribution of the spacer length; a k_{off} of 1 s^{-1} and x_β of 0.90 nm having been extracted. A comparison with the corresponding values extracted by the Bell–Evans model (see Table 6.2) indicates that the results from the two approaches are consistent. In addition, the authors have also carried out SPR measurements using a setup very similar to that of DFS by finding out a rather good agreement with the k_{off} values obtained by DFS. On such a basis, the authors deduced that DFS can be competitive with SPR, in terms of required efforts and time, when DFS measurements are carried out at a single loading rate value, and the related data analysis is performed by using Equation 6.1.

6.3.3 DFS STUDIES OF COMPLEXES WITH SHORT LIFETIME

By referring to Figure 6.1, complexes with dissociation rates higher than 10^{-1} s^{-1}, and then lifetimes shorter than 10 s are located at the opposite end with respect to avidin–biotin. These kinetic rates are consistent with a relatively high turnover and confer a transient character to these systems (Crowley and Ubbink, 2003). Biomolecules involved in transient complexes usually possess the additional capability to recognize different partners with a rather high degree of specificity. Furthermore, transient complexes are characterized by a high association rate (k_{on} in the range of 10^7–10^9 M^{-1}s^{-1}) and then affinity constants in the μM^{-1}–mM^{-1} range. All these features make transient complexes particularly intriguing from both fundamental and applicative points of view. For example, the study of transient complexes formed by biomolecules endowed with electron transfer capabilities could be enlightening about the molecular features and processes regulating the electron transfer in biological systems. In addition, a comparison of the biorecognition process occurring in these systems with that related to complexes characterized by longer lifetimes (such as avidin–biotin, antigen–antibody, etc.) could be insightful for the elucidation of the overall underlying a biorecognition process. On the other hand, DFS could be of remarkable help to study the interplay between the biorecognition and the electron transfer processes in transient complexes when they are integrated with electronic transducers to build up hybrid nanodevices for biosensing applications (Bonanni et al., 2007).

However, DFS has been only recently applied to investigate transient complexes; the relatively high dissociation rate, generally coupled with a relatively low affinity between the partners, make the experimental approach and the data analysis more challenging with respect to those formed by partners undergoing a tight binding. One of the first transient complexes studied by DFS is the one formed by the copper protein azurin and the heme-protein cytochrome c551 from the bacterium *Pseudomonas aeruginosa*, whose interaction drives the physiological electron transfer process occurring with high efficiency between them (Bonanni et al., 2005). Since the structure of this complex was not available, a computational docking study has been preliminarily carried out to search for the best steric fit between the partners.

It was found that the best complex involves a close contact between the hydrophobic regions of the two proteins (Cutruzzolá et al., 2002; De Grandis et al., 2007). According to this evidence, Azurin has been immobilized to the substrate through its native disulphide bridge that is located at the opposite end of the hydrophobic region; in such a way, the region containing the active sites was left available for the interaction. More specifically, two different immobilization strategies have been followed: In the first one, azurin was directly bound to bare gold (Bonanni et al., 2005), whereas in the second one, it was immobilized to gold through the introduction of alkanethiol as linker (Bonanni et al., 2006). In both the cases, cytochrome c551 was bound to the AFM tip through a PEG linker.

The unbinding forces obtained for these two systems (Table 6.3) are in the same range of the unbinding forces measured for avidin–biotin and for antigen–antibody pairs. These results confirm that the unbinding force value alone does not provide a quantitative evaluation of the complex strength. The unbinding force plotted as a function of the loading rate for the azurin–cytochrome c551 complex shows a linear trend for both the used immobilization strategies (see Figure 6.7), reflecting the existence of a single energy barrier. The extracted k_{off} values (Table 6.3) are indicative of a quite fast dissociation, consistently with the transient nature of this electron transfer complex (Crowley and Ubbink, 2003). The lower dissociation rate found for azurin anchored on functionalized gold with respect to that of azurin on bare gold suggests that when a direct interaction of the protein with the metal is prevented, its native configuration is better preserved, and in addition, the protein is endowed with a higher reorientational freedom for optimal interaction with its counterpart. These results provide some general hints on how to optimize the immobilization of biomolecules to a metal surface.

Dynamic force spectroscopy has been applied to study other complexes with short lifetimes, such as metal-containing partners, systems involved in neurological processes, or complexes relevant for cancer (see Table 6.3). Although the dissociation rate is generally higher than $1-10 \text{ s}^{-1}$, the corresponding unbinding force does not reveal any hint of the corresponding short lifetime, in agreement with the widely reported lack of correlation between the dissociation kinetics and the unbinding force values.

Yersin et al. have investigated by DFS the interaction between the glycoprotein transferrin and its receptor, which controls the level of Fe^{3+} (Yersin et al., 2008). Although Fe^{3+} is an important cofactor in many biological systems, free Fe^{3+} is both toxic for living cells and is insoluble. The transferrin receptor, located at the cell surface, binds iron-loaded transferrin and transports it to endosomes where Fe^{3+} can be released. To investigate the role played by Fe^{3+} on the formation of the complex, both the holo- and the apo-forms of transferrin have been used. The authors found that the unbinding force as a function of the logarithm of the loading rate is characterized by a single linear trend for the apo-form of transferrin, whereas two well-distinct linear regimes are observed for the holo-transferrin. The corresponding dissociation rates, reported in Table 6.3, indicate the occurrence of both a fast and a slow processes for the holo-complex, while a single process with an intermediate rate

TABLE 6.3

DFS Results for a Collection of Complexes with Short Lifetimes: Unbinding Force at a Given Loading Rate (in parentheses); Dissociation Rate k_{off} and Energy Barrier Width x_β (in parentheses) as Extracted by Applying the Bell–Evans Model for Each Linear Trend Observed in the F^* versus $\ln(r_F)$ plot (see Equation 6.2)

System: Substrate/ Tip	Immobilization: Substrate/ Tip	Cantilever Spring Constant k_{cant} (N/m)	Unbinding Force (pN) (Loading Rate (nN/s))	Loading Rate Range (nN/s)	k_{off} (s^{-1}) (x_β [nm])	Reference
Azurin/ Cytochrome c 551	Gold Si$_3$N$_4$ (PEG)	0.1–0.5	95 (10)	20–200	14 (0.14)	Bonanni et al., 2005;
Azurin/ Cytochrome c 551	Functionalized Gold Si$_3$N$_4$ (PEG)	0.022–0.040	140 (10)	30–150	6.7 (0.098)	Bonanni et al., 2006
Syntaxin/ Synaptobrevin	Ni–glass Ni–Si$_3$N$_4$ (PEG)	NA	220 (10)	1–20	6.3 (NA)	Liu et al., 2006
Syntaxin + SNAP25/ Synaptobrevin		NA	230 (10)	1–20	0.5 (NA)	
p53/ Azurin	Gold Si$_3$N$_4$ (PEG)	0.5	75 (10)	0.2–20	0.09 (0.52)	Taranta et al., 2008
Transferrin/ Tf receptor	Mica Si$_3$N$_4$ (PEG)	0.02	40 (10)	0.5–70	0.25 (0.81)	Yersin et al., 2008
p53/ Mdm2	Functionalized gold Si$_3$N$_4$ (PEG)	0.017–0.045	130 (10)	0.6–70	1.5 (0.17)	Funari et al., 2010

Note: Some details of the experimental setup: the immobilization strategies of the biomolecules to both the substrate and the AFM tip; the cantilever spring constant; the loading rate range. (NA = not available).

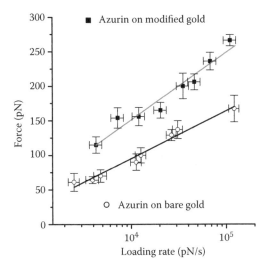

FIGURE 6.7 Unbinding force as a function of the logarithm of the loading rate measured by DFS for the azurin-cytochrome c551 complex using two different strategies for the immobilization of azurin on the substrate. (Adapted from Bonanni, B., Kamruzzahan, A. S. M., Bizzarri, A. R. et al. 2005. *Biophys. J.*, 89, 2783–2791. Adapted from Bonanni, B. et al. 2006. *J. Phys. Chem. B*, 110, 14574–14580.)

characterizes the apo-involving complex. In addition, at each analyzed loading rate, the average unbinding force was found to be lower for the complex involving the apo-form of transferrin. These results were considered indicative of the existence of two binding sites for the interaction between holo-transferrin and its receptor and of a single one for the apo-form, thus confirming a previously hypothesized model. By the same technique, the authors have investigated the interaction between isolated transferrin and the receptor endogenously expressed at the surface of living cells. They found substantially the same results as those obtained by using isolated biomolecules; thus inferring that for this complex the interaction does not depend on the molecular environment of the biomolecules.

Liu et al. have investigated the interactions between syntaxin and synaptobrevin in the presence of the synaptosome-associated protein (ternary complex) and in the absence of this protein (binary complex) (Liu et al., 2006). All these proteins belong to the superfamily of SNARE proteins, which play a primary role by mediating fusion of synaptic vesicles with the presynaptic membrane. In particular, the analyzed complex is involved in the exocytosis release of neurotransmitters. Measurements using isolated proteins have shown that the force necessary to dismantle a ternary complex is smaller than that necessary for the binary one. In addition, they found a different extension of the two systems (12 and 23 nm for the binary and ternary complexes, respectively) under the application of the force. They also found that the unbinding force as a function of the logarithm of the loading rate is characterized by a linear trend but with different slope and intercept parameters for the two systems (see Table 6.3). A dissociation rate of 6.3 s^{-1} observed for the binary complex is an order of magnitude lower than that of the ternary complex (0.5 s^{-1}). The higher stability shown by the ternary complex supports the important role of the synaptosome protein to allow positioning of vesicles at a maximal distance (about 12 nm) from the plasma membrane for a longer time than that corresponding to the binary complex. These findings concord with other results from other techniques and confirm a previous so-called zippering model taking into account the formation of coiled coils in the binary complex.

Successively, the analysis of the DFS data on the syntaxin-synaptobrevin binary complex has been extended within the framework of the very interesting Jarzyniski theoretical model (Liu et al., 2008). It allows to determine the equilibrium binding free energy of the complex from the work done by the applied force along several nonequilibrium paths connecting the initial and the final states of a reaction (see Chapter 3; Hummer and Szabo, 2005; Jarzynski, 1997). A simplified expression has been worked out for the binding free energy ΔG:

$$e^{-\Delta G/k_BT} = \sum_{i=1}^{N} \frac{1}{N} e^{-W_i/k_BT} \tag{6.4}$$

where N is the number of independent iterations of the unbinding process, and W_i is the work along the ith unbinding path done under the application of the external force. For a given force curve, the work done by the applied force during the unbinding process can be determined by calculating the integral from the beginning of the nonlinear course in the retraction curve up to the end of the jump-off event (see the inset of Figure 6.2). Accordingly, they have estimated a binding free energy for the syntaxin–synaptobrevin complex of about $(49 \pm 5)k_BT$. Such a value has been compared with that evaluated from the Arrhenius relationship ($k_{\text{off}} \propto e^{-(-\Delta G^*/k_BT)}$) at three different temperatures ($T = 277, 287, 297$ K) obtaining $\Delta G = (33 \pm 6)k_BT$. The discrepancy between the two ΔG values has been traced back to the fact that the applied force does not act along the bond axis, making in this case the application of the Bell–Evans model not completely appropriate. This approach, remarkably, endows DFS experiments with the capability to measure the binding free energy, besides determining k_{off} and x_β.

Recently, DFS has been applied to investigate complexes involving the human tumour suppressor p53, which is a protein known to play a crucial role in triggering cancer defense mechanisms. In the presence of different stress signals, p53 is stabilized through posttranslational modifications, its cellular levels increase, and it can induce the expression of target genes that, in turn, control the process of DNA-repair, the cell-cycle arrest, and the apoptotic cascade (Vogelstein et al., 2000). The activity of p53 is downregulated by the mdm2 oncogene that promotes its ubiquitin-dependent degradation through the formation of a complex with it (Chéne, 2004). On such a basis, the mdm2-p53 complex is a preferential target for anticancer drug design devoted to restore normal p53 function in tumour cells by preventing its mdm2-mediated inactivation. The DFS investigation of the interaction of p53 with mdm2 was done by using, for the first time, full-length proteins for both p53 and mdm2 (Funari et al., 2010); previous studies by bulk techniques having been carried out using only partial domains of both proteins (Schön et al., 2002). Funari et al. (2010) found a single linear trend in the unbinding force versus the logarithm of the loading rate; the corresponding dissociation rate k_{off} of 1.5 s^{-1} is indicative of a transient character. Interestingly, this result, obtained at single-molecule level and using full-length proteins, is in a good agreement with that obtained in bulk by partial chains of both the proteins ($1-2$ s^{-1}; Schön et al., 2002; Domenici et al., 2011). Notably, a comparison of measurements by DFS using full length or portions of biomolecules makes it possible to obtain information on which regions are involved in the interaction and even on the interplay among the different regions on the kinetic properties. Bizzarri and Cannistraro have analyzed the DFS data of the mdm2-p53 complex in the framework of the Jarzinski model by separately evaluating the contribution to the total binding free energy arising from both the stretching of the linker used in the setup and the complex unbinding process (Bizzarri and Cannistraro, 2011). The extracted unbinding free energy of -8.4 kcal/mol has been found to be in a good agreement with that measured in bulk by isothermal titration calorimetry again using partial domains of both the proteins (from -8.8 to -6.6 kcal/mol; Schön et al., 2002). Starting from the evidence that azurin was able to promote an anticancer activity through its binding to p53 *in vitro* and *in vivo*, the interaction between azurin and p53 was studied by DFS (Taranta et al., 2008). From the analysis of the unbinding force as a function of the logarithm of the loading rate, Taranta et al. found a single barrier in the energy landscape with a dissociation rate of 0.09 s^{-1} (see Table 6.3). This result, indicating a lifetime of about 10 s, is consistent with the formation of a relatively stable complex. With the aim to extract some information on the interaction sites between azurin and p53, the DFS study has been complemented by computational docking. The possible binding regions between azurin and two different partial domains of p53 (the DNA binding domain and the N-terminal domain) have been proposed and refined in connection with the available mutagenesis data (De Grandis et al., 2007; Taranta et al., 2009). These structural informations were found to be extremely insightful to design azurin-derived peptides retaining the same ability of azurin to penetrate cancerous cells and to extert a strong anticancer activity with minimal side effects (Yamada et al., 2009).

Although DFS has been mainly applied to investigate bimolecular complexes, the setup can be easily adapted to study ternary complexes or even to perform competitive binding experiments. Recently, a possible competition of azurin with mdm2 for the same binding site on p53 has been carried out by conceiving a competitive blocking experiments, according to the strategy sketched in Figure 6.8. First, the frequency of the unbinding events between p53 immobilized on the substrate and azurin anchored to the tip has been estimated before and after blocking the substrate with a solution of free azurin; a significant decrease in the unbinding frequency being observed upon azurin addition. The experiment has been then repeated by blocking the substrate with a solution of free mdm2. Successively, the experiment has been

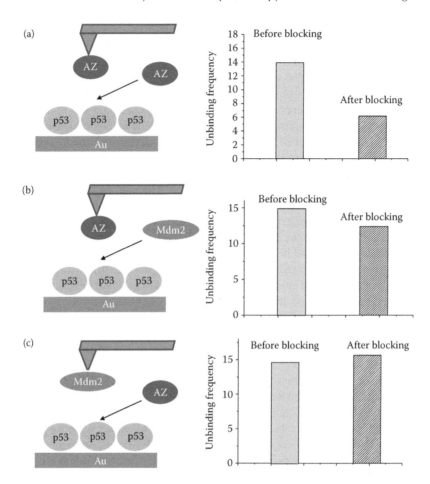

FIGURE 6.8 **(See color insert.)** Competitive blocking experiments on the p53–mdm2–azurin ternary complex. (a) Azurin is used as a competitor for the p53-azurin complex (left); unbinding frequencies before and after blocking the substrate with a solution of free azurin (right). (b) mdm2 is used as a competitor for the p53-azurin complex (left); unbinding frequencies before and after blocking the substrate with a solution of free mdm2 (right). (c) Azurin is used as a competitor for the p53-mdm2 complex (left); unbinding frequencies before and after blocking the substrate with a solution of free azurin (right). (Adapted from Bizzarri, A. R. and Cannistraro, S. 2009. *J. Phys. Chem. B*, 113, 16449–16464. Adapted from Funari, G. et al. 2010. *J. Mol. Recognit.*, 23, 343–351).

carried out by measuring the frequency of the unbinding events between p53 immobilized on the substrate and mdm2 anchored to the tip before and after blocking the substrate with a solution of free azurin. In both the latter cases, substantially no changes in the unbinding frequency has been observed. Such a finding has provided a demonstration that two proteins, azurin and mdm2, do not compete for the same binding site on p53.

6.3.4 DFS STUDIES OF COMPLEXES INVOLVING DNA OR APTAMERS

Deoxyribonucleic acid and ribonucleic acid (RNA) molecules drive and regulate the storage, transport, processing, and expression of genetic information in living organisms. DNA consists of two long polymers of nucleotides organized in complementary strands kept together by noncovalent bonds. RNA, which is derived from DNA through transcription process, differs from DNA by the substitution of the sugar deoxyribose with ribose, and it is usually single stranded. Since gene expression in eukaryotes is controlled by the specific binding of transcription factors to defined DNA sequences, the possibility to influence and control cell functions through modified synthetic transcriptional factors at single-molecule level offers fascinating prospects for molecular biology. Recently, a new class of molecules derived from DNA, called *aptamers*, have received a large attention, especially for applicative purposes (Cho et al., 2009). Aptamers are synthetic oligonucleotides derived from *in vitro* experiments in which single-stranded DNA or RNA molecules are isolated from a random sequence library according to their ability to selectively bind target molecules, which may be organic dyes, drugs, metal ions, amino acids, peptides, and proteins (Ellington and Szostak, 1990). With respect to antibodies, which, indeed, are widely used, aptamers present many advantages, including simpler synthesis, faster tissue penetration, easier modification and storage, and wider applicability; therefore, they may strongly rival antibodies in both diagnostic and therapeutic applications.

Table 6.4 reports the results from some DFS studies on (1) DNA molecules; (2) complexes involving DNA in interaction with peptides, proteins, drugs, and so on; and (3) aptamer–base complexes. It is interesting to remark that the unbinding force values for these systems are similar to those measured for protein pairs (see Tables 6.1 through 6.3). This suggests that the noncovalent interactions regulating biomolecular association possess general features, irrespective of the specific molecules involved.

The first DFS studies on DNA have been fully devoted to investigate the interaction between complementary DNA strands. With such an aim, complementary single strands of a DNA have been immobilized on an AFM tip and a surface via linkers (see Figure 6.9). Starting from the preliminary results of a previous work (Lee et al., 1994a), Strunz et al. have compared the unbinding force between DNA strands formed by 10, 20, and 30 base pairs (bp) (Strunz et al., 1999). For all these systems, the unbinding force plotted as a function of the logarithm of the loading rate has revealed a linear regime indicative of the existence of a single energy barrier to be overcome to break the double DNA strand. Accordingly, the kinetic mechanism regulating such a process has been described as a thermally driven dissociation, similarly to what happens for protein–protein interactions. In addition, they found that the width of the energy barrier linearly increases with the number of bp while the dissociation rate exhibits an exponential decrease. These effects have been interpreted in terms of a cooperative unbinding effect of the bp responsible for the

TABLE 6.4

DFS Results for a Collection of DNA-based Complexes Unbinding Force at a Given Loading Rate (in parentheses); Dissociation Rate k_{off} and Energy Barrier Width x_β (in parentheses) as Extracted by Applying the Bell-Evans Model for Each Linear Trend Observed in the F^* versus $\ln(r_F)$ plot (see Equation 6.2)

System: Substrate/ Tip	Immobilization: Substrate/ Tip	Cantilever Spring Constant k_{cant} (N/m)	Unbinding Force (pN) (Loading Rate (nN/s))	Loading Rate Range (nN/s)	k_{off} (s^{-1}) (x_β (nm))	Reference
DNA strand 10 bp	Glass (PEG)	0.012–0.017	20(1)	0.02–10	10^{-2} (1.0)	Strunz et al., 1999
DNA strand 20 bp	Si_3N_4 (PEG)		48 (1)		10^{-6} (1.4)	
DNA strand 30 bp			42(1)		10^{-10} (2.0)	
DNA strand 16 bp	Glass (PEG) Si_3N_4 (PEG)	0.02	$T = 27°C$ 46 (1)	0.05–10	$T = 11°C$ (0.18) $T = 27°C$ 0.5 (0.4) $T = 36°C$ 0.5 (0.9)	Schumakovitch et al., 2002
Lex A repressor/ DNA (rec A)	Aminoslide Si_3N_4	0.008–0.018	NA	0.5–9	3.1 (5.4)	Kühner et al., 2004
DNA fragment/ peptide	Mica Si_3N_4 (PEG)	NA	Wild-type NA mut1 61 (1) mut2 55 (1)	0.09–100	3.1 (0.68) 0.071 (0.93) 49.5 (0.72)	Eckel et al., 2003
DNA aptamers/ Immunoglobulin E	Silicon Si_3N_4	0.06–0.12	86–145 (NA)	80–210	NA	Jiang et al., 2003
DNA fragment/ Sfil protein	Mica Si_3N_4	0.04–0.07	100 (1)	2–500	38 (0.18) 70 (0.19) 248 (0.19)	Krasnoslobodtsev et al., 2007

Note: Some details of the experimental setup: the immobilization strategies of the biomolecules to both the substrate and the AFM tip; the cantilever spring constant; the loading rate range. (NA = not available).

(a)   (b)

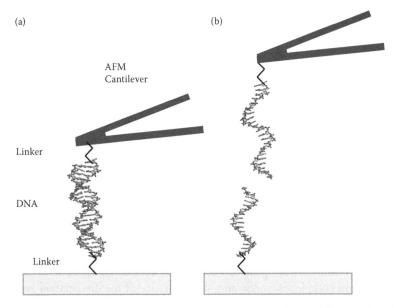

AFM
Cantilever

Linker

DNA

Linker

FIGURE 6.9 Sketch of the experimental setup used to measure by DFS the unbinding forces between complementary single DNA strands. (a) DNA was anchored to both the AFM tip and a substrate through a PEG linker. By approaching the tip to the substrate, a DNA duplex may be formed. (b) An unbinding process of DNA duplex occurs by retracting the tip from the substrate. (Adapted from Strunz, T. et al. 1999. *Proc. Natl. Acad. Sci. U.S.A,* 96:11277–11282.)

scaling of the unbinding forces. Schumakovitch et al. have successively extended the study of the unbinding process between DNA strands to include an analysis at different temperatures (Schumakovitch et al., 2002). They found that the dissociation rate increases with the temperature whereas the energy barrier width decreases (see Table 6.4). The temperature dependence of the dissociation rate has been observed to be in good agreement with the data obtained in bulk solution by the temperature jump technique. From the Arrhenius plot, they have extracted the activation enthalpy for the dissociation process by finding a value again consistent with that extracted by bulk experiments. Such an agreement has been put into relationship to the use of a slow pulling rate in DFS experiments, which allows to more correctly probe the energy barrier. At variance, the observed temperature dependence of the energy barrier width, which is assumed to be constant in the framework of the Bell–Evans model, has been tentatively ascribed to an entropic contribution to the process. In this respect, they have remarked that possible discrepancies among DFS data might arise from temperature effects that are not accounted for by the Bell–Evans model, and whose effective impact on the final results should be instead taken into account.

Eckel et al. have explored the DFS capabilities to investigate the interaction of a synthetic peptide with a double-stranded DNA sequence (Eckel et al., 2005). The peptide formed by 20 aminoacid residues has been derived from the DNA transcriptional activator PhoB involved in phosphate metabolism; such a peptide including the specific sequence responsible for the interaction with DNA. Besides the native form, two different single point mutated forms (mut1 and mut2) for the peptide have been analyzed. Preliminarily, the authors have conceived competition experiments using the native, or one of the mutated forms of the peptide, to investigate the specificity of the interactions with DNA. They found that the wild type and two of the three mutated forms resulted into many binding events that have been traced back to a specific binding to the DNA sequence. Instead, the other one yielded a few binding events, which statistically led to conclude that there is no interaction with DNA. In addition, the wild type and two mutated forms of the peptide exhibit a linear trend dependence for the unbinding force as a function of the logarithm of the loading rate when they interact with DNA (see Table 6.4). The extracted dissociation rate and the energy barrier width values are found drastically different in the various systems, indicating that even single point mutations of the peptide can significantly modulate its interaction properties with DNA. Furthermore, the dissociation rate and the energy barrier width values generally fall in the same range of those obtained for protein–protein complexes, confirming the similarity of the dissociation properties of these biomolecular systems. In a general way, the results put into evidence the high sensitivity of DFS to probe the binding specificity between these molecules, supporting even the possibility to combine DFS with chemical synthesis strategies for selection of drugs or ligands.

Krasnoslobodtsev et al. have probed the interaction between the tetrameric protein SfiI with a double-stranded oligonucleotide from DNA (Krasnoslobodtsev et al., 2007). SfiI belongs to a family of restriction enzymes that recognize and specifically bind to DNA for cleavage reaction. In particular, SfiI binds two DNA duplexes, each one containing 13 bp, separated by a 5 bp palindromic region that is apparently not essential for the recognition. To investigate the role of this nonspecific sequence, the authors have studied the unbinding force of SfiI interacting with one of the three different DNA fragments, each being formed by 40 bp including the two specific sequences and a different nonspecific sequence. For all the three systems, the unbinding force measured as a function of the logarithm of the loading rate has shown a linear trend with almost the same slope but with a different intercept. According to the Bell–Evans model, the existence of a single barrier characterized by both different dissociation rates and energy barrier widths has been derived (see Table 6.4). The rather high dissociation rate values observed for all the cases suggest that the complexes possess a transient character. Interestingly, the DFS measurements have been coupled with a topographic investigation by AFM imaging and it was found that the complexes are quite stable, irrespectively of the corresponding high dissociation rate value. To reconcile such an apparent inconsistency, the authors have suggested that the complex transiently dissociates while, however, the two biomolecules remain in close proximity. The significantly different dissociation rate values found for the three various complexes point out that the stability of each complex can be modulated by acting on a small portion of the DNA sequence that was assumed not to be directly involved in the biorecognition process. On such a basis, the authors have suggested a possible mechanism involving the formation of a dynamic complex coupled to a rolling process in which both the specific and the nonspecific sequences of DNA play some role. Jiang et al. have investigated the interaction between different DNA aptamers and the human immunoglobulin E, which plays a key role in allergic response (Jiang et al., 2003). They developed an immobilization strategy of the biomolecules, resulting in a rather high number of unbinding events (from 4 to 10) in each pull-off event. Under these conditions, the Poisson statistics provided a good description of the unbinding force distribution, allowing to single out the contribution from the rupture of a single pair of interacting biomolecules (see Chapter 4). They found an unbinding force of about 150 pN for a single specific interaction. The reliability and reproducibility of the results were demonstrated by repeating the experiments using different substrates and tips. Interestingly, they observed that the unbinding force value can be reduced by slightly increasing the concentration of NaCl; this is witnessing also the high sensitivity of DFS measurements to reveal even subtle changes in the physiological conditions. They have also compared the response of aptamers in competition with antibodies by carrying out control experiments in which aptamers were substituted with immunoglobulin E antibody. The finding that the affinity of aptamers to protein target matches or even surpasses that of antibodies lend significant grounds to the higher potentialities of aptamers for therapeutic and diagnostic applications.

Along the same direction, Basnar et al. have developed an approach exploiting aptamers for detection of thrombin, a protein bearing a high relevance in heart diseases (Bernhard et al., 2006). The interaction between a tip functionalized with aptamers and a gold substrate covered with thrombin molecules has resulted in a rather large number of rupture events (up to 15) during each retraction step. Application of the Poisson statistics has provided an unbinding force for each pair of about 5 pN; such a value, which is close to the instrumental limits of the AFM equipment, has been put into relationship to a melting of the quadruplex structure for the aptamers. These results may also be rewarding in light of extending the effective capability of DFS to the detection of extremely low interaction forces.

6.3.5 DFS of Complexes Involved in Adhesion and Aggregation Processes

A broad spectrum of biological processes requires controlled cell adhesion, that is, the binding of a cell to a substrate, which can be another cell, a surface, or an organic matrix. For example, cell-adhesion is involved in embryonic development, assembly of tissues, cellular communication, inflammation and wound healing, tumour metastasis, cell culturing, viral and bacterial infection, and so on. Cell adhesion is commonly regulated by specific cell-adhesion molecules (CAMs), which are typically transmembrane receptors, and comprise an intracellular domain interacting with cytoplasmic proteins and an extracellular domain that specifically binds to adhesion partners (Kemler, 1992). The major classes of CAMs in mammals are cadherins, selectins, integrins, and those belonging to the immunoglobulin superfamily. Some results from DFS studies on systems involved in adhesion processes are reported in Table 6.5.

TABLE 6.5

DFS Results for a Collection of Complexes or Systems Involved in Adhesion Processes: Unbinding Force at a Given Loading Rate (in Parentheses); Dissociation Rate k_{off} and Energy Barrier Width x_β (in Parentheses) as Extracted by Applying the Bell–Evans Model for Each Linear Trend Observed in the F^* versus $\ln(r_F)$ Plot (see Equation 6.2)

System: Substrate/Tip	Immobilization: Substrate/Tip	Cantilever spring constant k_{cant} (N/m)	Unbinding Force (pN) (Loading Rate (nN/s))	Loading rate range (nN/s)	k_{off} (s^{-1}) (x_β (nm))	Reference
P-selectin/ P-selectin ligand	Glass / Si$_3$N$_4$	0.06	120 (15)	12–240	0.022 (0.25)	Fritz et al., 1998
VE Cadherin/ VE Cadherin	Glass (PEG)/ Si$_3$N$_4$ (PEG)	0.03	32 (10)	10–120	1.8 (0.59)	Baumgartner et al., 2000
Platelet /	Glass	0.06	80 (12)	10–50	22 (0.1)	Lee and Marchant 2001
RGD-peptide P-selectin/ P-selectin ligand	Si$_3$N$_4$ Culture disk (cell) / Si$_3$N$_4$	0.01–0.04	150 (1)	0.1–10	0.2 (0.14)	Hanley et al., 2003
Fibronectin (wt)/ $\alpha_5\beta_1$ Integrin Fibronectin (mut.)/ $\alpha_5\beta_1$ Integrin	Culture disk / Si$_3$N$_4$ (cell)	10^{-5}–0.033	60 (1)	0.02–50	0.13 (0.4) 33.5(0.09) 0.012 (0.44) 29.1(0.91)	Li et al. 2003
VCAM-1 $\alpha_4\beta_1$ Integrin	Culture disk / Si$_3$N$_4$	0.01	50 (0.8)	0.1–100	59 (0.1) 0.13 (0.59)	Zhang et al., 2004
P-Selectin/ P-selectin ligand	Glass (lipid)/ Si$_3$N$_4$	0.004–0.013	20 (1)	3–300	0.077 (2.41) 33.6 (0.099)	Marshall et al., 2005
N-Cadherin/ N-Cadherin E-Cadherin/ E-Cadherin	Culture disk (cell)/ Si$_3$N$_4$ (cell) Culture disk (cell)/ Si$_3$N$_4$ (cell)	0.01	30 (1) 73 (1)	0.05–10	0.98 (0.77) 1.09 (0.32) 4.00 (0.10)	Panorchan et al., 2006
Erythrocyte / Fibrinogen Platelet/ Fibrinogen	Glass / Si$_3$N$_4$	0.019	73 (1) 60 (1)	1–30	0.023 (0.17) 60.9 (0.17) 0.034 (0.072) 0.026 (0.017)	Carvalho et al., 2010

Note: Some details of the experimental setup: the immobilization strategies of the biomolecules to both the substrate and the AFM tip; the cantilever spring constant; the loading rate range.

Although the first DFS studies on adhesion processes have been restricted to investigate the interaction between individual biomolecules, more recently the DFS capabilities have been extended to the study of the interactions between biomolecules embedded within the cell surface. In such a way, more reliable information on the behavior of biomolecules in physiological conditions can be extracted.

Baumgardner et al. have investigated the interaction between single recombinant cadherin proteins that are ubiquitous, calcium-dependent homophilic molecules involved in cell–cell adhesion (Baumgartner et al., 2000). Indeed, the cadherin–cadherin interaction plays a crucial role in a multitude of physiological and pathological processes, including embryogenesis, motility, differentiation, and carcinogenesis. They found that unbinding forces for a single cadherin–cadherin association are smaller than those previously obtained for other protein–protein complexes. Interestingly, they put into evidence that the unbinding force increases when a longer encounter time was applied; the encounter time being a delay time between the approaching and the retraction phases. Such an effect has been put into relationship to the occurrence of a cooperative interaction among cadherin molecules, likely responsible for multiple-adhesion processes. The authors have deduced some criteria to discriminate between specific and nonspecific unbinding events by taking into account the nonlinear extension in the retraction curve due to the stretching of the PEG linker used to bind cadherin to the AFM tip. More specifically, they have accepted only those force curves displaying such a nonlinear trend, whose corresponding unbinding length was consistent with that expected from the PEG stretching (see Figure 6.2 and Chapter 4). They found that the unbinding force as a function of the logarithm of the loading rate follows a linear trend indicative of the presence of a single barrier in the energy landscape, the corresponding parameters being reported in Table 6.5. The rather high dissociation rate value is indicative of a transient complex, consistently with the rapid remodeling of the cell shape to optimize the cellular adhesion at physiological conditions. Interestingly, the authors have developed a procedure to roughly estimate the association rate k_{on} from DFS data. In particular, they used the relationship $k_{on} = N_A V_{eff}/t^{0.5}$, where N_A is the Avogadro number, V_{eff} is the effective volume of a half-sphere, with radius R_{eff} around the tip, and $t^{0.5}$ is the time required for the half-maximal binding probability that can be evaluated from $t^{0.5} = 2R_{eff}/v$, where v is the approach speed of the cantilever. For the cadherin–cadherin association, they obtained a k_{on} in the range 10^3-10^4 M^{-1} s^{-1}. They have also derived an affinity constant $K_a = k_{on}/k_{off}$ in the range 10^3-10^5 M. Such an approach allowed them to analyze the affinity constant of the cadherin complexes as a function of the Ca^{2+} concentration. They found that the affinity drastically increased when the Ca^{2+} concentration exceeded a threshold at which the occurrence of some structural changes were able to modulate the intercellular adhesion processes.

Panorchan et al. have applied DFS to investigate complexes between different types of cadherin molecules (E- and N-cadherins), which were expressed on the surface of living cells that, in turn, were immobilized on both the tip and the substrate (see Figure 6.10; Panorchan et al., 2006). From the analysis of the unbinding force as a function of the logarithm of loading rate, they found two distinct linear

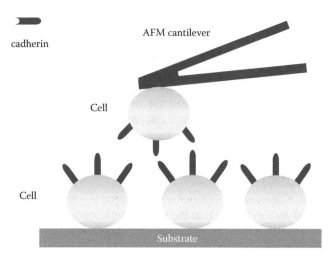

cadherin

AFM cantilever

Cell

Cell

Substrate

FIGURE 6.10   Sketch of the experimental setup used to measure by DFS the unbinding force between cadherin molecules expressed on the surface of living cells. (Adapted from Panorchan, P. et al. 2006. *J. Cell Sci.*, 119, 66–74.)

regimes for the E-cadherin–E-cadherin complex, and a single linear regime for the N-cadherin–N-cadherin one. Accordingly, the presence of one or two barriers in the energy landscape have been hypothesized in the unbinding process for these two systems, respectively. These differences have been traced back to the different biological function played by these two molecules. In particular, it has been suggested that the unbinding properties of E-cadherins could be related to the higher resistance to bond rupture under a mechanical stress with respect to N-cadherins. These results have led the authors to tentatively explain why metastatic cells express N-cadherin at their surfaces; transformed cells can easily break away from their neighbors in the primary tumor. It emerges that a detailed knowledge of the energy landscape, at single-molecule level, can c be useful to understand the physiological properties and the function of the biological systems. Finally, the authors, by using single recombinant molecules, have demonstrated that the unbinding properties of E-cadherins embedded in the cells are markedly different with respect to those extracted by using isolated biomolecules (see Table 6.4). This suggests, at the same time, some caution in the extrapolation to real systems of results obtained on single molecules. Indeed, the results on cadherin complexes show that the environment of biomolecules can regulate their kinetics, likely through a modulation of both the structural and the binding properties.

    The study of the interaction between cadherin molecules has recently taken advantage of the developments of AFM apparata capable of combining real-time topography and the characterization of their binding properties. Recognition imaging of VE-cadherin receptor binding sites both on cell surfaces and immobilized on a substrate has been performed by using a tip functionalized with VE-cadherin molecules in an instrumental modality called *Topography and Recognition* (TREC) (Chtcheglova et al., 2007). Such an approach has allowed Chtcheglova et al. to visualize, identify, and quantify local receptor binding sites at single-molecule level and to assign their locations to the topographical features of the cell surfaces. Such a combined approach is expected to become more and more followed in the study of biorecognition processes since it offers the possibility to extract simultaneous dynamical and spatial information on the systems at single-molecule level. Fritz et al. have investigated the interaction between P-selectin and P-selectin ligand (Fritz et al., 1998). P-selectin belongs to a family of glycoproteins that mediate adhesion with leukocytes or platelets and it is located within the endothelial cell wall, whereas the P-selectin ligand is expressed in the microvilli at the leukocyte surface. In particular, P-selectin supports leucocyte rolling under hydrodynamic flow via the interaction with its ligand. Interestingly, the authors have hypothesized that the application of the external force in the DFS experiments could mimic the physiological process. In other words, the binding and unbinding processes occurring during the DFS experiments could describe a rolling process on a surface under the action of a physiological fluid as depicted in Figure 6.11. Under this assumption, they have determined the dissociation rate from an analysis of the unbinding force trend versus the corresponding unbinding length in terms of a modified freely jointed chain model with a Monte Carlo simulation of the unbinding force distribution. They found a dissociation rate of 0.022 s^{-1}, which is indicative of a rather fast process, which could be due to the high chain-like elasticity of the system. The dissociation rate was also determined by SPR by finding a value of $3 \cdot 10^{-4}$ s^{-1}, which is markedly lower than that obtained by DFS. In general, it has been assumed that the SPR values, obtained without the application of external forces, could represent a lower limit for the DFS

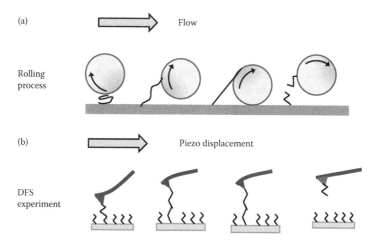

FIGURE 6.11 (a) Sketch of ligand-receptor interaction during leukocyte rolling. (b) Sketch of ligand–receptor unbinding during a DFS experiment. The retraction of the AFM cantilever from the substrate well mimics the physiological rolling process at the origin of the unbinding. (Adapted from Fritz, J. et al. 1998. *Proc. Natl. Acad Sci. U.S.A.*, 95:12283–12288.)

results. Moreover, they have analyzed the effects of the tip-to-substrate approaching speed on the unbinding process. Notably, they found that the unbinding force is practically unchanged while the adhesion probability increases as long as lower speed values are applied. Such a behavior has been attributed to the occurrence of a physiological mechanism finalized to prevent the rebinding during the rolling process. This experiment witnesses once more how DFS could be an extremely suitable tool for studying some aspects of biorecognition processes that cannot be investigated by standard techniques.

Hanley et al. have extended the work of Fritz et al. (1998) to include the investigation of the interaction of P-selectin with its ligand, which was embedded within the leukocyte cells, in turn immobilized on the substrate (Hanley et al., 2003). The authors found a dissociation rate value higher than that determined for isolated proteins (see Table 6.5); such an effect having been attributed to a structural modulation of the molecules when they are inserted in the cell surfaces. They have therefore drawn the attention to the crucial role played by the environment to establish the effective interaction properties of biomolecules with ligands. Furthermore, P-selectin molecules have been shown to display a different ability to bind ligands located on carcinoma cells with respect to normal ones. In this context, they have shown that an important contribution to the final binding affinity between the partners could arise from the protein density at the cell surface. Interestingly, these results have disclosed the possibility to evaluate the efficiency of therapeutic agents by monitoring the interaction properties of biomolecules upon changing the environmental conditions and under the action of mechanical stress.

Marshall et al. have also investigated the complex formed between the P-selectin and the P-selectin glycoprotein ligand (Marshall et al., 2003). They have observed a prolonged lifetime of the complex, followed by a shortening of it, upon the application of a mechanical force The observation of this phenomenon constitutes the first experimental evidence for the occurrence of "catch bonds" in a biological system. As previously mentioned, the application of a mechanical force to a system, instead of inducing a decrease of the energy barrier between the bound and the unbound states, sometimes could give rise to an increase of it with a concomitant rise of the bond lifetime (Prezhdo and Pereverzev, 2009). Catch bonds are expected to likely involve allosteric changes within the biomolecules, coupled with high-order fluctuations of the energy barrier and could have evolved in biological systems to fulfill specific functions. Moreover, the occurrence of catch bonds in the P-selectin-involved complex has been put into relationship to the capability of this system to support the rolling of leucocytes on a wall at low shear stresses. More generally, it has been suggested that transitions between catch and slip bonds might provide a general mechanism for the precise regulation of cell adhesion during mechanical stress. Notably, the discovery of catch bonds has been possible thanks to the DFS capability to follow a single biomolecular complex undergoing a controlled force. Successively, the same authors have focused the attention on the discrepancies among the results obtained by DFS for complexes involving P-selectin (Marshall et al., 2005). They deduced that the effective k_{off} value extracted from DFS experiments depends not only on the applied force on the system at a given instant but also on the entire "history" of the applied force. Accordingly, they suggested to revise the belief that the applied force could simply tilt the energy barrier, and they proposed to introduce a path-dependent analysis of the DFS data. From a general point of view, this means that the commonly used description of unbinding process in terms of the Bell–Evans model is an oversimplification which, in some cases, may give rise to some inconsistencies, as also observed for avidin–biotin pairs (see above). On such a basis, they solicited the scientific community to develop new theoretical tools, including force-history dependence, to analyze and interpret the DFS unbinding data of some biomolecular systems.

Li et al. (2003) have investigated the interaction between the transmembrane protein $\alpha_5\beta_1$ integrin with fibronectin, a glycoprotein largely present in the extracellular matrix such a complex plays a crucial role in cell differentiation and even in the adhesion of cells to the extracellular matrix. They have used a setup in which cell lines expressing only $\alpha_5\beta_1$ integrin have been immobilized on the AFM tip, whereas wild type or mutated fibronectin fragments have been immobilized on the substrate. They have focused the attention on the role played on the interaction properties by both the specific three aminoacid long sequence, named RGD, and the so-called synergy site of fibronectin. The presence of two linear trends in the unbinding force as a function of the logarithm loading rate for all the analyzed systems pointed out the existence of two energy barriers in the energy landscape (see Table 6.5). In particular, they suggested an inner and an outer energy barrier whose relative heights determine the overall kinetic properties of the system. The inner barrier has been found to depend only on the RGD sequence on fibronectin, whereas the outer one depends on both the RGD-specific sequence and the synergy site. On such a basis, it has been suggested that the presence of two barriers in the energy landscape could be a common characteristic of molecules involved in adhesion processes, since a modulation of their relative heights could give some flexibility in the response to external forces. Successively, the same authors have extended their approach to study the interaction of $\alpha_5\beta_1$ integrin with VCAM-1, an endothelial ligand essential for the extravasation of leukocytes, where they mediate the firm adhesion to surfaces (Zhang et al., 2004). The recombinant $\alpha_5\beta_1$ integrins (wild type and mutated forms) were immobilized on the AFM tip while VCAM-1 was anchored to a culture dish. Again, they found the existence of two barriers in the energy landscape for both wild type and mutant forms of VCAM-1 molecules. The inner barrier has been demonstrated to give a stronger resistance to the external forces through ionic interactions between Mg^{2+} and the N-terminal domain of a $\alpha_5\beta_1$ subunit. This provides a molecular basis to understand how the $\alpha_5\beta_1$ integrin–VCAM-1 interaction is able to resist a pulling force that is observed to occur in nature. This work witnesses the capability of DFS to provide remarkable insights into the mechanisms regulating the leukocyte ability to remain adherent to the surfaces in the presence of the shear force exterted by the bloodstream.

Dynamic force spectroscopy has been also applied to investigate complexes involved into platelet adhesion. A key role in these processes is played by fibrinogen, a soluble plasma glycoprotein converted into fibrin by thrombin during blood coagulation. Upon blood clotting, fibrinogen forms a fibrin network with the ability to entrap erythrocytes and platelets to form clots. An increased fibrinogen concentration is a factor that increases the risk of cardiovascular diseases. Lee and Marchant have studied the interaction between the human platelet $\alpha_{IIb}\beta_3$ receptor system with a ligand derived from fibrinogen (Lee and Marchant, 2001). The unbinding force plotted as a function of the logarithm of the loading rate shows a linear trend. The corresponding dissociation rate, reported in Table 6.5, results to be much higher than that measured in bulk solution by flow cytometry (0.2 s^{-1}). They suggested that such a discrepancy could arise either from some deformation of the platelet membrane or from conformational changes of the biomolecules during the DFS measurements. The presence of these changes likely due to the large flexibility displayed by the system suggests some caution in the use of the Bell–Evans model to more correctly analyze the corresponding data. Indeed, they proposed a revisitation of the Bell–Evans model to include possible deformations of the involved biological systems when the external force is applied.

Very recently, Carvalho et al. have applied DFS to elucidate the fibronogen-induced erythrocyte aggregation by studying the interaction of fibrinogen directly with two kinds of blood components, platelets and eryhrtocytes (Carvalho et al., 2010). Fibrinogen was immobilized on the AFM tip, while either the blood cells or the platelets were adsorbed on a glass substrate. They observed distinct peaks in the histograms of the unbinding force and attributed them to the occurrence of multiple binding events. Moreover, they find a smaller unbinding force for the fibrinogen–erythrocyte system with respect to that measured in fibrinogen–platelet interactions. In addition, for both the fibrinogen–platelet and the fibrinogen–erythrocyte interactions, the presence of two barriers in the energy landscape has been obtained. The corresponding dissociation rate values indicated the occurrence of a fast and a slow process for fibrinogen–erythrocyte and of two rather slow processes for fibrinogen–platelet (see Table 6.5). More specifically, they proposed that both the systems are characterized by a primary regime with similar dissociation rate and a secondary one which, however, is significantly different in the two systems. The observation of a fast process for the fibrinogen–erythrocyte was never observed before, probably due to the difficulty to follow, *in vivo*, a process with a lifetime of about a few milliseconds. On the basis of these results, the authors have hypothesized the existence of a novel receptor on the human eryhrtocyte membranes that can specifically bind fibrinogen. They have also ascertained that such a receptor is not significantly influenced by calcium as instead it occurs for the platelet receptor. As a negative control, they have tested a sample from a patient with a hereditary disease causing a deficiency in the fibrinogen receptor. The finding in this case of a drastic decrease in the unbinding force between fibrinogen and plateles, and also between fibrinogen and erythrocytes, has confirmed the existence of this new receptor. On such a basis, the authors have stressed that DFS is a highly sensitive nanotool for diagnostics of hematological diseases, offering new opportunities even in the functional evaluation of the disease severity.

6.4 DFS-BASED BIOSENSORS AND OTHER APPLICATIONS

The force detection capabilities of DFS have high potentialities in applicative fields, as well as in the development of nonstandard approaches to investigate biomolecular systems. For example, the use of a force-based discrimination with pN sensitivity has been recently applied for developing advanced bioanalytical devices and biosensors. However, for an extensive use in real analysis, DFS should be integrated within a high-throughput system able to explore a large number of interactions in short time. In this respect, the DFS sensing potentialities should be generally compared to those of other techniques and in particular with SPR, which is an emerging technique with a rapid expansion in the biomedical field. Although SPR is more suitable for an efficient routinary approach, DFS may have some advantages compared to it. Although SPR functions merely on metallic substrates, deposited as thin layers on glass, DFS measurements can be carried out on any substrate (metallic, semiconducting, or insulating). The detection sensitivity of SPR is limited to a few nanomolar of analytes with a molecular weight limit of a 2 kDa; at variance, DFS could be able to detect, in principle, the unbinding of a single ligand–receptor pair, irrespective of the size of the species involved in the recognition process. Indeed, DFS requires only a little amount of interacting species to probe the biological activity of the sensor surface; a few molecules, or even only one, onto the tips are sufficient to carry out reliable experiments. In the following part, same examples of promising approaches based on DFS to develop nanobiosensors or to study novel aspects of biological systems are discussed.

Blank et al. have developed a force-based biosensor to reveal a specific interactions between a ligand and a receptor by using a DNA zipper as molecular force sensor (Blank et al., 2003). Even if this method does not use an AFM equipment as such, it is susceptible to be implemented in a DFS experiment. The setup is sketched in Figure 6.12. One end of a complementary double DNA strand is bound to the marker (i.e., ligand) and the other one to a linker, which, in turn, is connected to the top surface, which could be replaced by the AFM tip; the DNA-bearing end being also marked with a fluorophore label. The receptor molecules are immobilized on the substrate using controlled arrays. By approaching the top surface to the substrate, a biorecognition process between the ligand and the receptor is promoted. During the retraction, the formed complexes whose interaction force overcomes the force between the DNA strands remain attached to the bottom plate with a concomitant deposition of the fluorophore; the substrate can be then scanned by a fluorescence equipment to reveal the position of the fluorescent signal. The authors remarked that a fluorescence-based detection of the substrates, coupled to an array organization of the biomolecules on the substrate, makes possible to extract the relevant information in a rapid and reliable way. Furthermore, they have demonstrated that the use of different DNA zippers allows to more use such a force-based approach to discriminate different complexes. Along the same direction, Kufer et al. have developed a DFS-based strategy for single-molecule cut-and-paste surface assembly (Kufer et al., 2008). Briefly, functional units (biotin labeled with a fluorophore) coupled to DNA

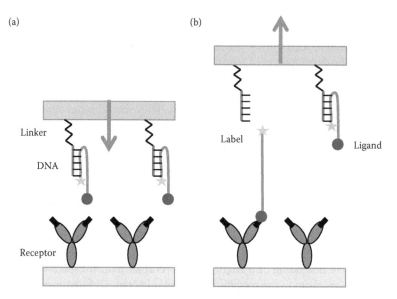

FIGURE 6.12 (See color insert.) Sketch of the experimental setup used for a force-based biosensor. A DNA strand is bound to the top plate through a linker. Both a fluorescent label and the ligand molecule are bound to DNA. The receptor molecules are immobilized on the bottom plate. (a) The top and the bottom plates approached each other to promote a biorecognition process between the partners. (b) During the retraction, one of the two DNA strands is expected to remain attached to the bottom plate, with a concomitant deposition of the fluorophore label upon the interaction between the ligand and the receptor overcomes that between the DNA strands. (Adapted from Blank, K. et al. 2003. *Proc. Nat. Acad. Sci. USA*, 100:11356–11360.)

oligomers were picked up from a depot area by means of a complementary DNA strand bound to an AFM tip. These units have been then transferred to and deposited on a target area to create basic geometric structures assembled from units with different functions. This process can be followed by monitoring the force curves; the spatial precision of the AFM coupled to the selectivity of the DNA biorecognition process among strands having been exploited for the bottom-up assembly of the biomolecular structures.

Miyachi et al. have exploited the biorecognition capability of DFS to screening DNA aptamers through a SELEX-based strategy (Miyachi et al., 2010). Single strands of DNA have been immobilized on the cantilever through the avidin–biotin pair while the target molecule, thrombin, has been anchored onto a gold substrate. During the approach of the tip to the substrate, a biorecognition between the aptamers and thrombin is promoted. Upon the retraction, if the binding force between the aptamer and the target molecule is higher than that between avidin and biotin, the latter interaction breaks and the DNA aptamer is deposited on the substrate; such a approach-retraction cycle being repeated to allow a further selection of DNA aptamers. Successively, the DNA molecules deposited on the substrate are recovered and amplified by polymerase chain reaction (PCR); these DNA aptamers could be subjected to other approach-retraction cycles to be further selected. They found that affinity force between DNA and thrombin gradually increases upon repeating selection runs. Such a method has allowed them to select aptamers with a stronger affinity for thrombin than those obtained with conventional strategies.

Very recently, Zhu et al. have developed a DFS-based approach to reveal methylation patterns in single-stranded DNA fragment (Zhu et al., 2010). The addition of a methyl group to cytosine plays a crucial role in the epigenetic gene regulation and therefore in human cancer research (Fazzari and Greally, 2004). However, the study of cytosine methylation cannot be easily carried on by conventional ensemble methods due to the weakness of the bp interaction. The used setup is sketched in Figure 6.13. Briefly, a monoclonal antibody specific for 5-methylcytosine was bound to the AFM tip through a PEG linker while 5-methylcytosine-containing single-stranded DNA was anchored to the substrate. Bringing the functionalized tip into contact with DNA promotes the formation of two distinct complexes, one for each

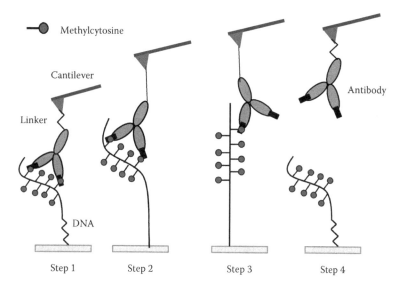

FIGURE 6.13 Sketch of the experimental setup used to reveal by DFS the molecular distance between two methylcytosine bases in a single-strand DNA. The antibody is anchored to the tip through a PEG linker, while the DNA is anchored to a glass substrate. The main stages of the retraction of the tip from the substrate, upon forming two distinct complexes between each of the two available arms of the antibody with a methylated DNA site are given in the following. Step 1: At the beginning the two arms of the antibody are bound to two different methylcytosine groups. Step 2: Successively, the linker is stretched, while both the free arms of the antibody remain bound to DNA. Step 3: By continuing the retraction, one of antibody-DNA bonds breaks. Step 4: Finally, the other antibody-DNA bond breaks and the linker goes back to its resting stage. (Adapted from Zhu, R. et al. 2010. *Nature Meth.*, 51:788–791.)

of the two available arms of the antibody interacting with a methylated DNA site. During the retraction process, the complexes involving the two arms of the antibody break in sequence from their corresponding ligands; the measurement of the PEG stretching allows to estimate the distance between the two methylated groups. Such a nanomechanical biorecognition-based approach made it possible to determine, therefore, a structural parameter of DNA whose characterization was not accessible by other methods. Moreover, the approach is suitable to be implemented in bionanodevices for high-throughput measurements of DNA chips to obtain posttranscriptional information on DNA.

The advent of AFM equipment with low-instrumental drift allows to follow the dynamical behavior of systems for a rather long time, opening, thus, the possibility to investigate aspects of biomolecular systems otherwise hidden or obscured by other processes. Very recently, Junker et al. have exploited the high sensitivity of DFS combined with high temporal stability to study conformational changes of calmodulin, which is one of the most prevalent Ca^{2+} signaling proteins in eukariotic cells (Junker et al., 2009). Notably, calmodulin can bind to more than 300 different target proteins to regulate numerous calcium-dependent functions. To study the conformational changes of calmodulin by DFS, the authors have sandwiched calmodulin, formed by two different domains, between the AFM tip and the substrate by using immunoglobulins. The application of a loading rate much lower in comparison to that usually used in standard DFS works has allowed them to follow the fluctuations of calmodulin in real time. An analysis of these fluctuations has shown that the calmodulin undergoes a folding–unfolding transition upon varying the Ca^{2+} concentration upon binding specific target peptides. These results witness that watching a single molecule at a time by DFS enable to follow molecular processes that would be lost in macroscopic measurements averaged over large numbers of molecules. On the other hand, the ability of DFS to watch folding or binding dynamics in real time could pave the way to direct observations of molecular reaction mechanisms as a sequence of structural events along microscopic pathways.

6.5 CONCLUSIONS

During the last years, DFS has become a progressively more rewarding and refined tool to investigate biomolecular systems. As it emerges from the examples discussed in the previous sections, DFS has been applied to a wide variety of biological complexes at the basis of many vital functions. Its capability to measure picoNewton unbinding forces between a single couple of biomolecules, in near-physiological conditions and without labeling, warrants enormous potentialities to deepen the knowledge of these systems, well-complementing information from traditional biochemical and spectroscopic techniques. It is now well ascertained that the knowledge of the unbinding force between two biomolecules does not directly provide a quantitative evaluation of the strength of the complex they may form. Indeed, quite different biomolecular systems, such as antigen–antibody, protein–ligand complexes, DNA-based systems, exhibit rather similar values for the unbinding force measured at the same loading rate. However, from a collection of unbinding force values, measured at different loading rates, information on the complex energy properties (number of barriers and related height and width) and on its dissociation rates can be extracted by applying suitable theoretical models. On the other hand, the application of an external force to unbind a biomolecular pair yields a decrease of the energy barrier and then a lowering of its lifetime making accessible the investigation of complexes characterized by extremely long lifetimes, whose study by standard approaches is rather difficult. Avidin–biotin and avidin-biotin like complexes have been extensively studied by many independent groups since 1994, and they constitute a benchmark for testing the capabilities of DFS to investigate the biorecognition processes. The wealth of related results appearing in the literature reports a significant variability in the values of both the dissociation rate and the energy barrier width. Careful and extensive DFS experiments, supported by appropriate data and theoretical analyses, have been devoted to reconcile the discrepancies between the results. These have been eventually attributed to (1) differences in the instrumental setups; (2) the different immobilization strategies; (3) the variability in the details of individual biomolecules, likely determined by local environmental heterogeneity; (4) intrinsic complexity of the biological systems arising from the existence of many local minima in the energy landscape, explored in a "history dependent" way during the DFS experiment.

Due to their enormous biological relevance, antigen–antibody complexes have also received large attention from the scientific community since the first DFS studies. DFS has put into evidence in these complexes the presence of more than one barrier in the energy landscape, whose height and width can be modulated by slight changes either in the environmental conditions or in the structural properties, leading to a fine tuning of the biological response of the systems. Different groups have compared DFS results on antigen–antibody pairs with those derived by SPR, showing that the results are in agreement when DFS experiments are carried out by applying low loading rate values; at higher values, the possibility of a deformation of the energy landscape should be taken into consideration, even from a theoretical point of view.

DFS has also demonstrated the capability to investigate complexes with short lifetimes, which are ubiquitous in cells where they play many important biological functions. In addition, DFS has been applied to investigate complexes involving DNA or aptamers, even in interaction with proteins. In general, these studies have been able to put into evidence a modulation of the energy barrier properties, and then of the kinetics, for partners differing even by single mutations, and this may be of considerable help to design more efficient therapeutic agents. Dynamic force spectroscopy at single-molecule level, can be applied to biomolecules embedded in the cell surface or in the membrane, making accessible the study of adhesion and aggregation processes between biomolecules in their physiological environment. In some cases, the application of an external force during DFS measurements could mimic the effect of the hydrodynamic flow in physiological conditions. More recently, the single-molecule detection capability of DFS together with the possibility to use only very small amount of samples to perform reliable experiments has become promising for the development of DFS-based innovative nanobiosensors.

Finally, the continuous improving of the AFM equipments is expected to have a high impact even on the field of biorecognition. A representative example is provided by the recent capability to follow the dynamical behavior of complexes for long time, thanks to the use of AFM with an extremely low drift. At the same time, some very recent development of AFM based on the combination of real-time visualization of the biomolecule topography and the characterization of their binding properties, offers the possibility to extract simultaneous kinetic and spatial information on these biological systems at single-molecule level. In summary, DFS is emerging as an extremely promising technique to investigate a large variety of biological systems in different conditions, with potentialities still partially unexplored.

ACKNOWLEDGMENT

This work has been partly supported by the AIRC (Associazione Italiana per la Ricerca sul Cancro) grant No. IG 10412.

REFERENCES

Allen, S., Chen, X., Davies, J., Davies, M. C., Dawkes, A. C., Edwards, J. C., Roberts, C. J., Sefton, J., Tendler, S. J. B., and Williams, P. M. 1997. Detection of antigene–antibody binding events with the atomic force microscope. *Biochemistry*, 36:7457–7463.

Baumgartner, W., Hinterdorfer, P., Ness, W., Raab, A., Vestweber, D., Schindler, H., and Drenckhahn, D. 2000. Cadherin interaction probed by atomic force microscopy. *Proc. Nat. Acad. Sci. USA*, 97:4005–4010.

Bayas, M. V., Schulten, K., and Leckband, D. 2003. Forced detachment of the CD2-CD58 complex. *Biophys. J.*, 84:2223–2233.

Bell, G. I. 1978. Models for the specific adhesion of cells to cells. *Science*, 200:618–627.

Bernhard, Elnathan, R., and Willner, I. 2006. Following aptamer–Thrombin binding by force measurements. *Analytical Chemistry*, 78:3638–3642.

Berquand, A., Xia, N., Castner, D. G., Clare, B. H., Abbott, N. L., Dupres, V., Adriaensen, Y., and Dufrêne, Y. F. 2005. Antigen binding forces of single antilysozyme Fv fragments explored by atomic force microscope. *Langmuir*, 21:5517–5523.

Bizzarri, A. R., Brunori, E., and Cannistraro, S. 2007. Docking and molecular dynamics simulation of the Azurin–Cytochrome c551 electron transfer complex. *J. Mol. Recognit.*, 20:122–131.

Bizzarri, A. R. and Cannistraro, S. 2009. Atomic force spectroscopy in biological complex formation: Strategies and perspectives. *J. Phys. Chem. B*, 113:16449–16464.

Bizzarri, A. R. and Cannistraro, S. 2010. The application of atomic force spectroscopy to the study of biological complexes undergoing a biorecognition process. *Chem. Soc. Rev. (Critical Reviews)*, 39:734–749.

Bizzarri, A. R. and Cannistraro, S. 2011. Free energy evaluation of the p53-Mdm2 complex from the unbinding work measured by dynamic force spectroscopy. *Phys. Chem. Chem. Phys.*, 13:2738–2743.

Bizzarri, A. R., DiAgostino, S., Andolfi, L., and Cannistraro, S. 2009. A combined atomic force microscope imaging and docking study to investigate the complex between p53 DNA binding domain and azurin. *J. Mol. Recognit.*, 22:506–515.

Blank, K., Mai, T., Gilbert, I., Schiffmann, S., Rankl, J., Zivin, R., Tackney, C., Nicolaus, T., Spinnler, K., Oesterhelt, F., Benoit, M., Clausen-Schaumann, H., and Gaub, H. E. 2003. A force-based protein biochip. *Proc. Nat. Acad. Sci. USA*, 100:11356–11360.

Bonanni, B., Andolfi, L., Bizzarri, A. R., and Cannistraro, S. 2007. Functional metalloproteins integrated with conductive substrates: detecting single molecules and sensing individual recognition events. *J. Phys. Chem. B*, 111:5062–5075.

Bonanni, B., Bizzarri, A. R., and Cannistraro, S. 2006. Optimized biorecognition of Cytochrome c551 and Azurin immobilized on thiol-terminated monolayers assembled on au(111) substrates. *J. Phys. Chem. B*, 110:14574–14580.

Bonanni, B., Kamruzzahan, A. S. M., Bizzarri, A. R., Rankl, C., Gruber, H. J., Hinterdorfer, P., and Cannistraro, S. 2005. Single molecule recognition between Cytochrome c551 and gold-immobilized azurin by force spectroscopy. *Biophys. J.*, 89:2783–2791.

Camacho, C. J. and Vajda, S. 2002. Protein–protein association kinetics and protein docking. *Curr. Opin. Struc. Biol.*, 12:36–40.

Carvalho, F. A., Connell, S., Miltenberger-Miltenyi, G., Pereira, S. V., Tavares, A., Ariens, R. A. S., and Santos, N. C. 2010. Atomic force microscope-based molecular recognition of a fibrinogen receptor on human erythrocytes. *ACS Nano*, 4:4609–4620.

Chéne, P. 2004. Inhibition of the p53-Mdm2 interaction: targeting a protein-protein interface. *Mol. Can. Res.*, 2:20–28.

Cho, E., Lee, J. W., and Ellington, A. D. 2009. Applications of aptamers as sensors. *Ann. Rev. Anal. Chem.*, 2:241–264.

Chtcheglova, L. A., Waschke, J., Wildling, L., Drenckhahn, D., and Hinterdorfer, P. 2007. Nano-scale dynamic recognition imaging on vascular endothelial cells. *Biophys. J.*, 93:L11–L13.

Crowley, P. B. and Ubbink, M. 2003. Close encounters of the transient kind: Protein interactions in the photosynthetic redox chain investigated by NMR spectroscopy. *Acc. Chem. Res.*, 36:723–730.

Cutruzzolá, F., Arese, M., Ranghino, G., van Pouderoyen, G., Canters, G., and Brunori, M. 2002. Pseudomonas aeruginosa cytochrome c551: probing the role of the hydrophobic patch in electron transfer. *J. Inorg. Biochem.*, 88:353–361.

De Grandis, V., Bizzarri, A., and Cannistraro, S. 2007. Docking study and free energy simulation of the complex between p53 DNA–binding domain and Azurin. *J. Mol. Recognit.*, 20:215–226.

de Odrowaz Piramowicz, M., Czuba P., Targosz, M., Burda, K., and Szymonski, M. 2006. Dynamic force measurements of avidin–biotin and streptavdin–biotin interactions using AFM. *Acta Biochim. Pol.*, 53:93–100.

De Paris, R., Strunz, T., Oroszlan, K., Guentherodt, H. J., and Hegner, M. 2000. Force spectroscopy and dynamics of the biotin-avidin bond studied by scanning force microscopy. *Single Mol.*, 1:285–290.

Domenici, F., Frasconi, M., Mazzei, F., D 'Orazi, G., Bizzarri, A. R., and Cannistraro, S. 2011. Azurin modulates the association of Mdm2 with p53: SPR evidence from interaction of the full-length proteins. *J. Mol. Recognit.*, 24:707–714.

Dudko, O. K., Filippov, A. E., Klafter, J., and Urbakh, M. 2003. Beyond the conventional description of dynamic force spectroscopy of adhesion bonds. *Proc. Natl. Acad. Sci. USA*, 100:11378–11381.

Eckel, R., Wilking, S. D., Becker, A., Sewald, N., Ros, R., and Anselmetti, D. 2005. Single-molecule experiments in synthetic biology: an approach to the affinity ranking of DNA–binding peptides. *Angew Chem. Int. Ed. Engl.*, 44:3921–3924.

Ellington, A. D. and Szostak, J. W. 1990. In vitro selection of RNA molecules that bind specific ligands. *Nature*, 346:812–822.

Evans, E. and Ritchie, K. 1997. Dynamic strength of molecular adhesion bonds. *Biophys. J.*, 72:1541–1555.

Fazzari, M. J. and Greally, J. M. 2004. Epigenomics: beyond CPG islands. *Nature Rev. Genet.*, 5:446–455.

Florin, E. L., Moy, V. T., and Gaub, H. E. 1994. Adhesion forces between individual ligand–receptor pairs. *Science*, 264:415–417.

Frauenfelder, H., Sligar, S. G., and Wolynes, P. G. 1991. The energy landscapes and motions of proteins. *Science*, 254:1598–1603.

Friddle, R. W. 2008. Unified model of dynamic forced barrier crossing in single molecules. *Phys. Rev. Lett.*, 100:138302–13805.

Fritz, J., Katopodis, A. G., Kolbinger, F., and Anselmetti, D. 1998. Force-mediated kinetics of single P-Selectin/ligand complexes observed by atomic force microscopy. *Proc. Natl. Acad. Sci. U.S.A.*, 95:12283–12288.

Funari, G., Domenici, F., Nardinocchi, L., Puca, R., D'Orazi, G., Bizzarri, A. R., and Cannistraro, S. 2010. Interaction of p53 with Mdm2 and Azurin as studied by atomic force spectroscopy. *J. Mol. Recognit.*, 23:343–351.

Grubmüller, H., Heymann, B., and Tavan, P. 1996. Ligand binding: molecular mechanics calculation of the Streptavidin–biotin rupture force. *Science*, 271:997–999.

Grunwald, C. 2008. Brief introduction to the streptavidin–biotin system and its usage in modern surface assays. *Z. Phys. Chem.*, 222:789–821.

Guo, S., Ray, C., Kirkpatrick, A., Lad, N., and Akhremitchev, B. 2008. Effects of multiple-bond ruptures on kinetic parameters extracted from force spectroscopy measurements: revisiting biotin-streptavidin interactions. *Biophys. J.*, 95:3964–3976.

Hanley, W., McCarty, O., Jadhav, S., Tseng, Y., Wirtz, D., and Konstantopoulos, K. 2003. Single molecule characterization of P-selectin/ligand binding. *J. Biol. Chem.*, 278:10556–10561.

Hinterdorfer, P., Baumgartner, W., Gruber, H. J., Schilcher, K., and Schindler, H. 1996. Detection and localization of individual antibody–antigen recognition events by atomic force microscope. *Proc. Natl. Acad. Sci. U.S.A*, 93:3477–3481.

Hinterdorfer, P. and Dufrêne, Y. F. 2006. Detection and localization of single molecular recognition events using Atomic Force Microscopy. *Nat. Methods*, 5:347–355.

Hummer, G. and Szabo, A. 2003. Kinetics from nonequilibrium single molecule pulling experiments. *Biophys. J.*, 85:5–15.

Hummer, G. and Szabo, A. 2005. Free energy surfaces from single-molecule force spectroscopy. *Acc. Chem. Res.*, 38:504–513.

Hyeon, C. B. and Thirumalai, D. 2003. Can energy landscape roughness of proteins and RNA be measured by using mechanical unfolding experiments? *Proc. Natl. Acad. Sci. U.S.A*, 38:10249–10253.

Izrailev, S., Stepaniants, S., Balsera, M., Oono, Y., and Schulten, K. 1997. Molecular dynamics study of unbinding of the avidin–biotin complex. *Biophys. J.*, 72:568–1581.

Janin, J. 1997. The kinetics of protein-protein recognition. *Proteins: Struct. Funct. Genet.*, 28:153–161.

Janshoff, A., Neitzert, M., Oberdörfer, Y., and Fuchs, H. 2000. Force spectroscopy of molecular systems–single molecule spectroscopy od polymers and biomolecules. *Angew. Chem. Int. Ed.*, 39:3212–3237.

Janshoff, A. and Steinem, C. 2001. Energy landscapes of ligand– receptor couples probed by dynamic force spectroscopy. *Chem. Phys. Chem.*, 2:577–579.

Jarzynski, C. 1997. A nonequilibrium equality for free energy differences. *Phys. Rev. Lett.*, 78:2690–2693.

Jena, B. P. and Hörber, H. J. K. 2002. *Atomic Force Microscope in cell biology. In Methods in Cell Biology*, volume 68. Academic Press: San Diego.

Jiang, Y., Zhu, C., Ling, L., Wan, L., Fang, X., and Bai, C. 2003. Specific aptamer-protein interaction studied by atomic force microscopy. *Anal. Chem.*, 75:2112–2116.

Jones, S. and Thornton, J. M. 1996. Principles of protein-protein interactions. *Proc. Natl. Acad. Sci. USA*, 93:13–20.

Junker, J. P., Ziegle, F., and M., R. 2009. Ligand-dependent equilibrium fluctuations of single Calmodulin molecules. *Science*, 323:633–637.

Junker et al., 2009; Kemler, R. 1992. Classical cadherins. *Semin. Cell Biol.* 3:149–155.

Krasnoslobodtsev, A. V., Shlyakhtenko, L. S., and Lyubchenko, Y. L. 2007. Probing interactions within the synaptic DNA–SfiI complex by AFM force spectroscopy. *J. Mol. Biol.*, 365:1407–1416.

Kufer, S. K., Puchner, E. M., Gumpp, H., Liedl, T., and Gaub, H. E. 2008. Single-molecule cut-and-paste surface assembly. *Science*, 319:594–596.

Kühner, F., Costa, L. T., Bisch, P. M., Thalhammer, S., Heckl, W. M., and Gaub, H. E. 2004. Lexa–DNA bond strength by single molecule force spectroscopy. *Biophys. J.*, 87:2683–2690.

Lauffenburger, D. and Linferman, J. J. 1993. *Receptors: Models for Binding, Trafficking and Signaling*. Oxford University Press, New York.

Lee, C. K., Wang, Y. M., Huang, L. S., and Lin, S. 2007. Atomic Force Microscope: Determination of unbinding force, off rate and energy barrier for protein–ligand interaction. *Micron*, 38:446–461.

Lee, G. U., Chrisey, L., and Colton, R. J. 1994a. Direct measurement of the forces between complementary strands of DNA. *Science*, 266:771–773.

Lee, G. U., Kidwell, D. A., and Colton, R. J. 1994b. Sensing discrete streptavidin–biotin interactions with atomic force microscopy. *Langmuir*, 10:354–357.

Lee, I. S. and Marchant, R. 2001. Force measurements on the molecular interactions between ligand (RDG) and human platelet $\alpha_{IIb}\beta_3$ receptor system. *Surf. Sci.*, 491:433–443.

Li, F., Redick, S., Erickson, H., and Moy, V. T. 2003. Force measurements of the $\alpha_5\beta_1$ Integrin–Fibronectin interaction. *Biophys. J.*, 84:1252–1262.

Lindorff-Larsen, K., Best, R. B., Depristo, M. A., Dobson, C. M., and Vendruscolo, M. 2005. Simultaneous determination of protein structure and dynamics. *Nature*, 433:128–132.

Liu, W., Montana, V., Bai, J., Chapman, E., Mohideen, U., and Parpura, V. 2006. Single molecule mechanical probing of the SNARE protein interactions. *Biophys. J.*, 91:744–758.

Liu, W., Montana, V., Parpura, V., and Mohideen, U. 2008. Comparative energy measurements in single molecule interactions. *Biophys. J.*, 95:419–426.

Lo, Y. S., Zhu, Y. J., and Beebe, T. P. 2001. Loading-rate dependence of individual ligand–receptor bond-rupture forces studied by atomic force microscope. *Langmuir*, 17:3741–3748.

Marshall, B. T., Long, M., Piper, J. W., Yago, T., McEver, R. P., and Zhu, C. 2003. Direct observation of catch bonds involving cell–adhesion molecules. *Nature*, 423:190–193.

Marshall, B. T., Sarangapani, K. K., Lou, J. H., Rodger, P., McEver, R. P., and Zhu, C. 2005. Force history dependence of receptor-ligand dissociation. *Biophys. J.*, 88:1458–1466.

Maynard, J. and Georgiou, G. 2000. Antibody engineering. *Annu. Rev. Biomed. Eng.*, 2:339–376.

Merkel, R., Nassoy, P., Leung, A., Ritchie, K., and Evans, E. 1999. Energy landscapes of receptor-ligand bonds explored with dynamic force spectroscopy. *Nature*, 397:50–53.

Miyachi, Y., Shimizu, N., Ogino, C., and Kondo, A. 2010. Selection of DNA aptamers using Atomic Force Microscope. *Nucl. Acid. Res.*, 38:e21–e28.

Morfill, J., Blank, K., Zahnd, C., Luginbuel, B., Kuehner, F., Gottschalk, K. E., PLueckthun, A., and Gaub, H. E. 2007. Affinity-matured recombinant antibody fragments analyzed by single molecule force spectroscopy. *Biophys. J.*, 93:3583–3590.

Morikis, D. and Lambris, J. D. 2004. Physical methods for structure, dynamics and binding in immunological research. *Trends Immunol.*, 25:700–707.

Moy, V. T., Florin, E. L., and Gaub, H. E. 1994a. Adhesive forces between ligand and receptor measured by AFM. *Science*, 264:415–417.

Moy, V. T., Florin, E. L., and Gaub, H. E. 1994b. Intermolecular forces and energies between ligands and receptors. *Science*, 266:257–259.

Neuert, G., Albrecht, C., Pamir, E., and Gaub, H. E. 2006. Dynamic force spectroscopy of the Digoxigenin-antibody complex. *FEBS Lett.*, 580:505–509.

Neuman, K. C. and Nagy, A. 2008. Single-molecule force spectroscopy: optical tweezers, magnetic tweezers and atomic force microscope. *Nature Meth.*, 5:491–505.

Nevo, R., Brumfeld, V., Kapon, R., Hinterdorfer, P., and Reich, Z. 2005. Direct measurement of protein energy landscape roughness. *EMBO Rep.*, 5:482–486.

Panorchan, P., Thompson, M. S., Davis, K. J., Tseng, Y., Konstantopoulos, K., and Wirtz, D. 2006. Single-molecule analysis of cadherin-mediated cell-cell adhesion. *J. Cell Sci.*, 119:66–74.

Patel, A. B., Allen, S., Davies, M. C., Roberts, C. J., Tendler, S. J. B., and Williams, P. M. 2004. Influence of architecture on the kinetic stability of molecular assemblies. *J. Am. Chem. Soc.*, 126:1318–1319.

Pincet, F. and Husson, J. 2005. The solution to the streptavidin-biotin paradox: the influence of history on the strength of single molecular bonds. *Biophys. J.*, 89:4374–4381.

Prezhdo, O. V. and Pereverzev, Y. V. 2009. Theoretical aspects of the biological catch bond. *Acc. Chem. Res.*, 42:693–703.

Rico, F. and Moy, V. T. 2007. Energy landscape roughness of the streptavidin–biotin interaction. *J. Mol. Recognit.*, 20:495–505.

Rief, M. and Grubmüller, H. 2002. Force spectroscopy of single biomolecules. *Chem. Phys. Chem.*, 3:255–261.

Ros, R., Schwesinger, F., Anselmetti, D., Kubon, M., Schafer, R., Pluckthun, A., and Tiefenauer, L. 1998. Antigen binding forces of individually addressed single-chain Fv antibody molecules. *J. Mol. Recognit.*, 20:7402–7405.

Schön, O., Friedler, A., Bycroft, M., Freund, S. M. V., and Fersht, A. R. 2002. Molecular mechansim of the interaction between Mdm2 and p53. *J. Mol. Biol.*, 323:491–501.

Schreiber, G. and Fersht, A. 1993. Interaction of Barnase and its polypeptide inhibitor Barstar studied by protein engineering. *Biochemistry*, 32:5145–5150.

Schreiber, G., Haran, G., and Zhou, H.-X. 2009. Fundamental aspects of protein–protein association kinetics. *Chemi. Rev.*, 109:83–860.

Schumakovitch, I., Grange, W., Strunz, T., Bertoncini, P., Guntherodt, H. J., and Hegner, M. 2002. Temperature dependence of unbinding forces between complementary DNA strands. *Biophys. J.*, 82:517–527.

Schwesinger, F., Ros, R., Strunz, T., Anselmetti, D., Güntherodt, H. J., Honegger, A., Jermutus, L., Tiefenauer, L., and Plückthun, A. 2000. Unbinding forces of single antibody–antigen complexes correlate with their thermal dissociation rates. *Proc. Natl. Acad. Sci. U.S.A*, 97:9972–9977.

Sekatskii, S. K., Favrea, M., Dietlera, G., Mikhailovb, A. G. Klinovb, D. V., Lukashb, S. V., and Deyevb, S. M. 2010. Force spectroscopy of Barnase–Barstar single molecule interaction. *J. Mol. Recognit.*, 23:583–588.

Strunz, T., Oroszlan, K., Schäfer, R., and Güntherodt, H. J. 1999. Dynamic force spectroscopy of single DNA molecules. *Proc. Natl. Acad. Sci. U.S.A*, 96:11277–11282.

Sulchek, T. A., Friddle, R. W., Langry, K., Lau, E. Y., Albrecht, H., Ratto, T. V., De Nardo, S. J., Colvin, M. E., and Noy, A. 2005. Dynamic force spectroscopy of parallel individual mucin1–antibody bonds. *Proc. Natl. Acad. Sci. U.S.A*, 102:16638–16643.

Taranta, M., Bizzarri, A. R., and Cannistraro, S. 2008. Probing the interaction between p53 and the bacterial protein Azurin by single molecule force spectroscopy. *J. Mol. Recognit.*, 21:63–70.

Taranta, M., Bizzarri, A. R., and Cannistraro, S. 2009. Modelling the interaction between the N-terminal domain of the tumor suppressor p53 and Azurin. *J. Mol. Recognit.*, 22:215–222.

Teulon, J. M., Delcuze, Y., Odorico, M., Chen, S. W., Parot, P., and Pellequer, J. L. 2011. Single and multiple bonds in (Strept)avidin–biotin interactions. *J. Mol. Recogn.*, 24:490–502.

Walton et al., 2008: Vogelstein, B., Lane, D., and Levine, A. J. 2000. Surfing the p53 network. *Nature* 408:307–310.

Walton, E. B., Lee, S., and Van Vliet, K. J. 2008. Extending bells model: how force transducer stiffness alters measured unbinding forces and kinetics of molecular complexes. *Biophys. J.*, 94:2621–2630.

Wilchek, M., Bayer, E. A., and Livnah, O. 2006. Essentials of biorecognition: the (Strept)avidin–biotin system as a model for protein–protein and protein–ligand interaction. *Immunology Lett.*, 103:27–32.

Williams, P. M., Moore, A., Stevens, M. M., Allen, S., Davies, M. C., Roberts, C. J., and Tendler, S. J. B. 2000. On the dynamic behaviour of the forced dissociation of ligand–receptor pairs. *J. Chem. Soc., Perkin Trans*, 2:5–8.

Wojcikiewicz, E. P., Abdulreda, M. H., Zhang, X., and Moy, V. T. 2006. Force spectroscopy of LFA–1 and its ligands, ICAM–1 and ICAM–2. *Biomacromolecules*, 7:3188–3195.

Yamada, T., Mehta, R. R., Lekmine, F., Christov, K., King, M. L., Majumdar, D., Shilkaitis, A., Green, A., Bratescu, L., Beattie, C. W., and Das Gupta, T. K. 2009. A peptide fragment of azurin induces a p53–mediated cell cycle arrest in human breast cancer cells. *Mol. Cancer Ther.*, 8:2947–2958.

Yersin, A., Osada, T., and Ikai, A. 2008. Exploring transferrin-receptor interactions at the single-molecule level. *Biopys. J.*, 94:230–240.

Yuan, C., Chen, A., Kolb, P., and Moy, V. T. 2000. Energy landscape of streptavidin-biotin complexes measured by atomic force microscope. *Biochemistry*, 39:10219–10223.

Zhang, X., Craig, S. E., Kirby, H., Humphries, M. J., and Moy, V. T. 2004. Molecular basis for the dynamic strength of the integrin $\alpha_4\beta_1$–VCAM-1 interaction. *Biophys. J.*, 87:3470–3478.

Zhu, R., Howorka, S., Pröll, J., Kienberger, F., Preiner, J., Hesse, J., Ebner, A., Pastushenko, V. P., Gruber, H. J., and Hinterdorfer, P. 2010. Nanomechanical recognition measurements of individual DNA molecules reveal epigenetic methylation patterns. *Nature Meth.*, 51:788–791.

Zlatanova, J., Lindsay, M., and Leuba, S. 2000. Single molecule force spectroscopy in biology using the atomic force microscope. *Prog. Biophys. Mol. Biol.*, 74:3761.

7 Conclusions and Perspectives

Anna Rita Bizzarri and Salvatore Cannistraro

The achievement of a complete and unified picture of biorecognition processes in living organisms requires a large effort based on the combination of standard experimental approaches with highly innovative techniques, with the support of adequate theoretical models (see Chapter 1). The traditional concepts of specificity, affinity, and rate constants, widely used to describe biorecognition, have to be updated also to take into account additional aspects, such as the distance and the orientation between the biomolecules, the eventual immobilization on the cell surface, the molecular density, and so on. In this context, single molecule techniques emerged as extremely useful tools to elucidate even subtle details of the biorecognition mechanisms. Dynamic force spectroscopy (DFS) has gained a prominent position among these techniques due to its ability to capture molecular events at the basis of the molecular interaction, well-complementing information coming from standard biomolecular and spectroscopic techniques operating in bulk. This essentially stems from the capability of DFS to measure unbinding forces with picoNewton sensitivity between single couples of biomolecules immobilized on suitable surfaces, under physiological conditions, without labeling and in real time.

The progressively higher relevance of DFS is witnessed by the continuous increase in the number of both scientific publications and atomic force microscopy (AFM) equipments devoted to DFS experiments in the worldwide scientific community. Indeed, DFS has been applied to investigate a variety of biomolecular systems playing many different biological functions (see Chapter 6), allowing also to elucidate the influence on the kinetic and thermodynamical properties of some important factors that are usually hidden when bulk techniques are used, such as punctual mutations within the partners, molecular heterogeneity, conformational changes, local environmental changes, molecular crowding, and so on. These investigations hold a remarkable interest since they offer also the possibility of tailoring the molecular structure and dynamics of biomolecules, ligands, drugs, and so on to optimize the function in which they are involved. DFS has made accessible the investigation of the energy landscape of interacting biomolecules evidencing the presence of many nearly isoenergetic local minima, whose exploration should be considered in order to achieve a consistent description of the kinetic response of biomolecular systems during the biorecognition process; this could also explain the "history" dependence of the unbinding data of the system. These results have moreover given new impetus to the development of theoretical models to interpret the mechanisms that govern the interactions between biomolecules. The use of suitable theoretical models is, on the other hand, essential to analyze DFS data since the unbinding properties of the partners are measured in nonequilibrium conditions under the application of an external force. It should be remarked that the application of the external force, yielding a lowering of the lifetime of the biological pairs, makes accessible the investigation of complexes characterized by extremely long lifetimes, such as avidin–biotin whose dissociation process does not take place spontaneously in a reasonable observation time. In general, the Bell–Evans model derived in the theoretical framework of a thermal escape over an energy barrier under the effect of an external force (see Chapter 3) has provided a good description of the unbinding force trend as a function of the logarithm of the loading rate, for the largest part of the analyzed biological systems, allowing to extract the dissociation rate and the barrier width of the energy landscape. However, the recurrent discrepancies found in the results from well-defined biological systems has recently stimulated the development of a more general theoretical description of the unbinding processes, even to reach a deeper understanding of biorecognition. In this respect, some important factors such as (1) the possible deformation of the energy landscape upon the application of the external force, (2) the relationship between the cantilever spring constant and the intermolecular potential stiffness, (3) the occurrence of rebinding processes between the partners, (4) the partial unfolding of the biomolecules upon the pulling force, (5) the applied force direction with respect to the reaction coordinate, should be taken into account by more comprehensive models. Along this direction, the Jarzynski theoretical model, which allows to evaluate the binding free energy, from the irreversible work done along nonequilibrium paths from the bound to the unbound state, constitutes an interesting novelty susceptible to enlarge the amount of information that can be obtained by DFS experiments on biological systems.

A crucial role is also played by the setup used to perform the DFS unbinding experiments. In particular, the methodologies followed to immobilize the biomolecules to the AFM surfaces (substrate and tip) should fulfill some important requirements: (1) Only two biomolecular partners should be possibly involved in the biorecognition process; (2) A reliable discrimination between specific and nonspecific unbinding events should be facilitated; and (3) The preservation of both the native structure and the functionality of the biomolecules has to be ensured (see Chapter 4). In general, an appropriate use of flexible linkers connecting the biomolecular partners to the inorganic surfaces, and undergoing a controlled stretching during the unbinding process, in most cases satisfy these conditions, especially when it is combined with suitable theoretical models or computer simulations. In spite of the large efforts made by the DFS community to continuously develop and implement new immobilization protocols, a definite strategy that could permit to standardize the experimental features and the corresponding data analysis is still lacking. It should, however, be kept in mind that the most appropriate procedure to immobilize biomolecules should be chosen according to the specific features of the system under analysis and to the information that should be extracted from the DFS data. For example, whole cells, or part of them, should be directly used in the DFS setup when studying the interaction between biomolecules which, in their native state, are located at the cell surface. In a DFS experiment, the application of an external force to a biomolecular system offers moreover the possibility to elucidate the molecular mechanisms in living organisms that operate, in physiological conditions, under the action of a mechanical force (such as in the presence of the bloodstream shear force of or in molecular motors).

The single molecule detection sensitivity coupled to the need of only a tiny amount of sample to carry out reliable experiments may endow DFS with the remarkable potentiality to drive the development of force/biorecognition-based nanobiosensors for applications in the field of early diagnostics. Either in this connection or for biorecognition research in general, the efforts toward implementation of efficient and reliable automatic procedures to analyze DFS results have enormously increased with some success (see Chapter 5).

More recent results witness that DFS applications to biorecognition processes are susceptible to undergo significant developments in the near future, even boosted by the improvements of AFM equipments (see Chapter 2). Quite recently developed low-drift AFM apparatus could lend higher potentialities to DFS experiments, making possible to perform measurements for longer times, allowing thus to follow processes at near equilibrium. On the other hand, new high-speed AFM equipments make accessible monitoring even faster biological events in real time. Furthermore, the combination of DFS with high-resolution AFM imaging allows both real-time topographical imaging and characterization of the binding properties of single biomolecular partners to visualize, identify, and quantify local receptor binding sites by assigning their locations to the topographical features of surfaces. Such a kind of combined approach is expected to become progressively more and more used since it offers the possibility to extract simultaneous dynamical and spatial information on the interacting systems at single molecule level. The use of conductive surfaces (tip and substrate) in DFS experiments would make feasible to combine DFS with conductive measurements (e.g., by scanning tunneling microscopy [STM] or conductive AFM) either to elucidate the interplay between electron transfer and biorecognition processes in electron transfer complexes, or to implement multisensing detection.

Finally, coupling DFS with ultra-sensitive optical techniques, such as advanced fluorescence and Raman–SERS (surface-enhanced Raman spectroscopy), could deserve a great promise for both a deeper study of biorecognition, enriched with chemical information, and the design of innovative nanobiosensors for early detection of biomarkers.

Index

T - #0369 - 101024 - C16 - 234/156/16 - PB - 9781138374522 - Gloss Lamination